Stetson University
3 4369 00420050 0

W9-AFP-645

WITHDRAWN

Undergraduate Texts in Mathematics

Editors
S. Axler
F.W. Gehring
K.A. Ribet

Springer
New York
Berlin
Heidelberg
Barcelona
Budapest
Hong Kong
London
Milan
Paris
Singapore
Tokyo

Undergraduate Texts in Mathematics

Anglin: Mathematics: A Concise History and Philosophy.
Readings in Mathematics.
Anglin/Lambek: The Heritage of Thales.
Readings in Mathematics.
Apostol: Introduction to Analytic Number Theory. Second edition.
Armstrong: Basic Topology.
Armstrong: Groups and Symmetry.
Axler: Linear Algebra Done Right. Second edition.
Beardon: Limits: A New Approach to Real Analysis.
Bak/Newman: Complex Analysis. Second edition.
Banchoff/Wermer: Linear Algebra Through Geometry. Second edition.
Berberian: A First Course in Real Analysis.
Bix: Conics and Cubics: A Concrete Introduction to Algebraic Curves.
Brémaud: An Introduction to Probabilistic Modeling.
Bressoud: Factorization and Primality Testing.
Bressoud: Second Year Calculus.
Readings in Mathematics.
Brickman: Mathematical Introduction to Linear Programming and Game Theory.
Browder: Mathematical Analysis: An Introduction.
Buskes/van Rooij: Topological Spaces: From Distance to Neighborhood.
Cederberg: A Course in Modern Geometries.
Childs: A Concrete Introduction to Higher Algebra. Second edition.
Chung: Elementary Probability Theory with Stochastic Processes. Third edition.
Cox/Little/O'Shea: Ideals, Varieties, and Algorithms. Second edition.
Croom: Basic Concepts of Algebraic Topology.
Curtis: Linear Algebra: An Introductory Approach. Fourth edition.
Devlin: The Joy of Sets: Fundamentals of Contemporary Set Theory. Second edition.
Dixmier: General Topology.
Driver: Why Math?
Ebbinghaus/Flum/Thomas: Mathematical Logic. Second edition.
Edgar: Measure, Topology, and Fractal Geometry.
Elaydi: Introduction to Difference Equations.
Exner: An Accompaniment to Higher Mathematics.
Fine/Rosenberger: The Fundamental Theory of Algebra.
Fischer: Intermediate Real Analysis.
Flanigan/Kazdan: Calculus Two: Linear and Nonlinear Functions. Second edition.
Fleming: Functions of Several Variables. Second edition.
Foulds: Combinatorial Optimization for Undergraduates.
Foulds: Optimization Techniques: An Introduction.
Franklin: Methods of Mathematical Economics.
Gordon: Discrete Probability.
Hairer/Wanner: Analysis by Its History.
Readings in Mathematics.
Halmos: Finite-Dimensional Vector Spaces. Second edition.
Halmos: Naive Set Theory.
Hämmerlin/Hoffmann: Numerical Mathematics.
Readings in Mathematics.
Hijab: Introduction to Calculus and Classical Analysis.
Hilton/Holton/Pedersen: Mathematical Reflections: In a Room with Many Mirrors.
Iooss/Joseph: Elementary Stability and Bifurcation Theory. Second edition.
Isaac: The Pleasures of Probability.
Readings in Mathematics.

(continued after index)

Robert Bix

Conics and Cubics

A Concrete Introduction to Algebraic Curves

With 148 Illustrations

Robert Bix
Department of Mathematics
The University of Michigan–Flint
Flint, MI 48502
USA

Editorial Board

S. Axler
Mathematics Department
San Francisco State University
San Francisco, CA 94132
USA

F.W. Gehring
Mathematics Department
East Hall
University of Michigan
Ann Arbor, MI 48109
USA

K.A. Ribet
Department of Mathematics
University of California at Berkeley
Berkeley, CA 94720-3840
USA

Mathematics Subject Classification (1991): 51-01, 14Hxx, 11Cxx

Library of Congress Cataloging-in-Publication Data
Bix, Robert, 1953–
Conics and cubics: a concrete introduction to algebraic curves / Robert Bix.
p. cm. — (Undergraduate texts in mathematics)
Includes bibliographical references and index.
ISBN 0-387-98401-1 (hc: alk. paper)
1. Curves, Algebraic. I. Title. II. Series.
QA567.B59 1998
516.3′52—dc21 97-46950

Printed on acid-free paper.

© 1998 Springer-Verlag New York, Inc.
All rights reserved. This work may not be translated or copied in whole or in part without the written permission of the publisher (Springer-Verlag New York, Inc., 175 Fifth Avenue, New York, NY 10010, USA), except for brief excerpts in connection with reviews or scholarly analysis. Use in connection with any form of information storage and retrieval, electronic adaptation, computer software, or by similar or dissimilar methodology now known or hereafter developed is forbidden.
The use of general descriptive names, trade names, trademarks, etc., in this publication, even if the former are not especially identified, is not to be taken as a sign that such names, as understood by the Trade Marks and Merchandise Marks Act, may accordingly be used freely by anyone.

Production managed by Karina Mikhli; manufacturing supervised by Jeffrey Taub.
Typeset by Asco Trade Typesetting Ltd., Hong Kong.
Printed and bound by R.R. Donnelley and Sons, Harrisonburg, VA.
Printed in the United States of America.

9 8 7 6 5 4 3 2 1

ISBN 0-387-98401-1 Springer-Verlag New York Berlin Heidelberg SPIN 10659128

To Peggy and Jonathan

Preface

Algebraic curves are the graphs of polynomial equations in two variables, such as $y^3 + 5xy^2 = x^3 + 2xy$. By focusing on curves of degree at most 3—lines, conics, and cubics—this book aims to fill the gap between the familiar subject of analytic geometry and the general study of algebraic curves. This text is designed for a one-semester class that serves both as a a geometry course for mathematics majors in general and as a sequel to college geometry for teachers of secondary school mathematics. The only prerequisite is first-year calculus.

On the one hand, this book can serve as a text for an undergraduate geometry course for all mathematics majors. Algebraic geometry unites algebra, geometry, topology, and analysis, and it is one of the most exciting areas of modern mathematics. Unfortunately, the subject is not easily accessible, and most introductory courses require a prohibitive amount of mathematical machinery. We avoid this problem by focusing on curves of degree at most 3. This keeps the results tangible and the proofs natural. It lets us emphasize the power of two fundamental ideas, homogeneous coordinates and intersection multiplicities.

On the other hand, an undergraduate or graduate course, based on this book, can be a sequel to college geometry for prospective or current teachers of secondary school mathematics. Because analytic geometry plays a crucial role in secondary school mathematics, teachers should see it taken beyond the level of mechanical computation. Because most college geometry courses focus on synthetic—that is, nonanalytic—geometry, it is desirable to devote a second geometry course to analytic geometry.

Every line can be transformed into the x-axis, and every conic can be transformed into the parabola $y = x^2$. We use these two basic facts to analyze the intersections of lines and conics with curves of all degrees, and to deduce special cases of Bezout's Theorem and Noether's Theorem. These results give Pascal's Theorem and its corollaries about polygons inscribed in conics, Brianchon's Theorem and its corollaries about polygons circumscribed about conics, and Pappus' Theorem about hexagons inscribed in lines. We give a remarkably simple proof of Bezout's Theorem for curves of all degrees by combining the result for lines with induction on the degrees of the curves in one of the variables. We use Bezout's Theorem to classify cubics. We introduce elliptic curves by proving that a cubic becomes an abelian group when we use collinearity to define addition of points; this fact plays a key role in number theory, and it is the starting point of the 1995 proof of Fermat's Last Theorem.

The exercises provide practice in using the results of the text, and they outline additional material. They can be homework problems when the book is used as a class text, and they are optional otherwise.

I am greatly indebted to Harry D'Souza for sharing his expertise, to Richard Alfaro for generating figures by computer, and to Michael Bix, Renate McLaughlin, and Kenneth Schilling for reviewing the manuscript. I am also grateful to the students at the University of Michigan-Flint who tried out the manuscript in class.

Flint, Michigan Robert Bix

Contents

CHAPTER I

Intersections of Curves

Introduction and History

Introduction

An algebraic curve is the graph of a polynomial equation in two variables x and y. Because we consider products of powers of both variables, the graphs can be intricate even for polynomials with low exponents. For example, Figure I.1 shows the graph of the equation

$$r^2 = \cos 2\theta$$

in polar coordinates. To convert this equation to rectangular coordinates and obtain a polynomial in two variables, we multiply both sides of the equation by r^2 and use the identity $\cos 2\theta = \cos^2\theta - \sin^2\theta$. This gives

$$r^4 = r^2\cos^2\theta - r^2\sin^2\theta. \tag{1}$$

We use the usual substitutions $r^2 = x^2 + y^2$, $r\cos\theta = x$, and $r\sin\theta = y$ to rewrite (1) as

$$(x^2 + y^2)^2 = x^2 - y^2.$$

Multiplying this polynomial out and collecting its terms on the left gives

$$x^4 + 2x^2y^2 + y^4 - x^2 + y^2 = 0. \tag{2}$$

Thus Figure I.1 is the graph of a polynomial in two variables, and so it is an algebraic curve.

We add two powerful tools for studying algebraic curves to the familiar techniques of precalculus and calculus. The first is the idea that

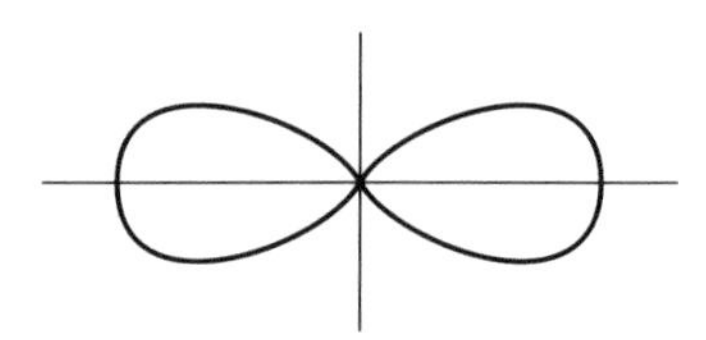

Figure I.1

curves can intersect repeatedly at a point. For example, it is natural to think that the curve in Figure I.1 intersects the x-axis twice at the origin because it passes through the origin twice. We develop algebraic techniques in Section 1 for computing the number of times that two algebraic curves intersect at the origin.

The second major tool for studying algebraic curves is the system of homogeneous coordinates, which we introduce in Section 2. This is a bookkeeping device that lets us study the behavior of algebraic curves at infinity in the same way as in the Euclidean plane. Erasing the distinction between points of the Euclidean plane and those at infinity simplifies our work greatly by eliminating special cases.

We combine the ideas of Sections 1 and 2 in Section 3. We use homogeneous coordinates to determine the number of times that two algebraic curves intersect at any point in the Euclidean plane or at infinity. We also introduce transformations, which are linear changes of coordinates. We use transformations throughout our work to simplify the equations of curves.

We focus on the intersections of lines and other curves in Section 4. If a line l is not contained in an algebraic curve F, we prove that the number of times that l intersects F, counting multiplicities, is at most the degree of F. This introduces one of the main themes of our work: the geometric significance of the degree of a curve. We also characterize tangent lines in terms of intersection multiplicities.

History

Greek mathematicians such as Euclid and Apollonius developed geometry to an extraordinary level in the third century B.C. Their algebra, however, was limited to vebral combinations of lengths, areas, and volumes. Algebraic symbols, which give algebraic work its power, arose only in the second half of the 1500s, most notably when François Vieta introduced the use of letters to represent unknowns and general coefficients.

Geometry and algebra were combined into analytic geometry in the first half of the 1600s by Pierre de Fermat and René Descartes. By asserting that any equation in two variables could be used to define a curve, they expanded the study of curves beyond those that could be constructed geometrically or mechanically.

Fermat found tangents and extreme points of graphs by using essentially the methods of present-day calculus. Calculus developed rapidly in the latter half of the 1600s, and its great power was demonstrated by Isaac Newton and Gottfried Leibniz. In particular, Newton used implicit differentiation to find tangents to curves, as we do after Theorem 4.10.

Apart from its role in calculus, analytic geometry developed gradually. Analytic geometers concentrated at first on giving analytic proofs of known results about lines and conics. Newton established analytic geometry as an important subject in its own right when he classified cubics, a task beyond the power of synthetic—that is, nonanalytic—geometry. We derive one of Newton's classifications of cubics in Chapter III.

While Fermat and Descartes were founding analytic geometry in the first half of the 1600s, Girard Desargues was developing a new branch of synthetic geometry called projective geometry. Renaissance artists and mathematicians had raised questions about drawing in perspective. These questions led Desargues to consider points at infinity and projections between planes, concepts we discuss at the start of Section 2. He used projections between planes to derive a remarkable number of theorems about lines and conics. His contemporary, Blaise Pascal, took up the projective study of conics, and their work was continued in the late 1600s by Philippe de la Hire.

Projective geometry languished in the 1700s as calculus and its applications dominated mathematics. Work on algebraic curves focused on their intersections, although multiple intersections were not analyzed systematically until the nineteenth century, as we discuss at the start of Chapter IV. We introduce intersection multiplicities in Section 1 so that we can automatically handle the special cases of theorems that arise from multiple intersections.

At the start of the 1800s, Gaspard Monge inspired a revival of synthetic geometry that focused particularly on the projective properties of conics. Mathematicians argued vigorously about the relative merits of synthetic and analytic geometry, although each subject actually drew strength from the other.

Analytic geometry was revolutionized when homogeneous coordinates were used to coordinatize the projective plane. Augustus Möbius introduced one system of homogeneous coordinates, barycentric coordinates, in 1827. He associated each point P in the projective plane with the triples of signed weights to be placed at the vertices of a fixed triangle so that P is the center of gravity. In 1830, Julius Plücker introduced the system of homogeneous coordinates that is currently used, which we introduce in Section 2.

Throughout the 1830s, Plücker used homogeneous coordinates to study curves. He obtained remarkable results, which we discuss at the start of Chapter II. Together with Riemann's work, which we discuss at the start of Chapter III, Plücker's results provided much of the inspiration for the subsequent development of algebraic geometry.

Möbius and Plücker also considered maps of the projective plane produced by invertible linear transformations of homogeneous coordinates. These are the transformations we discuss in Section 3. Much of nineteenth-century algebraic geometry was devoted to studying invariants, the algebraic combinations of coordinates of n-dimensional space that are preserved by invertible linear transformations. Founded by George Boole in 1841, invariant theory was developed in the latter half of the 1800s by such notable mathematicians as Arthur Cayley, James Sylvester, George Salmon, and Paul Gordan. Methods of abstract algebra came to dominate invariant theory when they were introduced by David Hilbert in the late 1800s and by Emmy Noether in the early 1900s.

§1. Intersections at the Origin

An important way to study a curve is to analyze its intersections with other curves. This analysis leads to the idea of two curves intersecting more than once at a point. We devote this section to studying multiple intersections at the origin, where the algebra is simplest.

A *polynomial* f or $f(x,y)$ in two variables is a finite sum of terms of the form ex^iy^j, where the coefficient e is a real number and the exponents i and j are nonnegative integers. We say that a term ex^ih^j has *degree* $i+j$ and that the *degree* of a nonzero polynomial is the maximum of the degrees of the terms with nonzero coefficients. For example, the six terms of the polynomial

$$y^3 - 2x^3y + 7xy - 3x^2 + 7x + 5$$

have respective degrees 3, 4, 2, 2, 1, and 0, and the degree of the polynomial is 4. *We work over the real numbers exclusively until we introduce the complex numbers in Section* 10.

We define an *algebraic curve* formally to be a polynomial $f(x,y)$ in two variables, and we picture the algbraic curve as the graph of the equation $f(x,y)=0$ in the plane. We abbreviate the term "algebraic curve" to "curve" because the only curves we consider are algebraic; that is, they are given by a polynomial equation in two variables. We refer both to the "curve $f(x,y)$" and to the "curve $f(x,y)=0$," and we even rewrite the equation $f(x,y)=0$ in algebraically equivalent forms. For example, we refer to the same curve as $y-x^2$, $y-x^2=0$, and $y=x^2$. Of course, we say that the curve $f(x,y)$ *contains* a point (a,b) and that the point *lies on*

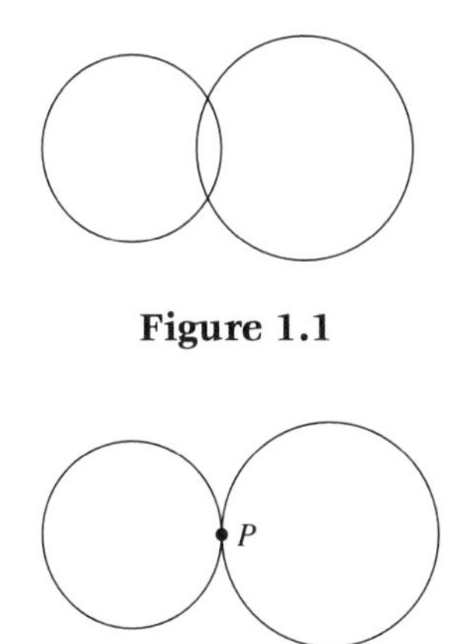

Figure 1.1

Figure 1.2

the curve when $f(a,b) = 0$. When the polynomial $f(x,y)$ is nonzero, we refer to its degree as the *degree* of the curve $f(x,y) = 0$.

One reason we define a curve formally to be a polynomial rather than its graph is to keep track of repeated factors. We imagine that the points of the graph that belong to repeated factors are themselves repeated. For example, we think of the curve

$$(y - x^2)^2(y - x)^3$$

as two copies of the parabola $y = x^2$ and three copies of the line $y = x$. This idea helps the geometry reflect the algebra.

We turn now to the idea that curves can intersect more than once at a point. As we noted in the chapter introduction, it is natural to think that the curve in Figure I.1 intersects the x-axis twice at the origin because the curve seems to pass through the origin twice.

For a different type of example, note that two circles with overlapping interiors intersect at two points (Figure 1.1). As the circles move apart, their two points of intersection draw closer together until they coalesce into a single point P (Figure 1.2). Accordingly, it seems natural to think that the circles in Figure 1.2 intersect twice at P.

Similarly, any line of positive slope through the origin intersects the graph of $y = x^3$ in three points (Figure 1.3). As the line rotates about the origin toward the x-axis, the three points of intersection move together at the origin, and they all coincide at the origin when the line reaches the x-axis. Accordingly, it is natural to think that the curve $y = x^3$ intersects the x-axis three times at the origin.

Let O be the origin $(0,0)$. We assign a value $I_O(f,g)$ to every pair of polynomials f and g. We call this value the *intersection multiplicity* of f and g at O, and we think of it as the number of times that the curves $f(x,y) = 0$ and $g(x,y) = 0$ intersect at the origin.

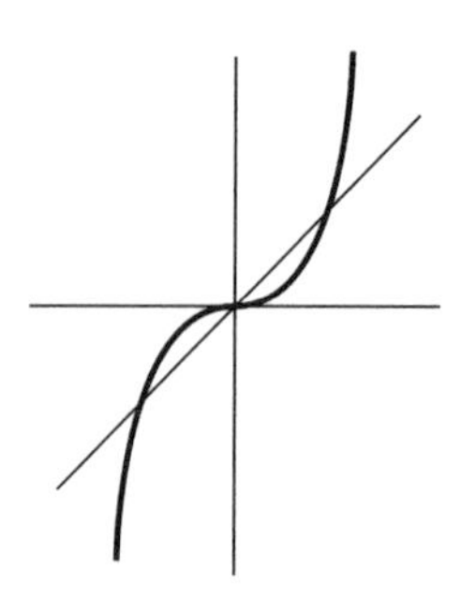

Figure 1.3

What properties should the assignment of the values $I_O(f,g)$ have? The proof of Theorem 1.7 will show that we need to allow for the possibility that curves intersect infinitely many times at the origin. We expect the following result, where the symbol ∞ denotes infinity:

Property 1.1
$I_O(f,g)$ is a nonnegative integer or ∞. $\square$

The order in which we consider two curves should not affect the number of times they intersect at the origin. This suggests the next property:

Property 1.2

$$I_O(f,g) = I_O(g,f). \qquad \square$$

If either of two curves fails to contain the origin, they do not intersect there, and their intersection multiplicity at the origin should be zero. On the other hand, if both curves contain the origin, they do intersect there, and their intersection multiplicity should be at least 1. Thus, we expect the following property to hold:

Property 1.3
$I_O(f,g) \geq 1$ if and only if f and g both contain the origin. $\square$

Of course, we consider ∞ to be greater than every integer, so that Property 1.3 allows for the possibility that $I_O(f,g) = \infty$ when f and g both contain the origin.

The y- and x-axes seem to intersect as simply as possible at the origin, and so we expect them to intersect only once there. Since the axes have equations $x = 0$ and $y = 0$, we anticipate the following property:

Property 1.4

$$I_O(x,y) = 1. \qquad \square$$

Let f, g, and h be three polynomials in two variables, and let (a,b) be a point. The equations

$$f(a,b) = 0 \qquad \text{and} \qquad g(a,b) = 0 \tag{1}$$

imply the equations

$$f(a,b) = 0 \qquad \text{and} \qquad g(a,b) + f(a,b)h(a,b) = 0 \tag{2}$$

Conversely, the equations in (2) imply the equations in (1). In short, f and g intersect at (a,b) if and only if f and $g + fh$ intersect there. Generalizing this to multiple intersections at the origin suggests the following:

Property 1.5

$$I_O(f,g) = I_O(f,g+fh). \qquad \square$$

One reason to expect that Property 1.5 holds for multiple as well as single intersections is the discussion accompanying Figures 1.1–1.3, which suggests that we can think of a multiple intersection of two curves as the coalescence of single intersections.

The equations $f(a,b) = 0$ and $g(a,b)h(a,b) = 0$ hold if and only if either $f(a,b) = 0 = g(a,b)$ or $f(a,b) = 0 = h(a,b)$. Thus, f and gh intersect at a point if and only if either f or g intersect there or f and h intersect there. That is, we get the points where f and gh intersect by combining the intersections of f and g with the intersections of f and h. As above, we expect this property to extend to multiple intersections because we think of a multiple intersection as the coalescence of single intersections. Thus, we expect the following:

Property 1.6

$$I_O(f,gh) = I_O(f,g) + I_O(f,h). \qquad \square$$

The value of $I_O(f,g)$ does not depend on the order of f and g (by Property 1.2). Thus, Property 1.5 states that the intersection multiplicity of two curves at the origin remains unchanged when we add a multiple of either curve to the other. Likewise, Property 1.6 shows that we can break up a product of two functions in either position of $I_O(\_,\_)$.

Property 1.6 reinforces the idea that repeated factors in a polynomial correspond to repeated parts of the graph. For example, Properties 1.2, 1.4, and 1.6 show that

$$I_O(x^2,y) = 2I_O(x,y) = 2.$$

When we think of $x^2 = 0$ as two copies of the line $x = 0$, it makes sense that $x^2 = 0$ intersects the line $y = 0$ twice at the origin, because each of the two copies of $x = 0$ intersects $y = 0$ once.

We use the term *intersection properties* to refer to Properties 1.1–1.6 and further properties introduced in Sections 3, 11, and 12. We must prove that we can assign values $I_O(f,g)$ for all pairs of curves f and g so that Properties 1.1–1.6 hold. We postpone this proof until Chapter IV so that we can proceed with our main task, using intersection properties to study curves. Of course, the results we obtain depend on our proving the intersection properties in Chapter IV.

In the rest of this section, we show how Properties 1.1–1.6 can be used to compute the intersection multiplicity of two curves at the origin. The discussion accompanying Figures 1.1–1.3 suggests that $I_O(f,g)$ measures how closely the curves f and g approach each other at the origin. When f is a factor of g, the graph of $g = 0$ contains the graph of $f = 0$. Thus, we are led to expect the following result:

Theorem 1.7

If f and g are polynomials such that f is a factor of g and the curve $f = 0$ contains the origin O, then $I_O(f,g)$ is ∞.

Proof

Consider first the case where g is the zero polynomial 0. (The theorem includes this case because the zero polynomial has every polynomial f as a factor, since $0 = f \cdot 0$.) Since $I_O(f,0) \geq 1$ (by Property 1.3), it follows for every positive integer n that

$$\begin{aligned} n \leq nI_O(f,0) &= I_O(f,0^n) \quad \text{(by Property 1.6)} \\ &= I_O(f,0). \end{aligned}$$

Because this holds for every positive integer $n, I_O(f,0)$ must be ∞.

In general, if g is any polynomial that has f as a factor, we can write $g = fh$ for a polynomial h. Then we have

$$\begin{aligned} I_O(f,g) &= I_O(f,fh) \\ &= I_O(f,fh - fh) \text{ (by Property 1.5)} \\ &= I_O(f,0) = \infty, \end{aligned}$$

by the previous paragraph. □

The proof of Theorem 1.7 shows why we needed to allow infinite intersection multiplicities in Property 1.1.

The following result shows that we can disregard factors that do not contain the origin when we compute intersection multiplicities at the origin:

Theorem 1.8

If f,g, and h are curves and g does not contain the origin, we have

$$I_O(f,gh) = I_O(f,h).$$

Proof

Properties 1.6, 1.3, and 1.1 show that

$$I_O(f,gh) = I_O(f,g) + I_O(f,h) = I_O(f,h),$$

since $I_O(f,g) = 0$ because g does not contain the origin. □

To illustrate the use of the intersection properties, we find the number of times that $y - x^2$ and $y^3 + 2xy + x^6$ intersect at the origin. We use Property 1.5 to eliminate y from the second polynomial by subtracting a suitable multiple of the first. To find this multiple, we use long division with respect to y to divide the first polynomial into the second, as follows:

$$\begin{array}{r@{}lllll}
 & y^2 + & x^2y & + 2x & + & x^4 \\
y - x^2 \,) & y^3 \quad + & 2xy & & + & x^6 \\
 & y^3 - x^2y^2 & & & & \\
\hline
 & x^2y^2 + & 2xy & & & \\
 & x^2y^2 - & x^4y & & & \\
\hline
 & & (2x + x^4)y & & + & x^6 \\
 & & (2x + x^4)y & - 2x^3 & - & x^6 \\
\hline
 & & & 2x^3 & + & 2x^6.
\end{array}$$

Each step of the division eliminates the highest remaining power of y until only a polynomial in x is left: the three steps of the division eliminate the y^3, y^2, and y terms. The division shows that

$$y^3 + 2xy + x^6 = (y - x^2)(y^2 + x^2y + 2x + x^4) + 2x^3 + 2x^6. \tag{3}$$

Thus, we are left with the remainder $2x^3 + 2x^6$, which does not contain y, when we subtract a multiple of $y - x^2$ from $y^3 + 2xy + x^6$. It follows that

$$I_O(y - x^2, y^3 + 2xy + x^6) = I_O(y - x^2, 2x^3 + 2x^6)$$

(by (3) and Property 1.5)

$$= I_O(y - x^2, x^3(2 + 2x^3))$$

$$= I_O(y - x^2, x^3) \quad \text{(by Theorem 1.8)}$$

$$= 3I_O(y - x^2, x) \quad \text{(by Property 1.6)}$$

$$= 3I_O(y, x)$$

(by Properties 1.2 and 1.5, since $y - x^2$ differs from y by a multiple of x)

$$= 3 \quad \text{(by Properties 1.2 and 1.4).}$$

Thus, $y = x^2$ intersects $y^3 + 2xy + x^6 = 0$ three times at the origin.

Of course, a polynomial $p(x)$ in one variable x is a finite sum of terms of the form ex^i, where e is a real number and i is a nonnegative integer. By generalizing the previous paragraph, we can find the number of times that a curve of the form $y = p(x)$ intersects any curve $g(x, y) = 0$ at the origin. This is easy to do because we do not need long division to find the remainder when $g(x, y)$ is divided by $y - p(x)$ with respect to y. The next theorem shows that the remainder is $g(x, p(x))$, the result of substituting $p(x)$ for y in $g(x, y)$. For example, we did not have to use long division above to find the remainder when $y^3 + 2xy + x^6$ is divided by $y - x^2$. All we needed to do was substitute x^2 for y in $y^3 + 2xy + x^6$ to find that the remainder is $(x^2)^3 + 2x(x^2) + x^6 = 2x^3 + 2x^6$, as before.

Theorem 1.9
Let $p(x)$ and $g(x, y)$ be polynomials.

(i) *If we use long division with respect to y to divide $g(x, y)$ by $y - p(x)$, the remainder is $g(x, p(x))$. This means that there is a polynomial $h(x, y)$ such that*

$$g(x, y) = (y - p(x))h(x, y) + g(x, p(x)). \tag{4}$$

(ii) *In particular, $y - p(x)$ is a factor of $g(x, y)$ if and only if $g(x, p(x))$ is the zero polynomial.*

Proof
(i) Let $h(x, y)$ be the quotient when we use long division with respect to y to divide $y - p(x)$ into $g(x, y)$. The remainder is a polynomial $r(x)$ in x because each step of the division eliminates the highest remaining power of y. We have

$$g(x, y) = (y - p(x))h(x, y) + r(x), \tag{5}$$

where $h(x, y)$ is the quotient. Substituting $p(x)$ for y in (5) makes $y - p(x)$ zero and shows that

$$g(x, p(x)) = r(x).$$

Together with (5), this gives (4).

(ii) If $g(x, p(x))$ is the zero polynomial, (4) shows that $y - p(x)$ is a factor of $g(x, y)$. Conversely, if $y - p(x)$ is a factor of $g(x, y)$, we can write

$$g(x, y) = (y - p(x))k(x, y)$$

for a polynomial $k(x, y)$. Substituting $p(x)$ for y shows that $g(x, p(x))$ is zero. □

We obtain a familiar result from Theorem 1.9 if we assume that x does not appear in p or g. Then p is a real number b, and g is a polynomial $g(y)$ in y. When we divide $g(y)$ by $y - b$, the quotient is a polynomial $h(y)$

in y, and the remainder is a real number r. This gives the following special case of Theorem 1.9, which we note for later reference:

Theorem 1.10
Let $g(y)$ be a polynomial in y, and let b be a real number.

(i) *The remainder when we divide $g(y)$ by $y-b$ is $g(b)$. This means that there is a polynomial $h(y)$ such that*

$$g(y) = (y-b)h(y) + g(b).$$

(ii) *In particular, $y-b$ is a factor of $g(y)$ if and only if $g(b) = 0$.* □

We can now find the intersection multiplicity at the origin of curves of the form $y = p(x)$ and $g(x,y) = 0$. By Theorem 1.9, we can eliminate all powers of y from $g(x,y)$ by subtracting a suitable multiple of $y - p(x)$, and we are left with $g(x, p(x))$. We can then use the intersection properties to find the intersection multiplicity. This gives the following result:

Theorem 1.11
Let $y = p(x)$ and $g(x,y) = 0$ be curves. Assume that $y = p(x)$ contains the origin and that $y - p(x)$ is not a factor of $g(x,y)$. Then the number of times that $y = p(x)$ and $g(x,y) = 0$ intersect at the origin is the smallest degree of any nonzero term of $g(x,p(x))$.

Proof
Since $y - p(x)$ is not a factor of $g(x,y)$, $g(x,p(x))$ is nonzero (by Theorem 1.9 (ii)). If s is the smallest degree of any nonzero term of $g(x,p(x))$, we can factor x^s out of every term of $g(x,p(x))$ and write

$$g(x,p(x)) = x^s q(x)$$

for a polynomial $q(x)$ whose constant term is nonzero.

Theorem 1.9 (i) shows that

$$g(x,y) = (y-p(x))h(x,y) + x^s q(x) \tag{6}$$

for a polynomial $h(x,y)$. Subtracting the product of $y-p(x)$ and $h(x,y)$ from $g(x,y)$ gives

$$I_O(y-p(x), g(x,y)) = I_O(y-p(x), x^s q(x))$$

(by (6) and Property 1.5)

$$= I_O(y-p(x), x^s)$$

(by Theorem 1.8, since the fact that $q(x)$ has nonzero constant term implies that the plane curve $q(x) = 0$ does not contain the origin)

$$= sI_O(y-p(x), x) \tag{7}$$

(by Property 1.6).

The assumption that $y = p(x)$ contains the origin means that $p(0) = 0$. Thus, the polynomial $p(x)$ has no constant term, and so we can factor x out of $p(x)$ and write

$$p(x) = xt(x) \tag{8}$$

for a polynomial $t(x)$. Adding x times $t(x)$ to $y - p(x)$ shows that

$$I_O(y - p(x), x) = I_O(y, x)$$

(by (8) and Properties 1.2 and 1.5)

$$= 1$$

(by Properties 1.2 and 1.4). Together with (7), this shows that $y = p(x)$ and $g(x,y) = 0$ intersect s times at the origin. □

After the proof of Theorem 1.8, it took some effort to find the number of times that $y - x^2$ and $y^3 + 2xy + x^6$ intersect at the origin. Theorem 1.11 makes it easy to do so.

EXAMPLE 1.12
How many times do the curves $y = x^2$ and $y^3 + 2xy + x^6 = 0$ intersect at the origin?

Solution
Substituting x^2 for y in $y^3 + 2xy + x^6$ gives

$$(x^2)^3 + 2x(x^2) + x^6 = 2x^3 + 2x^6.$$

Since this is nonzero, $y - x^2$ is not a factor of $y^3 + 2xy + x^6$ (by Theorem 1.9(ii)). Moreover, $y = x^2$ contains the origin, and so we can apply Theorem 1.11. The smallest power of x appearing in $2x^3 + 2x^6$ is x^3, and so the intersection multiplicity is 3, by Theorem 1.11. □

Theorem 1.11 makes it easy to determine the number of times that two curves intersect at the origin when the equation of one curve expresses y as a polynomial in x. This result enables us to determine the intersection multiplicities of lines and conics with other curves in Sections 4 and 5. Note that we can check the condition in Theorem 1.11 that $y - p(x)$ is not a factor of $g(x,y)$ by checking that $g(x, p(x))$ is nonzero (by Theorem 1.9(ii)).

Let $p(x)$ be a nonzero polynomial without a constant term. Since $p(0) = 0$, the curve $y = p(x)$ contains the origin. Since $p(x)$ is nonzero, $y - p(x)$ is not a factor of y. Thus, if we take $g(x,y)$ in Theorem 1.11 to be the polynomial y, we see that the intersection multiplicity of $y = p(x)$ and the x-axis $y = 0$ at the origin is the exponent of the smallest power of

x appearing in $p(x)$. For example, both of the curves

$$y = x^4 - 5x^3 + 7x^2 \qquad \text{and} \qquad y = 7x^2 \tag{9}$$

intersect the x-axis twice at the origin. It makes sense that these intersection multiplicities are equal because x^4 and x^3 approach zero faster than x^2 as x goes to zero, and so both curves in (9) approach the x-axis at the origin in essentially the same way.

The previous paragraph shows that, for any positive integer $n, y = x^n$ intersects the x-axis $y = 0$ n times at the origin. This reflects the fact that $y = x^n$ approaches the x-axis near the origin with increasing closeness as n grows. In particular, $y = x^3$ intersects the x-axis three times at the origin, which reflects the discussion accompanying Figure 1.3.

Theorem 1.11 determines the number of times that two curves intersect at the origin when the equation of one curve expresses y as a polynomial in x. On the other hand, we can find the number of times that any two curves intersect at the origin by applying Properties 1.1–1.6 and Theorems 1.7 and 1.8. The idea is to use Properties 1.5 and 1.6 to eliminate the highest power of y appearing in the equations of the curves. Repeating this until y has been eliminated from one of the equations gives the intersection multiplicity.

We illustrate this technique with an example. Note that the value of an intersection multiplicity remains unchanged if we add a multiple of one of the curves to the other (by Properties 1.2 and 1.5), but the intersection multiplicity can change if we multiply one of the two curves by a third (by Properties 1.2 and 1.6).

EXAMPLE 1.13
How many times do the curves $y^3 + 2x^5 = 0$ and $xy^2 + y - 3x^3 = 0$ intersect at the origin?

Solution
Although we can solve the first equation for y over the real numbers as $y = -2^{1/3}x^{5/3}$, this does not express y as a polynomial in x, and so we cannot apply Theorem 1.11. Instead, we repeatedly eliminate the highest power of y in the equations of the curves.

The highest power of y in the two given equations is y^3. We can eliminate the y^3 term by multiplying the first equation by x and subtracting y times the second equation. We use Properties 1.2 and 1.6 to evaluate the effect of multiplying the first equation by x:

$$I_O(y^3 + 2x^5, xy^2 + y - 3x^3)$$
$$= I_O(xy^3 + 2x^6, xy^2 + y - 3x^3) - I_O(x, xy^2 + y - 3x^3)$$

(by Properties 1.2 and 1.6)

$$= I_O(xy^3 + 2x^6, xy^2 + y - 3x^3) - I_O(x, y)$$

(multiplying x by $y^2 - 3x^2$ to get $xy^2 - 3x^3$, and subtracting this from $xy^2 + y - 3x^3$, by Property 1.5)

$$= I_O(xy^3 + 2x^6, xy^2 + y - 3x^3) - 1$$

(by Property 1.4). We can eliminate the y^3 term by subtracting y times the second polynomial from the first. By Properties 1.2 and 1.5, this gives

$$I_O(xy^3 + 2x^6 - y(xy^2 + y - 3x^3), xy^2 + y - 3x^3) - 1$$
$$= I_O(-y^2 + 3x^3y + 2x^6, xy^2 + y - 3x^3) - 1.$$

The next step is to eliminate one of the two y^2 terms. The easiest way to do this is to add x times the first polynomial to the second. This gives

$$I_O(-y^2 + 3x^3y + 2x^6, xy^2 + y - 3x^3 + x(-y^2 + 3x^3y + 2x^6)) - 1$$

(by Property 1.5)

$$= I_O(-y^2 + 3x^3y + 2x^6, (3x^4 + 1)y + 2x^7 - 3x^3) - 1.$$

We eliminate the remaining y^2 term by multiplying the first polynomial by $3x^4 + 1$ and adding y times the second polynomial. The curve $3x^4 + 1 = 0$ in the plane does not contain the origin (and is, in fact, empty). Thus, the value of the intersection multiplicity is unchanged if we multiply the first polynomial by $3x^4 + 1$ (by Property 1.2 and Theorem 1.8) and obtain

$$I_O(-(3x^4 + 1)y^2 + 3x^3(3x^4 + 1)y + 2x^6(3x^4 + 1),$$
$$(3x^4 + 1)y + 2x^7 - 3x^3) - 1$$
$$= I_O(-(3x^4 + 1)y^2 + (9x^7 + 3x^3)y + 6x^{10} + 2x^6,$$
$$(3x^4 + 1)y + 2x^7 - 3x^3) - 1.$$

Adding y times the second polynomial to the first eliminates the y^2 term, as desired, giving

$$I_O(11x^7y + 6x^{10} + 2x^6, (3x^4 + 1)y + 2x^7 - 3x^3) - 1 \tag{10}$$

(by Properties 1.2 and 1.5).

Factoring x^6 out of the first polynomial gives

$$I_O(x^6(11xy + 6x^4 + 2), (3x^4 + 1)y + 2x^7 - 3x^3) - 1$$
$$= I_O(x^6, (3x^4 + 1)y + 2x^7 - 3x^3) - 1$$

(by Property 1.2 and Theorem 1.8, since the curve $11xy + 6x^4 + 2 = 0$ does not contain he origin)

$$= 6I_O(x, (3x^4 + 1)y + 2x^7 - 3x^3) - 1$$

(by Properties 1.2 and 1.6). Using Property 1.5 to drop the terms $3x^4y + 2x^7 - 3x^3$, which are multiples of x, leaves $6I_O(x, y) - 1$, which equals 5 (by Property 1.4). The two given curves intersect five times at the origin. □

We can often simplify the work of computing intersection multiplicities by noticing that one of the polynomials factors and applying Property 1.6 or Theorem 1.8. For instance, by factoring x^6 out of the first polynomial in (10), we saved ourselves the work of using Property 1.5 to eliminate the y term. It is also worth noting that it is sometimes easier to work on eliminating powers of x rather than y.

The technique of eliminating a variable, which we illustrated in Example 1.13, lies at the heart of the study of algebraic curves. We use this technique to prove Bezout's Theorem 11.5, which determines how many times two curves intersect over the complex numbers.

We have not yet considered intersections of curves at points other than the origin. We postpone this until Section 3 so that we can use homogeneous coordinates to treat intersections at infinity at the same time as intersections in the Euclidean plane. We introduce homogeneous coordinates in the next section.

Exercises

1.1. How many times do the two given curves intersect at the origin?

(a) $y = x^3$ and $y^4 + 6x^3y + x^8 = 0$.
(b) $y = x^2 - 2x$ and $y^2 + 5y = 4x^3$.
(c) $y = x^2 + x$ and $y^2 = 3x^2y + x^2$.
(d) $x^3 + x + y = 0$ and $y^3 = 3x^2y + 2x^3$.
(e) $y^2 + x^2y - x^3 = 0$ and $y^2 + x^3y + 2x = 0$.
(f) $y^3 = x^2$ and $y^2 = x^3$.
(g) $y^4 = x^3$ and $x^2y^3 - y^2 + 2x^7 = 0$.
(h) $xy^2 + y - x^2 = 0$ and $y^3 = x^4$.
(i) $y^3 = x^2$ and $xy = y + x^2$.
(j) $y^3 = x^2$ and $xy^2 = 4y^2 + x^3$.
(k) $y^3 = x^2$ and $x^2y = 2y^2 + x^3$.
(l) $y^5 = x^7$ and $y^2 = x^3$.
(m) $y^2 = x^5$ and $y^3 - 4x^3y + x^4 = 0$.
(n) $y^3 = 2x^4$ and $x^2y^2 + y - x^2 = 0$.
(o) $xy^4 + y^3 = x^2$ and $y^5 + x^2 = xy$.

1.2. Consider the curve and the numbers s and t given in each part of this exercise. Show that there are s lines through the origin that intersect the curve more than t times there and that all other lines through the origin intersect the curve exactly t times there. Draw the curve and the s exceptional lines, showing that these are the lines through the origin that best approximate the curve there. In drawing the curve, it may be helpful to use polar coordinates or curve-sketching techniques from first-year calculus.

(a) $y = x^3 - 2x$, $s = 1$, $t = 1$.
(b) $y = x^3$, $s = 1$, $t = 1$.
(c) $y^2 = x^3$, $s = 1$, $t = 2$.
(d) $y^2 = x^4 + 4x^2$, $s = 2$, $t = 2$.
(e) $y^2 = x^4 - 4x^2$, $s = 0$, $t = 2$.
(f) $x^4 + x^2y^2 = y^2$, $s = 1$, $t = 2$.
(g) $x^2y^2 = x^2 - y^2$, $s = 2$, $t = 2$.
(h) $y^2 = x(x-1)^2$, $s = 1$, $t = 1$.
(i) $(x^2 + y^2)^2 = 2xy$, $s = 2$, $t = 2$.
(j) $(x^2 + y^2)^2 = xy^2$, $s = 2$, $t = 3$.
(k) $(x^2 + y^2)^2 = x^2(x + y)$, $s = 2$, $t = 3$.
(l) $(x^2 - y^2)^2 = xy$, $s = 2$, $t = 2$.
(m) $x^4 - y^4 = xy$, $s = 2$, $t = 2$.

1.3. Show that the graph of the equation $r = \sin(3\theta)$ in polar coordinates corresponds to a curve $f(x, y) = 0$ of degree 4. Follow the directions of Exercise 1.2 for this curve, with $s = 3$ and $t = 3$.

1.4. Let C and D be two different circles through the origin, and assume that the center of C lies on the x-axis. Prove that C and D intersect either twice or once at the origin, depending on whether or not the center of D lies on the x-axis. (This justifies the discussion accompanying Figures 1.1 and 1.2.)

1.5. Does Theorem 1.11 remain true without the assumption that $y = p(x)$ contains the origin? Justify your answer.

1.6. Let $f(x)$ and $g(x)$ be polynomials in one variable that have no common factors of positive degree. Prove that $f(x)y + g(x)$ does not factor as a product of two polynomials of positive degree.

1.7. Let $f(x, y)$ and $g(x, y)$ be polynomials in two variables, and let n be a nonnegative integer. Assume that every term in $f(x, y)$ has degree n and every term in $g(x, y)$ has degree $n + 1$. If $f(x, y)$ and $g(x, y)$ have no common factors of positive degree, prove that $f(x, y) + g(x, y)$ does not factor as a product of two polynomials of positive degree.

1.8. Let $f(x)$ be a polynomial in one variable. Prove that $y^2 + f(x)$ factors as a product of two polynomials of positive degree if and only if $f(x) = -g(x)^2$ for a polynomial $g(x)$.

1.9. Let $f(x)$ be a polynomial in one variable. Prove that $y^3 + f(x)$ factors as a product of two polynomials of positive degree if and only if $f(x) = g(x)^3$ for a polynomial $g(x)$.

1.10. Let $f(x, y)$ be a polynomial in two variables, and let $h(x)$ be a polynomial in one variable. Prove that f and h have intersection multiplicity ∞ at the origin if and only if x is a factor of both f and h.

(As in Example 1.13, one step in evaluating

$$I_O(f(x, y), g(x, y)) \tag{11}$$

for polynomials $f(x, y)$ and $g(x, y)$ is to replace it with

$$I_O(f(x, y), h(x)g(x, y)) - I_O(f(x, y), h(x)) \tag{12}$$

for a polynomial $h(x)$ in x alone. This replacement is justified by Property 1.6 unless $I_O(f(x,y),h(x)) = \infty$, which means that the quantity in (12) has indeterminate form $\infty - \infty$. In that case, this exercise shows that x is a factor of $f(x,y)$, and so we can use Properties 1.2 and 1.6 to evaluate (11). In this way, the techniques of Example 1.13 always apply.)

§2. Homogeneous Coordinates

The study of curves is greatly simplified by considering their behavior at infinity. This eliminates a number of special cases: for example, it enables us to study all conic sections—ellipses, parabolas, and hyperbolas—simultaneously in Chapter II.

We construct the "projective plane" in this section by adding "points at infinity" to the familiar Euclidean plane. We define a system of homogeneous coordinates for the projective plane, which lets us study curves at infinity in the same way as in the Euclidean plane. We focus on lines in the projective plane in this section, and we introduce curves of higher degree in Section 3.

We start with the familiar coordinate system on three-dimensional Euclidean space. Specifically, we choose a point O in Euclidean space to represent the origin (Figure 2.1). We select three mutually perpendicular lines through O to be the x-, y-, and z-axes. We associate the points on each axis with the real numbers in the usual way, so that O is the point 0 on each axis. We assign coordinates (a,b,c) to a point P in Euclidean space if the planes through P perpendicular to the x-, y-, and z-axes intersect them at the points a, b, and c, respectively. Of course, this gives the origin O coordinates $(0,0,0)$.

Projections suggest a way to study curves at infinity. Let $\mathscr{P}$ and $\mathscr{Q}$ be two planes in Euclidean space that do not contain the origin O. The *projection* from $\mathscr{P}$ to $\mathscr{Q}$ through O maps a point X on $\mathscr{P}$ to the point X' on $\mathscr{Q}$

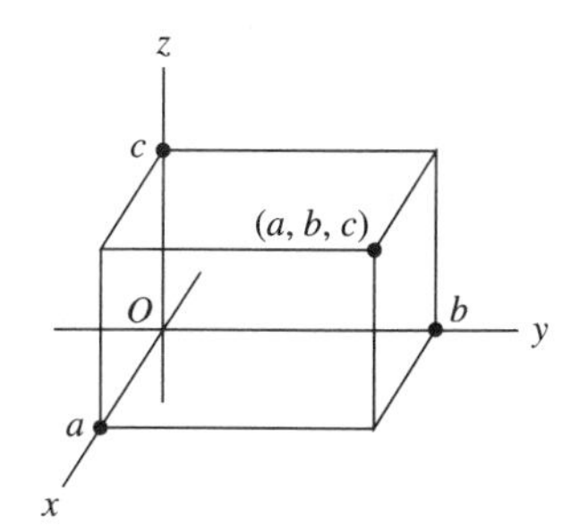

Figure 2.1

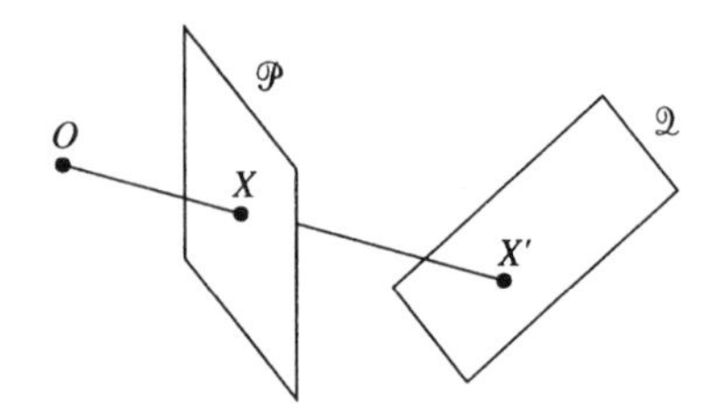

Figure 2.2

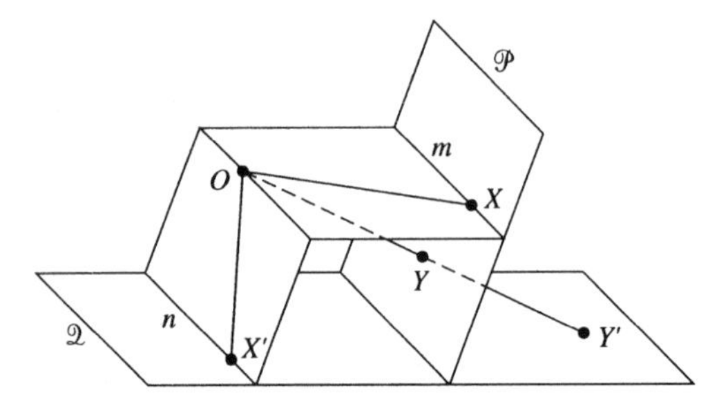

Figure 2.3

where the line through O and X intersects $\mathscr{Q}$ (Figure 2.2). Conversely, a point X' on $\mathscr{Q}$ is the image of the point X on $\mathscr{P}$ where the line through O and X' intersects $\mathscr{P}$. In this way, the projection matches up points X and X' on $\mathscr{P}$ and $\mathscr{Q}$ that lie on lines through O.

There are exceptions, however. When $\mathscr{P}$ and $\mathscr{Q}$ are not parallel, the plane through O parallel to $\mathscr{Q}$ intersects $\mathscr{P}$ in a line m (Figure 2.3). If X is any point of m, the line through O and X is parallel to $\mathscr{Q}$, and so X has no image on $\mathscr{Q}$. We call m the *vanishing line* on $\mathscr{P}$ because the points of m seem to vanish under the projection. In fact, as a point Y on $\mathscr{P}$ approaches m, its image Y' under the projection moves arbitrarily far away from the origin on $\mathscr{Q}$. This suggests that points on the vanishing line of $\mathscr{P}$ project to points at infinity on $\mathscr{Q}$.

Likewise, the plane through O parallel to $\mathscr{P}$ intersects $\mathscr{Q}$ in a line n, which we call the vanishing line on $\mathscr{Q}$. If X' is any point of n, the line through O and X' is parallel to $\mathscr{P}$, and we imagine that a point at infinity on $\mathscr{P}$ projects to X'.

In short, a projection between two planes that are not parallel matches up the points on the planes, except that points on the vanishing line of each plane seem to correspond to points at infinity on the other plane. This suggests that each plane has a line of points at infinity and that we can study these points by projecting them to ordinary points on another plane.

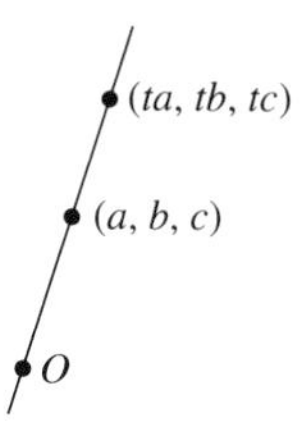

Figure 2.4

Accordingly, in order to study curves at infinity, we consider all points in Euclidean space except the origin. If X and X' are two of these points that lie on a line through the origin O, we think of X and X' as two representations of the same point under projection through O, as in Figure 2.2. That is, we think of all the points except O on each line in space through O as the same point.

Translating this into coordinates, we consider the triples (a,b,c) of real numbers except $O = (0,0,0)$. We think of all the triples (ta,tb,tc) as the same point as t varies over all nonzero real numbers; these are the triples except O on the line through O and (a,b,c) (Figure 2.4).

We make the following formal definition. The *projective plane* is the set of points determined by ordered triples of real numbers (a,b,c), where a, b, c are not all zero, and where the triples (ta,tb,tc) represent the same point as t varies over all nonzero real numbers (Figure 2.4). We call the ordered triples *homogeneous coordinates*. The term "homogeneous" indicates that all the triples (ta,tb,tc) represent the same point as t varies over all nonzero real numbers. For example, if we multiply the coordinates of $(1,-2,3)$ by 2, -3, and $\frac{1}{3}$, we see that the triples

$$(1,-2,3), \quad (2,-4,6), \quad (-3,6,-9), \quad (\tfrac{1}{3},-\tfrac{2}{3},1),$$

represent the same point.

It may seem odd to talk about a plane coordinatized by triples of real numbers, but the homogeneity of the coordinates effectively reduces the dimension by 1 from 3 to 2. For instance, if we consider points (a,b,c) with $c \neq 0$, dividing the coordinates by c gives $(a/c,b/c,1)$. Rewriting these points as $(d,e,1)$ for real numbers d and e shows that we are considering a two-dimensional set of points, although triples with last coordinate zero require separate consideration.

Geometrically, we relate the projective and Euclidean planes as follows. Triples of homogeneous coordinates correspond to lines in space through the origin O, as in Figure 2.4. Each line in space through O that does not lie in the plane $z = 0$ will be represented by the point where it intersects the plane $z = 1$ (Figure 2.5). We will identify the lines through O that lie in the plane $z = 0$ with the points at infinity of the plane $z = 1$.

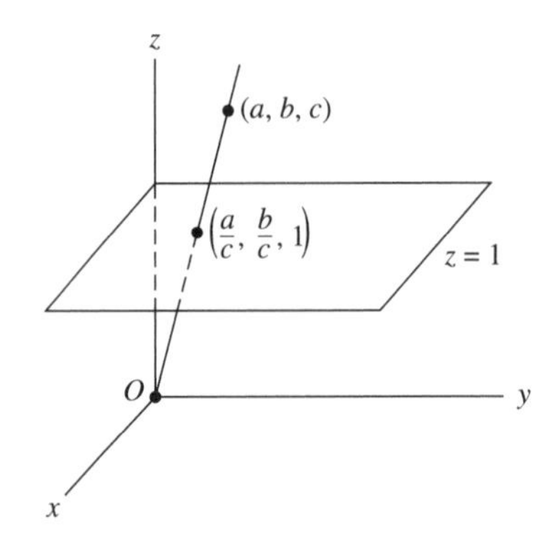

Figure 2.5

This will show that the projective plane consists of the Euclidean plane $z = 1$ together with additional points at infinity.

Algebraically, if $c \neq 0$, then $1/c$ is the one value of t such that the triple (ta, tb, tc) has last coordinate 1. Setting $t = 1/c$ gives the point $(a/c, b/c, 1)$ in the plane $z = 1$. Conversely, any point $(d, e, 1)$ in the plane $z = 1$ corresponds to a unique point in the projective plane, the point with homogeneous coordinates (td, te, t) for all nonzero numbers t. In this way, we have matched up the points in the projective plane whose last coordinate is nonzero with the points in the plane $z = 1$.

We think of the plane $z = 1$ as the Euclidean plane by identifying the points $(x, y, 1)$ and (x, y) of the two planes. Together with the last paragraph, this matches up the points in the projective plane whose last homogeneous coordinate is nonzero with the points of the Euclidean plane. A point in the projective plane that has homogeneous coordinates (a, b, c) for $c \neq 0$ is matched up with the point $(a/c, b/c)$ of the Euclidean plane. Conversely, a point (d, e) of the Euclidean plane is matched up with the point of the projective plane that has homogeneous coordinates $(d, e, 1)$ or, more generally, (td, te, t) for any nonzero number t.

We must still consider the points $(a, b, 0)$ in the projective plane whose last homogeneous coordinate is zero. We call these *points at infinity*. If $a \neq 0$, $1/a$ is the one value of t such that the triple $(ta, tb, 0)$ has first coordinate 1. Setting $t = 1/a$ gives the triple $(1, b/a, 0)$. We can choose $a \neq 0$ and b so that b/a is any real number s.

The only remaining point at infinity corresponds to the triples of homogeneous coordinates whose first and third coordinates are both zero. These triples are $(0, b, 0)$, where $b \neq 0$. Multiplying by $1/b$ gives the coordinates of the point in the unique form $(0, 1, 0)$,.

In short, every point in the projective plane can be written in exactly one way as one of the triples

$$(d, e, 1), \quad (1, s, 0), \quad (0, 1, 0), \tag{1}$$

as d, e, and s vary over all real numbers. The points in the projective plane whose last homogeneous coordinate is nonzero correspond to the triples $(d,e,1)$, which correspond in turn to the points (d,e) of the Euclidean plane. The points in the projective plane that have last homogeneous coordinate zero are the points at infinity, and they correspond to the triples $(1,s,0)$ and $(0,1,0)$.

We learn more about the points at infinity by relating them to the lines in the projective plane. A *line* in the projective plane is the set of points whose homogenous coordinates (x,y,z) satisfy an equation

$$px + qy + rz = 0, \tag{2}$$

where p, q, and r are real numbers that are not all zero. We call (2) the *equation* of the line.

It does not matter which triple of homogeneous coordinates of a point we substitute in (2). If a triple (x,y,z) satisfies (2), we can multiply the equation by a nonzero number t and obtain the equation

$$ptx + qty + rtz = 0, \tag{3}$$

which shows that the triple (tx,ty,tz) also satisfies (2).

We can also think of (3) as the result of multiplying the coefficients p,q,r of (2) by a nonzero number t. Thus, the equivalence of (2) and (3) shows that a line stays unchanged when we multiply the coefficients in its equation by a nonzero number.

To understand the lines in the projective plane, first consider the lines given by (2) with $q \neq 0$. Dividing this equation by q and solving for y gives the equivalent equation

$$y = \left(-\frac{p}{q}\right)x + \left(-\frac{r}{q}\right)z.$$

As p,q, and r vary over all real numbers with $q \neq 0$, we obtain the equations

$$y = mx + nz \tag{4}$$

for all real numbers m and n. The corresponding lines in the Euclidean plane consist of all points (x,y) such that the triple $(x,y,1)$ satisfies (4). This gives the lines

$$y = mx + n \tag{5}$$

in the Euclidean plane. As m and n vary over all real numbers, (5) gives all lines in the Euclidean plane that are not vertical. In short, the lines in the projective plane given by (2) for $q \neq 0$ correspond to the lines in the Euclidean plane that are not vertical.

Consider next the lines given by (2) with $q = 0$ and $p \neq 0$. Dividing the equation $px + rz = 0$ by p and solving for x gives the equation

$$x = \left(-\frac{r}{p}\right)z.$$

As r and p vary over all real numbers with $p \neq 0$, we obtain the equations

$$x = hz \tag{6}$$

for all real numbers h. The corresponding lines in the Euclidean plane consist of the points (x, y) such that $(x, y, 1)$ satisfies (6). This gives the lines

$$x = h \tag{7}$$

in the Euclidean plane. As h varies over all real numbers, (7) gives all vertical lines in the Euclidean plane. Thus, the lines in the projective plane given by (2) with $q = 0$ and $p \neq 0$ correspond to the vertical lines in the Euclidean plane.

The last two paragraphs show that the lines in the projective plane given by (2) when p or q is nonzero correspond to the lines of the Euclidean plane. The only other line in the projective plane is given by (2) with $p = 0 = q$ and $r \neq 0$ (since the coefficients p, q, r in (2) are not all zero). Then (2) becomes $rz = 0$, and dividing this equation by r gives $z = 0$. We call the line $z = 0$ in the projective plane the *line at infinity*. Of course, the points (a, b, c) of the projective plane that lie on the line $z = 0$ are exactly those whose last coordinate c is zero. Thus the line at infinity consists exactly of the points at infinity.

In short, *the lines of the projective plane are the lines of the Euclidean plane plus the line at infinity, which consists of the points at infinity.*

We can now relate the points at infinity with the lines of the Euclidean plane. As we saw in the discussion before (1), each point at infinity can be written in exactly one way as

$$(1, s, 0) \quad \text{or} \quad (0, 1, 0) \tag{8}$$

for a real number s. The lines $y = mx + n$ and $x = h$ correspond to the lines $y = mx + nz$ and $x = hz$ (by the discussions relating (4) to (5) and (6) to (7)). For any real number s, the point at infinity $(1, s, 0)$ lies on the line $y = mx + nz$ if and only if m equals s, and it does not lie on any of the lines $x = hz$. The point at infinity $(0, 1, 0)$ lies on all the lines $x = hz$ and on none of the lines $y = mx + nz$. In short, *each point at infinity lies on exactly those lines of the Euclidean plane that form a family of parallel lines: the point at infinity $(1, s, 0)$ lies on the lines $y = sx + n$ of slope s for all real numbers n, and the point at infinity $(0, 1, 0)$ lies on the vertical lines $x = h$ for all real numbers h. In this way, we match up the points at infinity with the families of parallel lines in the Euclidean plane.*

We now know that the projective plane consists of the points and lines of the Euclidean plane, additional points at infinity, and one added line at infinity. The line at infinity contains all the points at infinity and no points of the Euclidean plane. Each point at infinity lies on exactly those lines in the Euclidean plane that form a family of parallel lines, and there is exactly one point at infinity for each family of parallel lines.

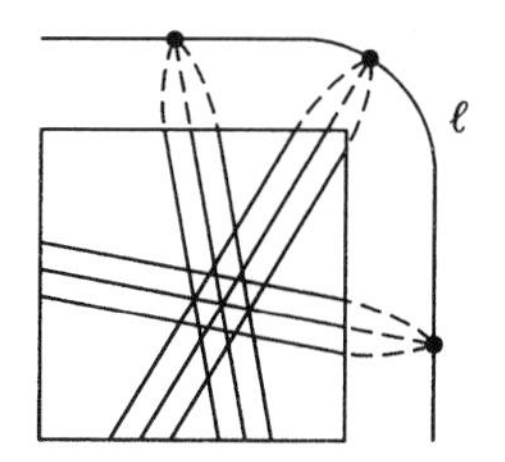

Figure 2.6

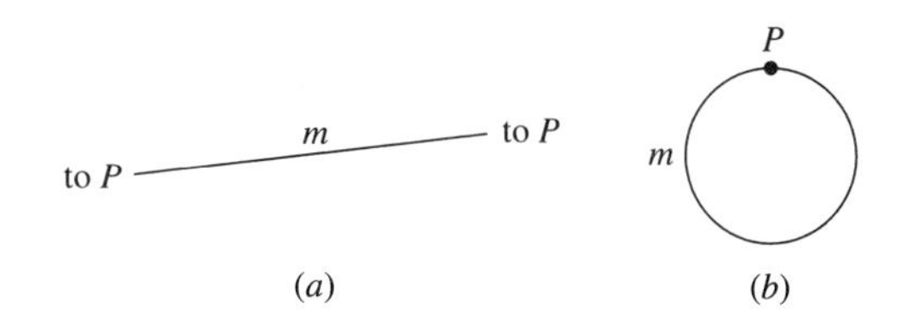

Figure 2.7

Figure 2.6 suggests the form of the projective plane. The square represents the Euclidean plane, and the line l represents the line at infinity. Dotted lines connect points at infinity with parallel lines in the Euclidean plane that contain them.

Let P be the point at infinity on a line m in the Euclidean plane. We imagine that we can reach P by proceeding infinitely far along m in either direction (Figure 2.7(a)). This suggests that the two "ends" of m in the Euclidean plane are joined at infinity by the point P so that m forms a closed curve (Figure 2.7(b)).

An important consequence of adding the points at infinity is that we no longer need to consider special cases created by parallel lines. In the Euclidean plane, two lines intersect in a point unless they are parallel. On the other hand, any two lines in the projective plane intersect in a point: parallel lines in the Euclidean plane intersect at infinity in the projective plane (Figure 2.6).

Theorem 2.1

Any two lines intersect at a unique point in the projective plane.

Proof

Two parallel lines in the Euclidean plane do not intersect in the Euclidean plane, and they contain the same point P at infinity; thus, P is their unique point of intersection (Figure 2.8). Two lines in the Euclidean plane that are not parallel intersect exactly once in the projective plane because they intersect exactly once in the Euclidean plane and contain

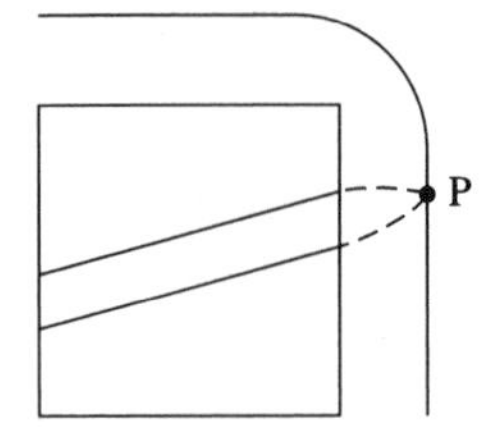

Figure 2.8

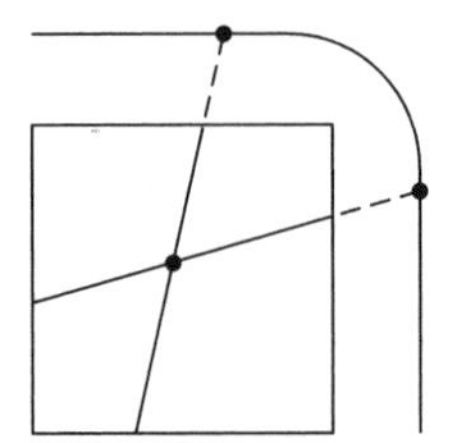

Figure 2.9

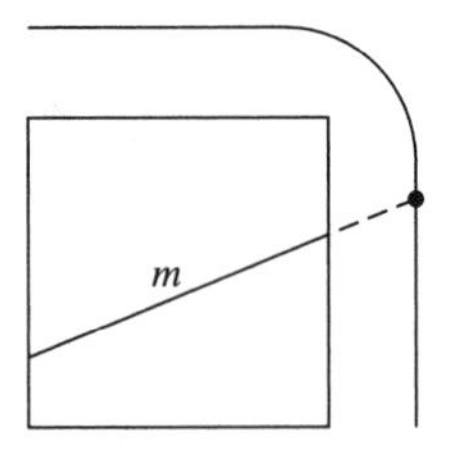

Figure 2.10

different points at infinity (Figure 2.9). A line m of the Euclidean plane intersects the line at infinity at the unique point at infinity that lies on m and all lines parallel to it (Figure 2.10). These three cases include all possibilities for two lines in the projective plane. □

In analogy with Theorem 2.1, we prove that any two points lie on a unique line in the projective plane. Unlike Theorem 2.1, this property already holds in the Euclidean plane, and so we need only show that it still holds when we add the points and the line at infinity.

Theorem 2.2
Any two points lie on a unique line in the projective plane.

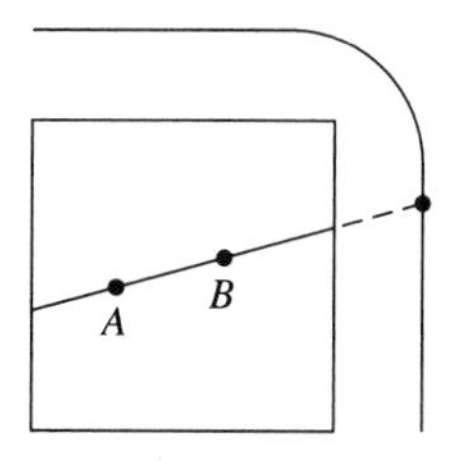

Figure 2.11

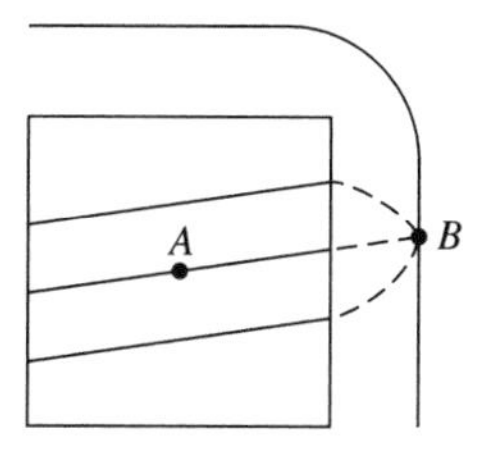

Figure 2.12

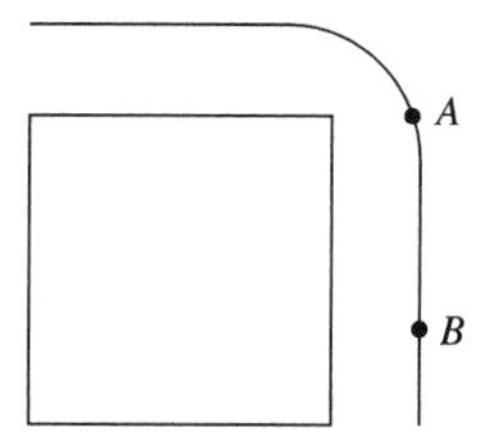

Figure 2.13

Proof

Two points A and B in the Euclidean plane lie on a unique line in the Euclidean plane; this is the unique line of the projective plane through A and B because the line at infinity contains only points at infinity (Figure 2.11). The unique line through a point A of the Euclidean plane and a point B at infinity is the line through A in the Euclidean plane that belongs to the family of parallel lines containing B (Figure 2.12). The unique line through two points A and B at infinity is the line at infinity (Figure 2.13), since each line of the Euclidean plane contains only one point at infinity. These three cases cover all possibilities for two points in the projective plane. □

By Theorem 2.1, any two lines l and m intersect at a unique point in the projective plane, which we write as $l \cap m$. By Theorem 2.2, any two

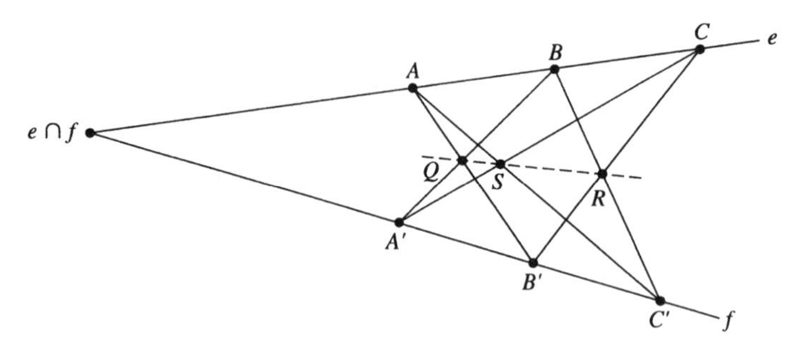

Figure 2.14

points A and B lie on a unique line in the projective plane, which we write as AB. We call points *collinear* if they all lie on one line, and we call lines *concurrent* if they all lie on one point. This notation makes it easy to state the following result, which we prove in Section 6 as Theorem 6.5:

Theorem 2.3 (Pappus' Theorem)
Let e and f be two lines in the projective plane. Let A, B, and C be three points of e other than $e \cap f$, and let A', B', and C' be three points of f other than $e \cap f$. Then the points $Q = AB' \cap A'B$, $R = BC' \cap B'C$, and $S = CA' \cap C'A$ are collinear (Figure 2.14). □

Note that Pappus' Theorem is a result about the collinearity of points. The projective plane is well suited to such results: by Theorem 2.1, any two lines in the projective plane intersect at a point, without the exceptions created in the Euclidean plane by parallel lines. On the other hand, because distances and angles are undefined at infinity, results about these concepts do not readily extend from the Euclidean to the projective plane.

Because the position of the line at infinity is unspecified in Pappus' Theorem, we can obtain a number of different results about the Euclidean plane from Pappus' Theorem by taking the line at infinity in various positions. The points at infinity vanish, and the lines of the Euclidean plane that intersect at a point at infinity are parallel.

For example, suppose we take the line BC' in Pappus' Theoerem to be the line at infinity. Because B is now at infinity, $A'B$ is the line g through A' parallel to e, and we have $Q = AB' \cap A'B = AB' \cap g$ (Figure 2.15). Because C' is now at infinity, $C'A$ is the line h through A parallel to f, and we have $S = CA' \cap C'A = CA' \cap h$. The conclusion of Pappus' Theorem is equivalent to the assertion that the lines BC', $B'C$, and QS lie on a common point R. Because BC' is now the line at infinity, the conclusion asserts that $B'C$ and QS meet at a point R at infinity, which means that the lines $B'C$ and QS are parallel. The lines e and f are not parallel because

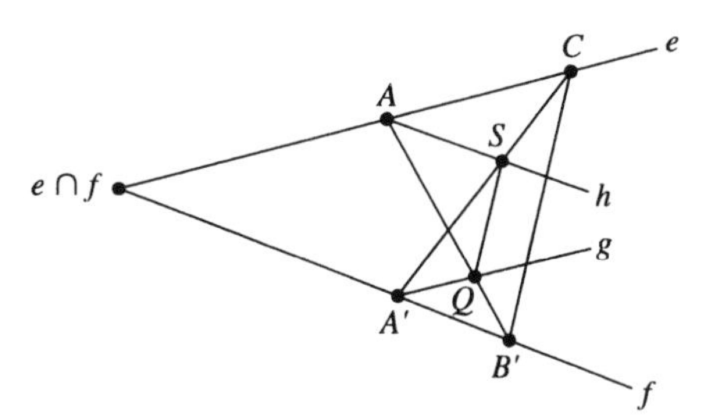

Figure 2.15

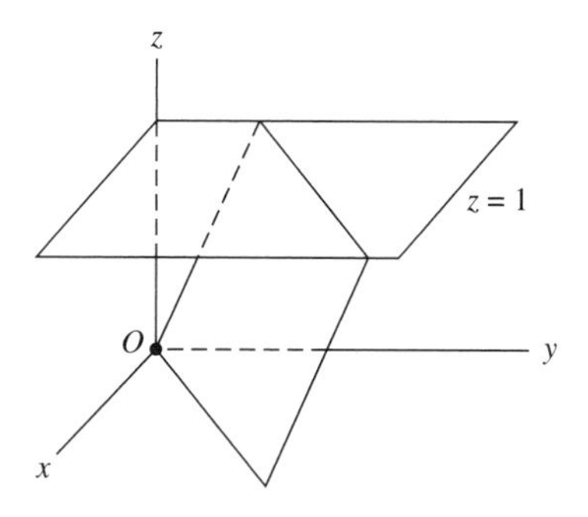

Figure 2.16

their intersection $e \cap f$ does not lie on the line at infinity BC'. Thus, we obtain the following result from Pappus' Theorem by taking BC' to be the line at infinity:

Theorem 2.4

In the Euclidean plane, let e and f be two lines that are not parallel. Let A and C be two points of e other than $e \cap f$, and let A' and B' be two points of f other than $e \cap f$. Let Q be the point where AB' intersects the line g through A' parallel to e, and let S be the point where CA' intersects the line h through A parallel of f. Then the lines QS and $B'C$ are parallel (Figure 2.15). □

We defined a line in the projective plane to be the set of points in the projective plane whose homogeneous coordinates (x, y, z) satisfy (2), where the coefficients p, q, r in (2) are real numbers that are not all zero. We justified this definition algebraically by showing that the lines it gives correspond to the lines of the Euclidean plane plus the line at infinity. We can also justify the definition geometrically, as follows.

If we take (x, y, z) to be the usual three-dimensional coordinates in Euclidean space, as in the discussion accompanying Figure 2.1, (2) is the general equation of a plane through the origin in Euclidean space. Thus, using homogeneous coordinates, we can identify the lines of the projective plane with the planes through the origin in Euclidean space. Just as we picture a line through the origin in Euclidean space as a point by intersecting it with the plane $z = 1$ (Figure 2.5), we picture a plane through the origin in Euclidean space as a line by intersecting it with the plane $z = 1$ (Figure 2.16). The plane $z = 0$, which does not intersect the plane $z = 1$, corresponds to the line at infinity.

Exercises

2.1. Homogeneous coordinates of a point in the projective plane are given in each part of this exercise. Determine whether the point lies in the Euclidean plane or at infinity. If the point lies in the Euclidean plane, determine its usual (x, y) coordinates. If the point lies at infinity, determine the slope of the lines in the Euclidean plane that contain the point.
(a) $(4,2,-3)$. (b) $(1,-2,4)$.
(c) $(0,5,2)$. (d) $(3,0,-5)$.
(e) $(-2,5,0)$. (f) $(6,2,0)$.
(g) $(-1,3,-4)$. (h) $(5,0,0)$.
(i) $(0,3,0)$. (j) $(0,0,-2)$.

2.2. A point of the projective plane is given in each part of this exercise. Determine homogeneous coordinates of the point in one of the forms listed in (1).
(a) The point $(2,5)$ in the Euclidean plane.
(b) The point $(0,-3)$ in the Euclidean plane.
(c) The point $(1,4)$ in the Euclidean plane.
(d) The point at infinity on lines of slope 3.
(e) The point at infinity on lines of slope $-\frac{2}{3}$.
(f) The point at infinity on vertical lines.
(g) The point at infinity on horizontal lines.

2.3. In each part of this exercise, the equation of a line in the projective plane is given in the form of (2). Determine whether the equation represents a line of the Euclidean plane or the line at infinity. In the first case, write the equation of the line as $y = mx + n$ or $x = h$ in the usual (x, y) coordinates of the Euclidean plane.
(a) $6x - 2y + 3z = 0$. (b) $2x + 5z = 0$.
(c) $x + 3y + 4z = 0$. (d) $7z = 0$.
(e) $3x + 2y = 0$. (f) $4y - 2z = 0$.
(g) $x - 4z = 0$. (h) $-2x + 4y + z = 0$.

2.4. A line of the projective plane is given in each part of this exercise. Write the equation of the line in homogeneous coordinates in the form of (2). In parts (a)–(e) write the point at infinity on the line in one of the forms in (8).

(a) The line $y = 2x - 3$ in the Euclidean plane.
(b) The line $y = -x/3$ in the Euclidean plane.
(c) The line $x = 2$ in the Euclidean plane.
(d) The line $y = 4$ in the Euclidean plane.
(e) The line $y = x + 2$ in the Euclidean plane.
(f) The line at infinity.

2.5. In each part of this exercise, two lines in the projective plane are given in homogeneous coordinates in the form of (2). The lines intersect at a unique point P (by Theorem 2.1). Find homogeneous coordinates for P in one of the forms in (1). If P is a point of the Euclidean plane, find its usual (x, y) coordinates. If P lies at infinity, find the slope of the lines in the Euclidean plane that contain P.
(a) $x + 2y - 6z = 0$ and $3x + 4y - 15z = 0$.
(b) $-2x + 4y - z = 0$ and $x - 2y + 3z = 0$.
(c) $3x + y + 5z = 0$ and $z = 0$.
(d) $2x + 3y - 6z = 0$ and $-x + y + 3z = 0$.
(e) $6x - 2y + 4z = 0$ and $3x - z = 0$.
(f) $3x + y - 2z = 0$ and $6x + 2y + 5z = 0$.
(g) $4x + 3y + 16z = 0$ and $3x + 2y + 10z = 0$.

2.6. In each part of this exercise, homogeneous coordinates are given for two points in the projective plane. The points lie on a unique line l (by Theorem 2.2). Find an equation for l in homogeneous coordinates in the form of (2). Determine whether l is a line of the Euclidean plane and, if so, write its equation in (x, y) coordinates in one of the forms $y = mx + n$ or $x = h$.
(a) $(4, -1, 3)$ and $(2, 5, 1)$. (b) $(4, 3, 2)$ and $(-2, 5, 1)$.
(c) $(2, 5, 1)$ and $(6, 1, 3)$. (d) $(-4, 5, 6)$ and $(2, 3, -3)$.
(e) $(4, 5, 0)$ and $(1, -3, 0)$. (f) $(0, 1, -2)$ and $(-3, 2, -4)$.
(g) $(3, 5, 2)$ and $(4, 1, 0)$. (h) $(4, 6, -2)$ and $(5, 0, 0)$.

2.7. State the version of Pappus' Theorem 2.3 that holds in the Euclidean plane in the following cases. Illustrate each version with a figure in the Euclidean plane.
(a) C is the only point at infinity named.
(b) Q is the only point at infinity named.
(c) QR is the line at infinity, and it does not contain $e \cap f$.
(d) QR is the line at infinity, and it contains $e \cap f$.
(e) f is the line at infinity.
(f) $(e \cap f)S$ is the line at infinity, and it does not contain Q.
(g) $B'S$ is the line at infinity, and it does not contain B.
(h) BB' is the line at infinity, and it does not contain S.
(i) BB' is the line at infinity, and it contains S.
(j) None of the points named lies at infinity.

2.8. The following theorem is proved in Exercise 3.21 (Figure 2.17):

Theorem

In the projective plane, let e and f be two lines on a point P. Let A, B, C be three points of e other than P, and let A', B', C' be three points of f other than P. Assume that the lines AA', BB', CC' are concurrent at a point T. Set $Q = AB' \cap A'B$ and $R = BC' \cap B'C$. Then the points P, Q, R are collinear.

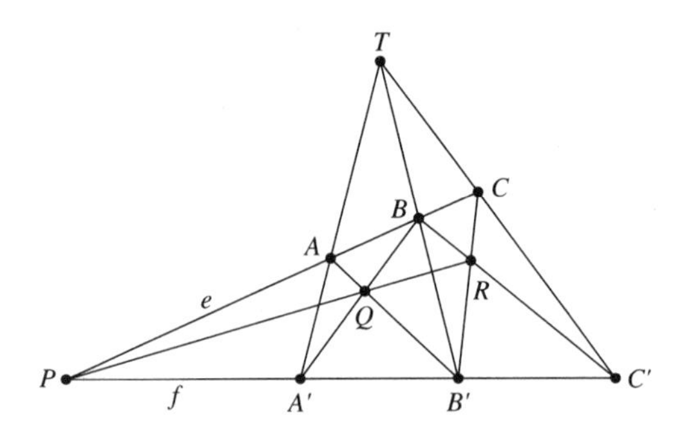

Figure 2.17

State the version of this theorem that holds in the Euclidean plane in the following cases. Draw a figure in the Euclidean plane to illustrate each version.

(a) Q is the only point at infinity named.
(b) C' is the only point at infinity named.
(c) B' is the only point at infinity named.
(d) P is the only point at infinity named.
(e) T is the only point at infinity named.
(f) f is the line at infinity.
(g) $B'C$ is the line at infinity.
(h) $A'C$ is the line at infinity.
(i) PR is the line at infinity.
(j) PT is the line at infinity.
(k) QT is the line at infinity, and it does not contain C.
(l) CC' is the line at infinity, and it does not contain Q.
(m) CC' is the line at infinity, and it contains Q.
(n) CQ is the line at infinity, and it does not contain C'.

2.9. The following theorem is proved in Exercise 3.21 (Figure 2.17). It is the converse of the theorem in Exercise 2.8.

Theorem
In the projective plane, let e and f be two lines on a point P. Let A, B, C be three points of e other than P, and let A', B', C' be three points of f other than P. Set $Q = AB' \cap A'B$ and $R = BC' \cap B'C$. Assume that the points P, Q, R are collinear. Then the lines AA', BB', CC' are concurrent at a point T.

State the version of this theorem that holds in the Euclidean plane in the cases in Exercise 2.8. Draw a figure in the Euclidean plane to illustrate each version.

2.10. The following theorem is proved in Exercise 5.21. (Figure 2.18):

Theorem
In the projective plane, let e and f be two lines on a point P. Let A and A' be two points of e other then P, and let B, B', C be three points of f other than P. Set $G = AB \cap A'B'$, $H = AB' \cap A'B$, $I = AB \cap A'C$, and $J = AC \cap A'B$. Then the lines GH, IJ, and e are concurrent.

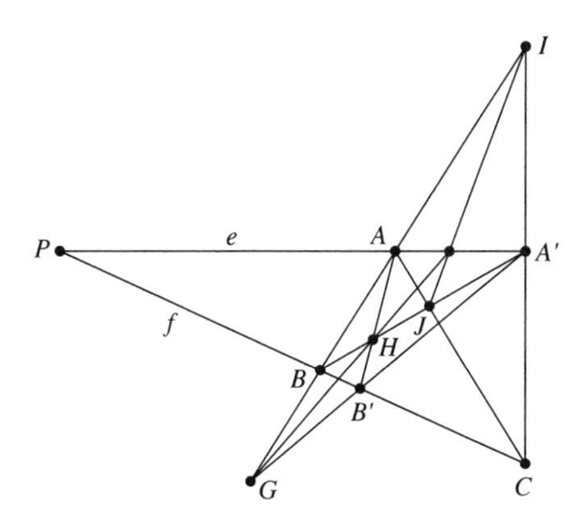

Figure 2.18

State the version of this theorem that holds in the Euclidean plane in the following cases. Draw a figure in the Euclidean plane to illustrate each version.

(a) A' is the only point at infinity named.
(b) B is the only point at infinity named.
(c) G is the only point at infinity named.
(d) e is the line at infinity.
(e) f is the line at infinity.
(f) GH is the line at infinity.
(g) HI is the line at infinity.
(h) GI is the line at infinity.
(i) $A'B$ is the line at infinity.
(j) AB' is the line at infinity.

§3. Intersections in Homogeneous Coordinates

We considered intersections of curves at the origin in Section 1, and we enlarged the Euclidean plane to the projective plane in Section 2. We combine these ideas in this section and consider intersections of curves at all points in the projective plane.

We start by extending algebraic curves from the Euclidean to the projective plane by homogenizing polynomials. We then consider intersection multiplicities at any point in the projective plane. We introduce transformations, which are linear changes of variables in homogeneous coordinates. We show that we can transform any four points, no three of which are collinear, into any other four such points. Because transformations preserve intersection multiplicities, we can find the number of times that two curves intersect at any point in the projective plane by transforming that point to the origin.

We start by extending algebraic curves from the Euclidean to the projective plane. Some care is required, because a polynomial equation $g(x,y,z) = 0$ in three variables does not generally define a curve in the projective plane. In fact, g must have the property that

$$g(a,b,c) = 0 \quad \text{if and only if} \quad g(ta,tb,tc) = 0$$

for any $t \neq 0$ and $(a,b,c) \neq (0,0,0)$, so that the choice of the homogeneous coordinates for a point is irrelevant. For example, the equation $x = 1$ does not define a curve in the projective plane because $x = 1$ does not imply that $tx = 1$ for $t \neq 1$.

Let d be a nonnegative integer. A *homogeneous polynomial* $F(x,y,z)$ of *degree* d in variables x,y,z is an expression

$$F(x,y,z) = \sum e_{ij}x^i y^j z^{d-i-j}, \tag{1}$$

where the sigma represents summation, the coefficients e_{ij} are real numbers that are not all zero, and i and j vary over pairs of nonnegative integers whose sum is at most d. In short, a homogeneous polynomial of degree d is a nonzero polynomial such that the exponents of the variables in every term sum to d. We use capital letters to designate homogeneous polynomials.

Multiplying x,y,z in (1) by a nonzero number t gives

$$F(tx,ty,tz) = \sum e_{ij}(tx)^i(ty)^j(tz)^{d-i-j}.$$

Because t is raised to the power $i + j + (d - i - j) = d$ in every term, we can factor out t^d and obtain

$$F(tx,ty,tz) = t^d \sum e_{ij}x^i y^j z^{d-i-j} = t^d F(x,y,z).$$

It follows that $F(ta,tb,tc) = 0$ if and only if $F(a,b,c) = 0$ for any $t \neq 0$ and any point (a,b,c). In other words, if one choice of homogeneous coordinates for a point satisfies the equation $F = 0$, they all do.

In homogeneous coordinates, an *algebraic curve*—or, simply, a *curve*—is a homogeneous polynomial $F(x,y,z)$. We imagine that the curve consists of all points in the projective plane that satisfy the equation $F(x,y,z) = 0$, where points corresponding to repeated factors of F are repeated as many times as the factor. We have seen that the choice of homogeneous coordinates for each point is immaterial. We often refer to the curve F by the equation $F(x,y,z) = 0$ or its algebraic equivalents. We call the degree of F the *degree* of the curve.

For any homogeneous polynomial $F(x,y,z)$, set $f(x,y) = F(x,y,1)$. Setting $z = 1$ in (1) gives

$$f(x,y) = \sum e_{ij}x^i y^j.$$

A point (x,y) of the Euclidean plane lies on the graph of $f(x,y)=0$ if and only if the corresponding point $(x,y,1)$ lies on the graph of $F(x,y,z)=0$. Thus, the curves $f=0$ and $F=0$ contain the same points of the Euclidean plane, and we call f the *restriction* of F to the Euclidean plane.

Conversely, if $f(x,y)$ is a nonzero polynomial of degree d in two variables, we extend the curve $f(x,y)=0$ from the Euclidean to the projective plane as follows. The *homogenization* $F(x,y,z)$ of f is the homogeneous polynomial obtained by multiplying each term of f by the power of z needed to produce a term of degree d. That is, if

$$f(x,y)=\sum e_{ij}x^iy^j, \tag{2}$$

we get

$$F(x,y,z)=\sum e_{ij}x^iy^jz^{d-i-j}, \tag{3}$$

so that F is homogeneous of the same degree d as f. Setting $z=1$ in the right-hand side of (3) gives the right-hand side of (2). This shows that

$$F(x,y,1)=f(x,y), \tag{4}$$

and so $F=0$ and $f=0$ contain the same points of the Euclidean plane. We call the curve $F=0$ the *extension* of the curve $f=0$ to the projective plane. We obtain the graph of F from the graph of f by adding points at infinity, namely, the points $(x,y,0)$ such that $F(x,y,0)=0$. Each point at infinity can be written in exactly one way as $(1,s,0)$ or $(0,1,0)$ for a real number s, as in (8) in Section 2.

For example, suppose we consider the hyperbola $xy=1$ in the Euclidean plane (Figure 3.1). The polynomial $xy-1$ has degree 2, and so we multiply each term by the power of z needed to raise the degree to 2. Thus, the homogenization is $xy-z^2$, and the curve $xy=1$ in the Euclidean plane extends to the curve $xy=z^2$ in the projective plane. The points $(x,y,1)$ on $xy=z^2$ are exactly the points (x,y) on $xy=1$, and so both curves contain the same points of the Euclidean plane.

Which of the points $(1,s,0)$ and $(0,1,0)$ at infinity lie on $xy=z^2$? Substituting $(1,s,0)$ gives $s=0$, and substituting $(0,1,0)$ gives the true statement $0=0$. Thus, $xy=z^2$ contains exactly two points at infinity, $(1,0,0)$

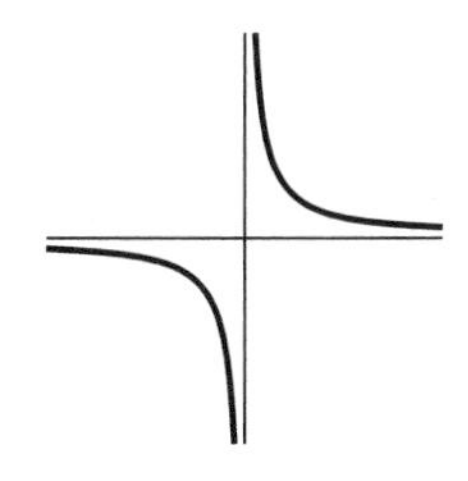

Figure 3.1

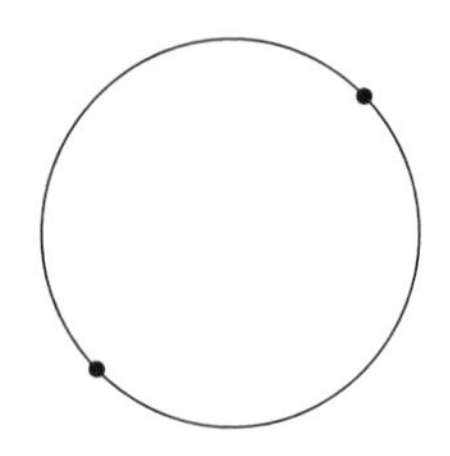

Figure 3.2

and $(0,1,0)$. As in the discussion after (8) of Section 2, $(1,0,0)$ is the point at infinity on the lines of slope 0—the horizontal lines—of the Euclidean plane, and $(0,1,0)$ is the point at infinity on the vertical lines. We imagine that the two ends of the hyperbola in Figure 3.1 that approach the y-axis meet at the point at infinity on vertical lines, and that the two ends that approach the x-axis meet at the point at infinity on horizontal lines. Adding these two points at infinity joins the two pieces of the hyperbola into a simple closed curve, as in Figure 3.2. The fact that Figure 3.2 is simpler than Figure 3.1 suggests that working in the projective plane may simplify the study of curves.

Lines in the projective plane, which we defined before (2) of Section 2, are exactly the curves of degree 1. Homogenization gives the same relationship that we introduced in (4)–(7) of Section 2 between lines of the Euclidean and projective planes. The lines $y = mx + n$ and $x = h$ of the Euclidean plane extend to the lines $y = mx + nz$ and $x = hz$ of the projective plane. The line at infinity $z = 0$ is not the extension of any line of the Euclidean plane because the polynomial z is not the homogenization of any polynomial in x and y: the polynomial 1 has degree 0 and is its own homogenization.

Let $f(x,y)$ be a nonzero polynomial, and let $F(x,y,z)$ be its homogenization. We often refer to the curve F as "the curve f in the projective plane" because f is more familiar than F. In effect, we automatically extend curves to the projective plane by homogenizing them. For example, "the curve $xy = 1$ in the projective plane" is the curve $xy = z^2$ in homogeneous coordinates.

Now that we have defined curves in the projective plane, it is natural to consider their intersection multiplicities. We assume that the *intersection multiplicity* $I_P(F,G)$ is an integer associated with every pair of homogeneous polynomials $F(x,y,z)$ and $G(x,y,z)$ and every point P of the projective plane. We think of $I_P(F,G)$ as the number of times that the curves F and G intersect at P.

The number of times that two curves intersect at the origin should not change when we restrict the curves from the projective to the Euclidean plane and replace homogeneous coordinates with the usual (x,y) coordinates. We formalize this as the following property, which we establish in Chapter IV along with the other intersection properties:

Property 3.1
Let $F(x,y,z)$ and $G(x,y,z)$ be homogeneous polynomials, and set $f(x,y) = F(x,y,1)$ and $g(x,y) = G(x,y,1)$. Then we have

$$I_O(F(x,y,z),G(x,y,z)) = I_O(f(x,y),g(x,y)),$$

where O is the origin. □

In Section 1, we considered intersections only at the origin. We can now define the intersection multiplicity of two curves in the Euclidean plane at any point of the plane.

Definition 3.2
Let $f(x,y)$ and $g(x,y)$ be nonzero polynomials, and let $F(x,y,z)$ and $G(x,y,z)$ be their homogenizations. Let (a,b) be a point of the Euclidean plane. Then we define the *intersection multiplicity* $I_{(a,b)}(f,g)$ of the curves $f(x,y) = 0$ and $g(x,y) = 0$ at the point (a,b) in the Euclidean plane to be the intersection multiplicity $I_{(a,b,1)}(F,G)$ of the curves $F(x,y,z) = 0$ and $G(x,y,z) = 0$ at the point $(a,b,1)$ in the projective plane. □

We think of the quantity $I_{(a,b)}(f,g)$ in Definition 3.2 as the number of times that the curves $f = 0$ and $g = 0$ in the Euclidean plane intersect at the point (a,b). Definition 3.2 and the discussion before Property 1.1 give two ways to assign intersection multiplicities at the origin, but Property 3.1 and (4) show that these two ways agree.

We saw in Section 2 that we can identify the points and lines of the projective plane with the lines and planes through the origin in Euclidean space. We introduce transformations—linear changes of variables in homogeneous coordinates—to take advantage of the symmetry of Euclidean space and transfer it to the projective plane. We use transformations in two key ways. First, we compute the intersection multiplicity of two curves at any point in the projective plane by transforming that point to the origin and using the techniques of Section 1. Second, we use transformations to simplify the equations of curves.

Definition 3.3
A *transformation* is a map from the projective plane to itself that takes any point (x,y,z) to the point (x',y',z') determined by the equations

$$\begin{aligned} x' &= ax + by + cz, \\ y' &= dx + ey + fz, \\ z' &= gx + hy + iz, \end{aligned} \tag{5}$$

where a–i are real numbers such that the equations in (5) are equivalent to equations of the form

$$\begin{aligned} x &= Ax' + By' + Cz', \\ y &= Dx' + Ey' + Fz', \\ z &= Gx' + Hy' + Iz', \end{aligned} \tag{6}$$

that express x, y, z in terms of x', y', z' for real numbers A–I. □

If x, y, z are not all zero, the equations in (6) imply that the corresponding values of x', y', z' are not all zero. Moreover, if we replace x, y, z in (5) with tx, ty, tz for a nonzero number t, the corresponding values of x', y', z' are also multiplied by t. Thus, the equations in (5) map each point (x, y, z) in the projective plane to a well-defined point (x', y', z'), as Definition 3.3 asserts.

We consider several examples of transformations. Translating the Euclidean plane h units horizontally and k units vertically maps any point (x, y) to the point $(x+h, y+k)$. The corresponding map of the projective plane sends (x, y, z) to (x', y', z'), where

$$\begin{aligned} x' &= x + hz, \\ y' &= y + kz, \\ z' &= z. \end{aligned} \tag{7}$$

Note that we have made the right-hand sides of these equations homogeneous of degree 1 by multiplying the constants h and k by z. These equations give a transformation of the projective plane because we can solve them for x, y, z in terms of x', y', z', as Definition 3.3 requires:

$$\begin{aligned} x &= x' - hz', \\ y &= y' - kz', \\ z &= z'. \end{aligned}$$

Setting $z = 1$ in (7) shows that the transformation maps $(x, y, 1)$ to $(x+h, y+k, 1)$, and so it extends to the projective plane the translation of the Euclidean plane taking (x, y) to $(x+h, y+k)$. The equations in (7) map each point $(x, y, 0)$ at infinity to itself, which makes sense because a translation does not change the slopes of lines.

Another way to exploit the symmetry of the projective plane is to interchange coordinates. For example, interchanging the first and third coordinates maps (x, y, z) to (x', y', z'), where

$$x' = z, \qquad y' = y, \qquad z' = x. \tag{8}$$

These equations have the form of both (5) and (6), and so they give a transformation. Likewise, any permutation—that is, any rearrangement—of the coordinates is a transformation. We use these transformations to eliminate distinctions between points at infinity and points of the Euclidean plane. For example, the transformation in (8) maps the points on the line at infinity $z = 0$ to the points on the y-axis $x' = 0$.

The third basic type of transformation multiplies coordinates by nonzero constants. If r, s, t are nonzero numbers, we can solve the equations

$$x' = rx, \qquad y' = sy, \qquad z' = tz, \tag{9}$$

for x, y, z and obtain

$$x = \frac{1}{r}x', \qquad y = \frac{1}{s}y', \qquad z = \frac{1}{t}z'.$$

Thus, there is a transformation that maps (x, y, z) to (rx, sy, tz).

We show next that we can obtain new transformations from given ones by reversing them or performing them in sequence. In this way, we obtain a wide range of transformations from the three basic types we have introduced.

Because the systems of equations in (5) and (6) are equivalent, if there is a transformation mapping (x, y, z) to (x', y', z'), there is also a transformation mapping (x', y', z') to (x, y, z). Thus, we can reverse any transformation.

Suppose that we are given the transformation in (5) mapping (x, y, z) to (x', y', z'). Suppose that we are also given a transformation mapping (x', y', z') to (x'', y'', z''), where

$$\begin{aligned} x'' &= jx' + ky' + lz', \\ y'' &= mx' + ny' + oz', \\ z'' &= px' + qy' + rz'. \end{aligned} \tag{10}$$

Substituting the equations in (5) into these equations gives

$$\begin{aligned} x'' &= j(ax + by + cz) + k(dx + ey + fz) + l(gx + hy + iz), \\ y'' &= m(ax + by + cz) + n(dx + ey + fz) + o(gx + hy + iz), \\ z'' &= p(ax + by + cz) + q(dx + ey + fz) + r(gx + hy + iz). \end{aligned}$$

Collecting terms gives

$$\begin{aligned} x'' &= (ja + kd + lg)x + (jb + ke + lh)y + (jc + kf + li)z, \\ y'' &= (ma + nd + og)x + (mb + ne + oh)y + (mc + nf + oi)z, \\ z'' &= (pa + qd + rg)x + (pb + qe + rh)y + (pc + qf + ri)z, \end{aligned} \tag{11}$$

which has the form of (5). Moreover, because the equations in (10) give a transformation, we can solve them for x', y', z' in terms of x'', y'', z'' and obtain

$$x' = Jx'' + Ky'' + Lz'',$$
$$y' = Mx'' + Ny'' + Oz'',$$
$$z' = Px'' + Qy'' + Rz'',$$

for real numbers J–R. Substituting these expressions into (6) expresses x,y,z in terms of x'',y'',z''. Thus the equations in (11) give a transformation mapping (x,y,z) to (x'',y'',z''). This is the net result of following the transformation taking (x,y,z) to (x',y',z') with the transformation taking (x',y',z') to (x'',y'',z''). In short, we can combine two transformations into a third one by performing them in sequence.

How much latitude do we have in constructing transformations? We note that any transformation must preserve lines: points are collinear if and only if their images under the transformation are collinear. To see this, let the transformation taking (x,y,z) to (x',y',z') be given by the equations in (5). A line in the projective plane has equation

$$px + qy + rz = 0, \tag{12}$$

where p,q,r are constants that are not all zero. Substituting the expressions for x,y,z in (6) into (12) gives

$$p(Ax' + By' + Cz') + q(Dx' + Ey' + Fz') + r(Gx' + Hy' + Iz') = 0.$$

Collecting terms gives

$$(pA + qD + rG)x' + (pB + qE + rH)y' + (pC + qF + rI)z' = 0. \tag{13}$$

Substituting the expressions for x',y',z' in (5) turns (13) back into (12). Since the coefficients in (12) are not all zero, the same holds for (13), and so (13) represents a line. A point (x,y,z) lies on the line in (12) if and only if its image (x',y',z') lies on the line in (13). Because a transformation is reversible, it follows that points are collinear if and only if their images are collinear.

We can produce a wide range of transformations by combining the three kinds of transformations in the discussions accompanying (7)–(9). In fact, we can transform any four points, no three of which are collinear, into any other four points, no three of which are collinear. We say that a transformation *fixes* a point if it maps the point to itself. We call points *distinct* when no two of them are equal.

Theorem 3.4
In the projective plane, let A,B,C,D be four points, no three of which are collinear, and let A',B',C',D' be four points, no three of which are collinear. Then there is a transformation that maps A,B,C,D to A',B',C',D', respectively.

Proof

We start by proving that there is a transformation that maps A,B,C,D to $(1,0,0)$, $(0,1,0)$, $(0,0,1)$, $(1,1,1)$. At least one coordinate of A is nonzero. Because we can use a transformation to interchange the coordinates of A, we can assume that the last coordinate is nonzero. Because the coordinates of A are homogeneous, we can divide them all by the last one, so that we have $A = (r,s,1)$ for numbers r and s. Then the transformation

$$x' = x - rz, \qquad y' = y - sz, \qquad z' = z,$$

maps A to $(0,0,1)$. Following this with the transformation that interchanges the first and third coordinates gives a transformation that maps A to $(1,0,0)$. Let B_1 be the image of B under this transformation.

Because transformations are reversible, they map distinct points to distinct points. Accordingly, since $B \neq A$, we have $B_1 \neq (1,0,0)$. Thus, either the second or third coordinate of B_1 is nonzero. Interchanging these coordinates fixes $(1,0,0)$, and so we can assume that the last coordinate of B_1 is nonzero. Dividing through by this coordinate gives B_1 homogeneous coordinates $(t,u,1)$ for real numbers t and u. The transformation

$$x' = x - tz, \qquad y' = y - uz, \qquad z' = z,$$

maps B_1 to $(0,0,1)$ and fixes $(1,0,0)$. Following this with the transformation that interchanges the last two coordinates gives a transformation that maps B_1 to $(0,1,0)$ and fixes $(1,0,0)$. Applying this transformation after the one at the end of the previous paragraph gives a transformation that maps A to $(1,0,0)$ and B to $(0,1,0)$. Let C_1 be the image of C under this transformation.

We are given that C does not lie on line AB. Since transformations preserve collinearity, C_1 does not lie on the line through $(1,0,0)$ and $(0,1,0)$. This is the line $z = 0$, and so the last coordinate of C_1 is nonzero. Dividing the coordinates of C_1 by this number gives C_1 homogeneous coordinates $(v,w,1)$ for numbers v and w. The transformation

$$x' = x - vz, \qquad y' = y - wz, \qquad z' = z,$$

fixes $(1,0,0)$ and $(0,1,0)$ and maps C_1 to $(0,0,1)$. Applying this transformation after the one at the end of the previous paragraph gives a transformation that maps A,B,C to $(1,0,0)$, $(0,1,0)$, $(0,0,1)$. Let D_1 be the image of D under this transformation.

D does not lie on any of the lines AB, BC, CA. Because transformations preserve collinearity, D_1 does not lie on the line $z = 0$ through $(1,0,0)$ and $(0,1,0)$, the line $x = 0$ through $(0,1,0)$ and $(0,0,1)$, or the line $y = 0$ through $(1,0,0)$ and $(0,0,1)$. Thus, every coordinate of D_1 is nonzero, and we write D_1 as (h,k,l) for nonzero numbers h,k,l. The transformation

$$x' = \frac{x}{h}, \qquad y' = \frac{y}{k}, \qquad z' = \frac{z}{l},$$

maps D_1 to $(1,1,1)$ and fixes $(1,0,0)$, $(0,1,0)$, and $(0,0,1)$. (For example, the transformation maps $(1,0,0)$ to $(1/h,0,0)$, which equals $(1,0,0)$ in homogeneous coordinates.) Applying this transformation after the one at the end of the previous paragraph gives a transformation that maps A,B,C,D to $(1,0,0)$, $(0,1,0)$, $(0,0,1)$, $(1,1,1)$.

By symmetry, there is also a transformation that maps A',B',C',D' to $(1,0,0)$, $(0,1,0)$, $(0,0,1)$, $(1,1,1)$. Reversing this transformation gives a transformation that maps $(1,0,0)$, $(0,1,0)$, $(0,0,1)$, $(1,1,1)$ to A',B',C',D'. Applying this transformation after the one at the end of the previous paragraph gives a transformation that maps A,B,C,D to A',B',C',D'. □

Let $F(x,y,z)$ be a homogeneous polynomial of degree d. We can write

$$F(x,y,z) = \sum e_{ij} x^i y^j z^{d-i-j} \tag{14}$$

for constants e_{ij} not all zero. Substituting the expressions for x,y,z in (6) into F gives a polynomial

$$\begin{aligned} &F'(x',y',z') \\ &\quad = \sum e_{ij}(Ax'+By'+Cz')^i(Dx'+Ey'+Fz')^j(Gx'+Hy'+Iz')^{d-i-j}. \end{aligned} \tag{15}$$

Expanding the right-hand side of (15) shows that every term of F' has the same degree d as F. The reversability of the transformation and the fact that F is nonzero implies that F' is also nonzero. Thus, F' is homogeneous of the same degree d as F. Because the right-hand sides of (14) and (15) are related by the substitutions in (6), we see that

$$F(s,t,u) = F'(s',t',u'), \tag{16}$$

for any point (s,t,u) in the projective plane, where (s',t',u') is the image of (s,t,u) under the transformation in (5). Because the transformation in (5) is reversible, it matches up the points of the curve $F(x,y,z)=0$ and $F'(x',y',z')=0$ (by (16)), and every curve F' of degree d arises in this way from a unique curve of degree d. We call F' the *image* of F under the transformation. We have shown that transformations map curves of each degree d among themselves. The case $d=1$ shows that transformations preserve lines, as we saw in the discussion accompanying (12) and (13).

Note that *the transformation taking* (x,y,z) *to* (x',y',z') *given by* (5) *acts on curves by substituting the expressions in* (6) *for* x,y,z. For example, consider the transformation

$$x' = x+2z, \qquad y' = y-3z, \qquad z' = z, \tag{17}$$

that translates points in the Euclidean plane 2 units horizontally and -3 units vertically. Solving these equations for x,y,z in terms of x',y',z' gives

$$x = x'-2z', \qquad y = y'+3z', \qquad z = z'. \tag{18}$$

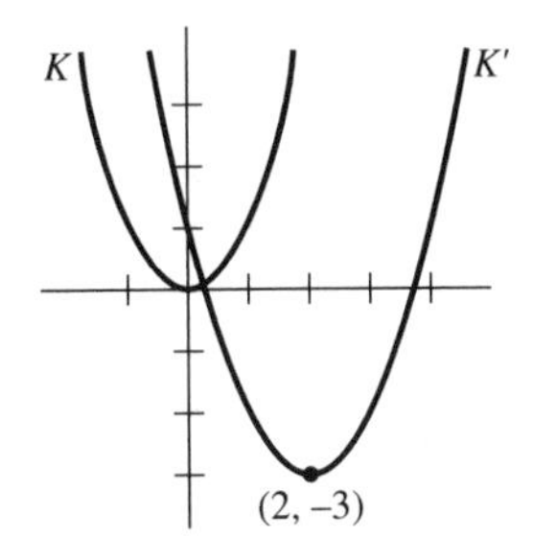

Figure 3.3

To determine the image of the curve

$$yz = x^2 \tag{19}$$

under the transformation in (17), substitute the expressions for x, y, z from (18) into (19) to produce

$$(y' + 3z')z' = (x' - 2z')^2. \tag{20}$$

Multiplying this equation out and collecting terms gives

$$y'z' = x'^2 - 4x'z' + z'^2. \tag{21}$$

Thus, the transformation in (17) maps (19) to (21). Setting $z = 1$ in (19) and $z' = 1$ in (20) gives the familiar result that the parabola K with equation $y = x^2$ and vertex $(0, 0)$ can be translated 2 units to the right and 3 units down to give the parabola K' with equation $y' + 3 = (x' - 2)^2$ and vertex $(2, -3)$ (Figure 3.3).

We use transformations to study curves of degree at most 3 by simplifying their equations. We have just noted that transformations preserve the degree of a curve. We also need to know that transformations preserve intersection multiplicities. We prove this result, which we now state formally, in Chapter IV along with the other intersection properties.

Property 3.5
Let a transformation of the projective plane map (x, y, z) to (x', y', z'). Let P be any point of the projective plane, and let P' be its image under the transformation. Let $F(x, y, z) = 0$ and $G(x, y, z) = 0$ be curves, and let $F'(x', y', z') = 0$ and let $G'(x', y', z')$ be their images under the transformation. Then we have

$$I_P(F(x, y, z), G(x, y, z)) = I_{P'}(F'(x', y', z'), G'(x', y', z')). \qquad \square$$

If the transformation in Property 3.5 is given by the equations in (5), we obtain F' and G' by substituting the expressions in (6) for x, y, z in F and G, as discussed after the proof of Theorem 3.4.

We can now generalize the intersection properties in Section 1 from intersections at the origin to intersections at any point. We use Theorem 3.4 and the fact that transformations preserve intersection multiplicities to transform any point of intersection of two curves to the origin.

Theorem 3.6

In the projective plane, let $F(x,y,z)=0$, $G(x,y,z)=0$, and $H(x,y,z)=0$ be curves, and let P be a point. Then the following results hold:

(i) $I_P(F,G)$ *is a nonnegative integer or* ∞.
(ii) $I_P(F,G)=I_P(G,F)$.
(iii) $I_P(F,G)\geq 1$ *if and only if F and G both contain P.*
(iv) $I_P(F,G)=I_P(F,G+FH)$ *if $G+FH$ is homogeneous.*
(v) $I_P(F,GH)=I_P(F,G)+I_P(F,H)$.
(vi) $I_P(F,G)=\infty$ *if F is a factor of G and contains P.*

Proof

There is a transformation taking P to the origin, by Theorem 3.4. The intersection multiplicity of two curves at P equals the intersection multiplicity of their images at the origin (by Property 3.5). We can compute the intersection multiplicities of curves in the projective plane at the origin by restricting the curves to the Euclidean plane (by Property 3.1). Thus, statements (i)–(vi) follow from Properties 1.1–1.3, 1.5, 1.6, and Theorem 1.7. □

Parts (v) and (iii) of Theorem 3.6 show that

$$I_P(F,kG)=I_P(F,k)+I_P(F,G)=I_P(F,G)$$

for any real number $k\neq 0$. That is, multiplying a curve G by a nonzero constant k does not change its intersection multiplicities with other curves. Accordingly *we consider kG to be the same curve as G for all real numbers $k\neq 0$*. That is, *we consider two homogeneous polynomials to be the same curve exactly when they are scalar multiples of each other.*

It is natural to think of the polynomials kG as the same curve for all real numbers $k\neq 0$ because the equations $kG(x,y,z)=0$ and $G(x,y,z)=0$ have the same solutions in the projective plane. We identified lines differing by nonzero constant multiples when we observed that every line in the projective plane except $z=0$ is given by (4) or (6) of Section 2. We also identified each line with its nonzero scalar multiples when we proved in Theorems 2.1 and 2.2 that two lines intersect at a unique point and two points lie on a unique line.

We need one more basic result relating intersections and transformations. Part (ii) of the next theorem states that translations of the Euclidean plane preserve intersection multiplicities. This holds because translations extend to transformations of the projective plane, and transformations preserve intersection multiplicities. Part (iii) states that restricting curves from the projective to the Euclidean plane preserves intersection multiplicities. This generalizes Property 3.1 by replacing the origin with any point in the Euclidean plane. It is a companion result to Definition 3.2, which shows that extending curves from the Euclidean to the projective plane preserves intersection multiplicities; Theorem 3.7(iii) is slightly more general than Definition 3.2, since there are curves in the projective plane such as $xz = 0$ that are not extensions of curves in the Euclidean plane because they have z as a factor.

Theorem 3.7
Let a and b be real numbers.

(i) *Let $F(x, y, z)$ and $G(x, y, z)$ be homogeneous polynomials, and let their restrictions to the Euclidean plane be*

$$f(x, y) = F(x, y, 1) \qquad \text{and} \qquad g(x, y) = G(x, y, 1). \tag{22}$$

Then

$$I_{(a,b,1)}(F(x, y, z), G(x, y, z)) \tag{23}$$

equals

$$I_{(0,0)}(f(x + a, y + b), g(x + a, y + b)). \tag{24}$$

(ii) *If $f(x, y)$ and $g(x, y)$ are nonzero polynomials, we have*

$$I_{(a,b)}(f(x, y), g(x, y)) = I_{(0,0)}(f(x + a, y + b), g(x + a, y + b)). \tag{25}$$

(iii) *Let F and G be homogeneous polynomials, and define f and g by the equations in (22). Then we have*

$$I_{(a,b,1)}(F, G) = I_{(a,b)}(f, g). \tag{26}$$

Proof
(i) As in the discussion accompanying (7), the equations

$$x' = x - az, \qquad y' = y - bz, \qquad z' = z, \tag{27}$$

represent a transformation because they can be solved for x, y, z in terms of x', y', z', as follows:

$$x = x' + az', \qquad y = y' + bz', \qquad z = z'. \tag{28}$$

The transformation in (27) maps $(a, b, 1)$ to $(0, 0, 1)$. Accordingly, Property 3.5, which states that transformations preserve intersection multiplicities, shows that the quantity in (23) equals

$$I_{(0,0,1)}(F(x' + az', y' + bz', z'), G(x' + az', y' + bz', z')), \tag{29}$$

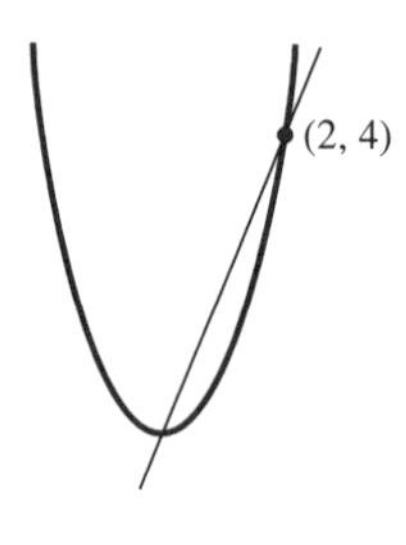

Figure 3.4

where we have substituted the expressions for x, y, z from (28) into (23). Setting $z' = 1$ in the polynomials in (29) gives the polynomials in (24) (by (22)). Thus, the intersection multiplicities in (29) and (24) are equal (by Property 3.1). (The primes in (29) are immaterial, since they merely show that this expression arises from a transformation.) In short, the quantities in (23) and (24) are equal because they both equal the quantity in (29).

(ii) Let $F(x, y, z)$ and $G(x, y, z)$ be the homogenizations of $f(x, y)$ and $g(x, y)$. Equation (4) shows that the equations in (22) hold. Thus, part (i) shows that the quantities in (23) and (24) are equal. The quantity in (23) equals the left-hand side of (25) (by Definition 3.2). Hence, (25) holds.

Part (iii) follows by combining parts (i) and (ii). □

Theorem 3.7(ii) makes it easy to compute the number of times that two curves intersect at any point in the Euclidean plane: we translate the point to the origin and then apply the techniques of Section 1. For example, suppose we want to compute the number of times that $y = x^2$ and $y = 2x$ intersect at $(2, 4)$ (Figure 3.4). Theorem 3.7(ii) shows that

$$\begin{aligned} I_{(2,4)}(y - x^2, y - 2x) &= I_{(0,0)}(y + 4 - (x + 2)^2, y + 4 - 2(x + 2)) \\ &= I_{(0,0)}(y - x^2 - 4x, y - 2x). \end{aligned}$$

By Theorems 1.9(ii) and 1.11, this intersection multiplicity is the smallest degree of any nonzero term produced by substituting $2x$ for y in $y - x^2 - 4x$ and collecting terms, which gives $-x^2 - 2x$. This degree is 1, and so $y = x^2$ intersects $y = 2x$ once at $(2, 4)$.

To find the number of times that two curves intersect at a point P at infinity, transform P to a point of the Euclidean plane by interchanging coordinates and then apply Theorem 3.7. For example, suppose that we want to find the number of times that the hyperbola $x^2 - y^2 = 1$ intersects its asymptote $y = x$ at infinity (Figure 3.5). Converting to homogeneous coordinates, we want the intersection multiplicity of $x^2 - y^2 - z^2 = 0$ and $y - x = 0$ at $(1, 1, 0)$. We interchange x and z to move the point of intersection into the Euclidean plane. This gives

$$I_{(0,1,1)}(z^2 - y^2 - x^2, y - z) \tag{30}$$

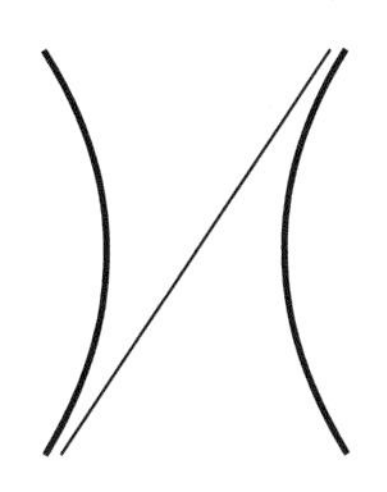

Figure 3.5

(by Property 3.5). Taking $a = 0$ and $b = 1$ in Theorem 3.7(i) shows that the quantity in (30) equals

$$I_O(1 - (y+1)^2 - x^2, (y+1) - 1) = I_O(-y^2 - 2y - x^2, y),$$

where O is the origin. This intersection multiplicity is 2, by Theorems 1.9(ii) and 1.11, since setting $y = 0$ in $-y^2 - 2y - x^2$ gives $-x^2$. Thus, the hyperbola intersects its asymptote twice at infinity.

We end this section with a remark for readers familiar with linear algebra. Transformations of the projective plane correspond to invertible linear transformations of R^3, by Definition 3.3. Because linear transformations of R^3 are determined by the images of three linearly independent vectors, it may seem surprising that Theorem 3.4 shows that there are four degrees of freedom in defining transformations of the projective plane. In fact, the fourth degree of freedom arises from the homogeneity of coordinates in the projective plane, as the second-to-last paragraph of the proof of Theorem 3.4 shows.

Exercises

3.1. A curve $f(x, y) = 0$ in the Euclidean plane is given in each part of this exercise. Determine the extension $F(x, y, z) = 0$ of the curve to the projective plane, where F is the homogenization of f. Determine the points at infinity on the extension, writing each one as in (8) of Section 2, and determine the slope of the lines in the Euclidean plane on each of these points.
(a) $x^4 + 3x^2y = 4y^4 - 5y^3 - y^2 + 2y + 6$.
(b) $y^2 - 3xy + 5x - 2y = 21$.
(c) $y^2 = x^3 + 5x$.
(d) $y^3 = 4x^2y + 8x + 12$.
(e) $x^3 - 3x^2y + 2xy - 4y = 10$.

3.2. A curve $F(x, y, z) = 0$ in the projective plane is given in each part of the exercise. Determine the equation $f(x, y) = 0$ of the curve's restriction to the Euclidean plane.

(a) $z^3 = x^2z - 2xy^2 + 3y^3$.
(b) $8x^3 + 2x^2z - xyz + y^3 + 3yz^2 + 4z^3 = 0$.
(c) $x^4 - 2x^2yz + xyz^2 + 3y^4 + 5yz^3 - 2z^4 = 0$.
(d) $2xyz^2 + x^3z - yz^3 + 3x^3y = 0$.

3.3. In each part of Exercise 3.2, determine the points at infinity on the given curve. Write each point at infinity in one of the forms in (8) of Section 2, and specify the slope of the lines in the Euclidean plane on each point.

3.4. Two curves and a point in the Euclidean plane are given in each part of this exercise. Use Theorem 3.7(ii) to find the number of times that the curves intersect at the point.
(a) $x^3 + y^2 = 4$, $x^3y + y^2 = 4$, $(0, 2)$.
(b) $x^3 + y^2 = 5$, $x^2 + 2y^2 = 9$, $(1, -2)$.
(c) $xy = 2$, $xy^2 = 4$, $(1, 2)$.
(d) $x^2 + 2xy = 4$, $x^2 + y^2 = 4y + 4$, $(-2, 0)$.
(e) $x^2 + 2xy - 2y = 1$, $x^3 = y^2 + 2y + 2$, $(1, -1)$.
(f) $x^2 + xy + y = 1$, $xy^2 + 4 = 0$, $(-1, 2)$.

3.5. Each part of this exercise gives two curves in homogeneous coordinates and a point at infinity in the projective plane. Find the number of times that the curves intersect at the point as in the discussion accompanying Figure 3.5.
(a) $xy = 2x^2 + z^2$, $y^2 + yz = 4x^2$, $(1, 2, 0)$.
(b) $3y^2 + xy + 2z^2 = 0$, $z^3 = xy^2 + 3y^3$, $(3, -1, 0)$.
(c) $3y = x + 2z$, $3y^3 + xz^2 = xy^2$, $(3, 1, 0)$.
(d) $xy + y^2 = z^2$, $x^2 - y^2 = 2z^2$, $(1, -1, 0)$.
(e) $xz^2 + x^2y = 4y^3$, $x^2z^2 + 3xy^3 = 6y^4$, $(2, 1, 0)$.

3.6. Consider the equations

$$x' = 2x, \qquad y' = 4x - y, \qquad z' = x - 3y + z. \tag{31}$$

(a) Show that these equations give a transformation by solving them for x, y, z in terms of x', y', z', as in (6).
(b) Determine the image of the line $y = 3x - 2z$ under the transformation in (31).
(c) Determine the image of the curve $x^2 - y^2 = z^2$ under the transformation in (31).

3.7. Do parts (a)–(c) of Exercise 3.6 for the equations

$$x' = 3y, \qquad y' = x + 2z, \qquad z' = 2x - z. \tag{32}$$

3.8. Do parts (a)–(c) of Exercise 3.6 for the equations

$$x' = 3x + 2y - z, \qquad y' = x + 3y, \qquad z' = x + 2y. \tag{33}$$

3.9. Compute the combined effect of performing the following sequences of transformations:
(a) Following the transformation in (31) with that in (32).
(b) Following the transformation in (32) with that in (31).
(c) Following the transformation in (31) with that in (33).
(d) Following the transformation in (33) with that in (31).

(e) Following the transformation in (32) with that in (33).
(f) Following the transformation in (33) with that in (32).

3.10. Let A, B, C be three collinear points, and let A', B', C' be three collinear points. Use Theorem 3.4 to prove that there is a transformation that maps A, B, C to A', B', C', respectively. (We use this result in Exercises 3.11, 3.15, and 7.14.)

3.11. (a) Consider any transformation that fixes the origin, the point $(1, 0)$ in the Euclidean plane, and the point at infinity on horizontal lines. Prove that there are real numbers a, b, e, h such that $a \neq 0, e \neq 0$, and the transformation maps

$$(x, y, z) \to (ax + by, ey, hy + az).$$

Conclude that the transformation fixes every point on the x-axis.
(b) Let A, B, C be three points on a line l. Use part (a) and Exercise 3.10 to prove that every transformation that fixes A, B, C also fixes every point of l. (We use this exercise in Exercises 5.18–5.22, 6.17–6.20, and 13.6–13.12.)

3.12. Let l and m be two lines that do not contain a point T. Prove that there is a transformation that maps X to $TX \cap m$ for each point X of l. (*Hint*: One possible approach is to use Theorem 3.4 to reduce to the case where l is the x-axis, m is the y-axis, and T is the point at infinity on lines of slope -1.)

3.13. Consider a transformation that maps a line l to a line $m \neq l$. Prove that the transformation fixes $l \cap m$ if and only if there is a point T lying on neither l nor m such that the transformation maps X to $TX \cap m$ for every point X of l.

(*Hint*: If the given transformation fixes $l \cap m$, why is there a point T such that the transformation in Exercise 3.12 agrees with the given transformation on $l \cap m$ and two other points of l? Why does it follow from Exercise 3.11(b) that the two transformations agree on every point of l?)

3.14. Let A, B, C, D be four points, no three of which are collinear, in the projective plane. Prove that $AB \cap CD$, $AC \cap BD$, and $AD \cap BC$ are three noncollinear points. (This exercise is used in Exercises 5.22, 6.18, and 13.11. One possible approach to this exercise is to use Theorem 3.4 to reduce to the case where A–D are particular points and direct computation can be used.)

3.15. Let A, B, C, D be four collinear points. Prove that there is a transformation that interchanges A with C and B with D. (This exercise is used in Exercise 5.22. It may be helpful in doing this exercise to use Exercise 3.10 to transform A, B, C into three particular collinear points.)

3.16. Consider a curve of the form $y = f(x)$, where $f(x)$ is a polynomial in x of positive degree n. Prove that the curve has exactly one point P at infinity, that it intersects every vertical line exactly $n - 1$ times at P, and that it intersects the line at infinity exactly n times at P.

(The case $n = 1$ may require separate consideration.)

3.17. Four points, no three of which are collinear, are given in each part of this exercise. There is a transformation that maps the points $(1,0,0)$, $(0,1,0)$, $(0,0,1)$, $(1,1,1)$ to the four given points, by Theorem 3.4. Find equations as in (5) that give such a transformation. Recall that the homogeneous coordinates of a point can be multiplied by a nonzero number t without affecting the point.

(a) $(0,2,1)$, $(1,2,-1)$, $(0,1,0)$, $(1,3,2)$.
(b) $(1,1,0)$, $(1,2,0)$, $(0,1,1)$, $(0,1,-1)$.
(c) $(3,0,5)$, $(0,1,2)$, $(1,0,-1)$, $(3,-1,4)$.
(d) $(1,1,1)$, $(1,0,1)$, $(0,1,2)$, $(1,0,0)$.

3.18. (a) If a transformation fixes each of the points $(1,0,0)$, $(0,1,0)$, $(0,0,1)$, and $(1,1,1)$, prove that the transformation has the form $x' = tx$, $y' = ty$, $z' = tz$ for a nonzero number t. Conclude that the transformation fixes every point.

(b) In the projective plane, let A, B, C, D be four points, no three of which are collinear, and let A', B', C', D' be four points, no three of which are collinear. Theorem 3.4 states that there is a transformation that maps A, B, C, D to A', B', C', D'. Use part (a) and Theorem 3.4 to prove that this transformation is unique; that is, if two transformations map A, B, C, D to A', B', C', D', prove that every point has the same image under both transformations.

3.19. Consider the following result:

Theorem

In the projective plane, let N, A, A' be three collinear points, and let l be a line that does not contain A or A'. Then there is a transformation that fixes N and every point of l, maps A to A', and sends each point X to a point X' collinear with X and N.

(a) Prove the theorem when N is the origin and l is the line at infinity by considering the transformations $x' = rx$, $y' = ry$, $z' = z$ for nonzero numbers r.

(b) Prove the theorem when l is the line at infinity and N is the point at infinity on vertical lines by considering the transformations $x' = x$, $y' = y + kz$, $z' = z$ for nonzero numbers k.

(c) Prove the theorem in general by combining parts (a) and (b) with Theorem 3.4.

3.20. This exercise contains the proof of the following result (Figure 3.6):

Desargues' Theorem

Let A, C, E, A', C', E' be distinct points such that no two of the lines $AC, CE, AE, A'C', C'E', A'E', AA', CC', EE'$ are equal. Set $P = AC \cap A'C'$, $Q = AE \cap A'E'$, and $R = CE \cap C'E'$. Then the lines AA', CC', EE' are concurrent if and only if the points P, Q, R are collinear.

(a) Prove that $P \neq Q$. Set $l = PQ$ and prove that neither A nor A' lies on l. Set $N = AA' \cap CC'$, and prove that neither A nor A' equals N.

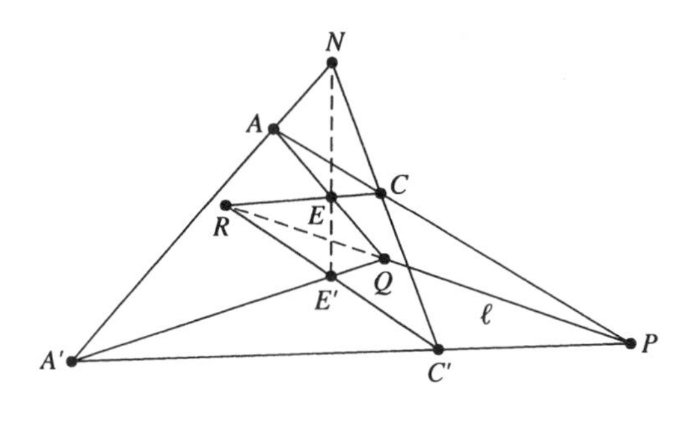

Figure 3.6

(b) If AA', CC', EE' are concurrent, prove that the transformation in the theorem in Exercise 3.19 maps C to C' and E to E', and deduce that P, Q, R are collinear.

(c) If P, Q, R are collinear, prove that the transformation in the theorem in Exercise 3.19 maps C to C' and E to E', and conclude that AA', CC', EE' are concurrent.

3.21. Use Desargues' Theorem from Exercise 3.20 to prove the following results:
(a) The theorem in Exercise 2.8.
(b) The theorem in Exercise 2.9.

3.22. Let $F(x, y, z)$ be a homogeneous polynomial. Prove that $F(x, y, z)$ is the homogenization of a polynomial in x and y if and only if $F(x, y, z)$ does not have z as a factor.

3.23. Let $F(x, y, z)$ be a homogeneous polynomial, let $f(x, y) = F(x, y, 1)$ be the restriction of F to the Euclidean plane, and let $F_1(x, y, z)$ be the homogenization of f. Prove that $F = z^s F_1$, where z^s is the highest power of z that can be factored out of F.

3.24. Prove that any factor of a homogeneous polynomial is itself homogeneous.

3.25. If $f(x, y)$ is a nonzero polynomial in two variables, prove that the curves $f(x, y) = 0$ and $f(x + h, y + k) = 0$ have the same points at infinity for any real numbers h and k. (Thus, translating a curve does not affect its points at infinity.)

3.26. Prove that a transformation fixes every point at infinity if and only if there are numbers s, h, k such that s is nonzero and the transformation maps (x, y) to $(sx + h, sy + k)$ for each point (x, y) in the Euclidean plane. Show that such a transformation exists for any numbers s, h, k such that $s \neq 0$.

§4. Lines and Tangents

We take the general results of the first three sections and use them now to analyze the intersections of lines and curves. We prove that a line intersects a curve of degree n that does not contain it at most n times, counting multiplicities; this means that the sum of the multiplicities of the intersections is at most n. This is the first of many results about the geometric significance of the degree of a curve.

In the second half of this section, we analyze the number of times that a line and a curve intersect at a point. If P is a point of a curve F, and if there is a unique line that intersects F more than once at P; we call this line the tangent to F at P. We show that this is equivalent to using implicit differentiation to find tangents to curves, as in first-year calculus. We end the section by using tangents to characterize pairs of curves that intersect more than once at a point.

We start by observing that any two lines in the projective plane should intersect with multiplicity 1 because their intersection is as simple as possible. We give a formal proof by transforming the lines to the x- and y-axes and applying Property 1.4.

Theorem 4.1
Any two lines in the projective plane intersect with multiplicity 1 *at their unique point of intersection.*

Proof
Let l and m be the two given lines. They intersect at a unique point P (by Theorem 2.1). Let Q be a second point on l, and let R be a second point on m. There is a transformation that maps P to the origin O, Q to a second point on the y-axis $x = 0$, and R to a second point on the x-axis $y = 0$ (by Theorem 3.4). This transformation maps l and m to $x = 0$ and $y = 0$, and so we have $I_P(l, m) = I_O(x, y) = 1$ (by Properties 3.5, 3.1, and 1.4). □

Our goal in the first half of this section is to generalize Theorem 4.1 by determining the number of times, counting multiplicities, that a line intersects any algebraic curve in the projective plane. Theorem 1.11 determines the number of times that a curve of the form $y = f(x)$ intersects a curve $g(x, y) = 0$ at the origin when the second curve does not contain the first. We now find the number of times that these curves intersect at any point $(a, f(a))$ on $y = f(x)$ in the Euclidean plane. We do so by using Theorem 3.7(ii) to translate $(a, f(a))$ to the origin and applying Theorem 1.11.

Theorem 4.2
Let $y = f(x)$ and $g(x, y) = 0$ be curves, and let a be a real number. If

$$g(x, f(x)) = (x - a)^s h(x) \tag{1}$$

for an integer $s \geq 0$ and a polynomial $h(x)$ such that $h(a) \neq 0$, then s is the number of times that $y = f(x)$ and $g(x, y) = 0$ intersect at the point $(a, f(a))$.

Proof

Theorem 3.7(ii) shows that the intersection multiplicity of the curves $y - f(x)$ and $g(x, y)$ at $(a, f(a))$ equals the intersection multiplicity of the curves

$$y + f(a) - f(x + a) \qquad \text{and} \qquad g(x + a, y + f(a)) \tag{2}$$

at the origin. We think of the first polynomial in (2) as y minus the quantity $f(x + a) - f(a)$. Substituting this quantity for y in the second polynomial in (2) gives

$$g(x + a, f(x + a)). \tag{3}$$

This polynomial is nonzero, because it becomes $g(x, f(x))$ if we substitute $x - a$ for x, and (1) and the assumption that $h(a) \neq 0$ show that $g(x, f(x))$ is nonzero. Thus, the first polynomial in (2) is not a factor of the second (by Theorem 1.9(ii)). Moreover, the first polynomial in (2) takes the value zero when $x = 0$ and $y = 0$. Hence, Theorem 1.11 shows that the number of times that the curves in (2) intersect at the origin is the smallest degree of any nonzero term in (3).

Substituting $x + a$ for x in (1) shows that

$$g(x + a, f(x + a)) = x^s k(x),$$

where $k(x) = h(x + a)$ is a polynomial such that $k(0) = h(a) \neq 0$. It follows that s is the smallest degree of any nonzero term of (3), since the fact that $k(0) \neq 0$ means that the constant term of $k(x)$ is nonzero. Together with the first and last sentences of the previous paragraphs, this shows that $y = f(x)$ and $g(x, y) = 0$ intersect s times at $(a, f(a))$. □

To find the points in the Euclidean plane where curves $y = f(x)$ and $g(x, y) = 0$ intersect, we naturally substitute $f(x)$ for y in $g(x, y) = 0$ and take the roots of $g(x, f(x)) = 0$. This commonsense procedure works for multiple intersections as well: the number of times that $y = f(x)$ intersects $g(x, y) = 0$ at a point $(a, f(a))$ is the number of times that $x - a$ is a factor of $g(x, f(x))$. This is the gist of Theorem 4.2, which we restate as follows. We use the next result to study the intersections of lines and curves in Theorems 4.4 and 4.5 and conics and curves in Theorems 5.8 and 5.9.

Theorem 4.3

Let $y = f(x)$ and $g(x, y) = 0$ be curves in the Euclidean plane. If $y - f(x)$ is not a factor of $g(x, y)$, we can write

$$g(x, f(x)) = (x - a_1)^{s_1} \cdots (x - a_v)^{s_v} r(x), \tag{4}$$

where the a_i are distinct real numbers, the s_i are positive integers, and $r(x)$ is a polynomial that has no real roots. Then $y = f(x)$ and $g(x, y) = 0$ intersect

s_i times at the point $(a_i, f(a_i))$ for $i = 1, \ldots, v$, and these are the only points of intersection in the Euclidean plane.

Proof

Since $y - f(x)$ is not a factor of $g(x, y)$, the polynomial $g(x, f(x))$ is nonzero (by Theorem 1.9(ii)). Factor as many polynomials of degree 1 as possible out of $g(x, f(x))$. The number of factors cannot exceed the degree of $g(x, f(x))$ because $g(x, f(x))$ is nonzero. When the process of factorization ends, the remaining factor $r(x)$ has no factors of degree 1, and so it has no real roots (by Theorem 1.10(ii)). Thus, we can factor $g(x, f(x))$ as in (4).

Since $a_i \neq a_j$ for $i \neq j$, (4) shows that

$$g(x, f(x)) = (x - a_i)^{s_i} h(x),$$

where $h(x)$ is a polynomial such that $h(a_i) \neq 0$. Thus, s_i is the number of times that $y = f(x)$ and $g(x, y) = 0$ intersect at the point $(a_i, f(a_i))$ (by Theorem 4.2). If a is any real number other than $a_1, \ldots, a_v$, (4) shows that $g(a, f(a)) \neq 0$, and so the curve $g(x, y) = 0$ does not contain the point $(a, f(a))$ and does not intersect $y = f(x)$ there (by Theorem 3.6(iii)) and Definition 3.2). Likewise, $y = f(x)$ does not intersect $g(x, y) = 0$ at any point (a, b) in the Euclidean plane with $b \neq f(a)$, since these points do not lie on $y = f(x)$. □

In short, if a curve has the special form $y = f(x)$, Theorem 4.3 gives the multiplicities of all of its intersections in the Euclidean plane with any curve $g(x, y) = 0$ that does not contain it. We simply substitute $f(x)$ for y in $g(x, y)$ and factor the resulting polynomial $g(x, f(x))$. In order to apply Theorem 4.3, we must check that $y - f(x)$ is not a factor of $g(x, y)$, but we can do so simply by checking that $g(x, f(x))$ is nonzero (by Theorem 1.9(ii)).

In either the Euclidean or the projective plane, we say that two curves *intersect d times, counting multiplicities*, if d is the sum of the intersection multiplicities of the curves at all points in the plane. Let $g(x, y) = 0$ be a curve of degree n, and let $y = mx + b$ be a nonvertical line that does not lie entirely on the curve. Because the degree of $g(x, mx + b)$ is at most the degree n of g, Theorem 4.3 shows that the line intersects the curve at most n times, counting multiplicities, in the Euclidean plane. We extend this result to the projective plane in Theorem 4.5. We start with the special case where the line is the x-axis $y = 0$. We single this case out so that we can return to it in the proofs of Theorems 5.2, 6.4, 9.1, and 11.1.

Theorem 4.4

Let $G(x, y, z)$ be a homogeneous polynomial of degree n that does not have y as a factor. If we set $g(x, y) = G(x, y, 1)$, we can write

$$g(x, 0) = (x - a_1)^{s_1} \cdots (x - a_v)^{s_v} r(x), \tag{5}$$

for distinct real numbers a_i, positive integers s_i, and a polynomial $r(x)$ that has no real roots. Then the number of times, counting multiplicities, that the curve $G = 0$ intersects the x-axis $y = 0$ in the projective plane is the degree n of G minus the degree of $r(x)$.

Proof

Since G is homogeneous of degree n, we can write

$$G(x, y, z) = \sum e_{ij}x^i y^j z^{n-i-j}. \tag{6}$$

The terms without y's are those of the form $e_{i0}x^i z^{n-i}$, where $j = 0$. At least one of the coefficients e_{i0} is nonzero because y is not a factor of G. The corresponding term $e_{i0}x^i$ in

$$g(x, y) = G(x, y, 1) = \sum e_{ij}x^i y^j \tag{7}$$

is nonzero, and so y is not a factor of $g(x, y)$. Thus, by Theorem 4.3, we can factor $g(x, 0)$ as in (5), and the total number of times, counting multiplicities, that $g(x, y)$ intersects the x-axis $y = 0$ in the Euclidean plane is $s_1 + \cdots + s_v$. This is also the total number of times, counting multiplicities, that $G(x, y, z) = 0$ intersects $y = 0$ in the Euclidean plane (by Theorem 3.7(iii)).

Equation (5) shows that $s_1 + \cdots + s_v$ is the degree of $g(x, 0)$ minus the degree of $r(x)$. Setting $y = 0$ in (7) gives

$$g(x, 0) = \sum e_{i0}x^i,$$

the sum of the terms in (7) that do not contain y. Thus, the degree of $g(x, 0)$ is the largest integer d such that $e_{d0} \neq 0$; such an integer exists because the e_{i0} are not all zero. In short, the number of times, counting multiplicities, that $G = 0$ and $y = 0$ intersect in the Euclidean plane is d minus the degree of $r(x)$, where d is the largest integer such that $e_{d0} \neq 0$.

How many times do $G = 0$ and $y = 0$ intersect at infinity? The only possible point of intersection at infinity is $(1, 0, 0)$ (by Theorem 3.6(iii) and the fact that this is the only point at infinity on $y = 0$). Interchanging x and z shows that the number of times that $G(x, y, z) = 0$ and $y = 0$ intersect at $(1, 0, 0)$ equals the number of times that $G(z, y, x) = 0$ and $y = 0$ intersect at $(0, 0, 1)$ (by Property 3.5 and the discussion accompanying (8) of Section 3). Setting $z = 1$ shows that this is the number of times that the curves $G(1, y, x) = 0$ and $y = 0$ in the Euclidean plane intersect at the origin (by Property 3.1). Replacing x by 1 and z by x in (6) shows that

$$G(1, y, x) = \sum e_{ij}y^j x^{n-i-j}. \tag{8}$$

At least one of the terms $e_{i0}x^{n-i}$ with $j = 0$ has nonzero coefficient, and so y is not a factor of $G(1, y, x)$. Thus, the number of times that $G(1, y, x) = 0$ and $y = 0$ intersect at the origin is the smallest degree of any nonzero term of

$$G(1,0,x) = \sum e_{i0}x^{n-i}$$

(by Theorem 1.11 and (8)). This exponent is $n-d$, because $n-i$ decreases as i increases and d is the largest integer such that $e_{d0} \neq 0$. In short, $G(x,y,z) = 0$ and $y = 0$ intersect $n-d$ times, counting multiplicities, at infinity.

We obtain the total number of intersections, counting multiplicities, of $G = 0$ and $y = 0$ in the projective plane by summing the number of intersections in the Euclidean plane and the number at infinity. We have seen that the number in the Euclidean plane is d minus the degree of $r(x)$, and the number at infinity is $n-d$, and so the sum is n minus the degree of $r(x)$. □

We can now prove that any line intersects any curve of degree n that does not contain it at most n times in the projective plane, counting multiplicities. We need one preliminary observation. If a transformation mapping (x,y,z) to (x',y',z') takes homogeneous polynomials $F(x,y,z)$ and $G(x,y,z)$ to $F'(x',y',z')$ and $G'(x',y',z')$, then F is a factor of G if and only if F' is a factor of G'. In fact, using the equations in (6) of Section 3 to substitute for x, y, z changes an equation

$$G(x,y,z) = F(x,y,z)H(x,y,z)$$

into an equation

$$G'(x',y',z') = F'(x',y',z')H'(x',y',z'),$$

where H and H' are homogeneous polynomials, and this process can be reversed because transformations can be reversed.

Theorem 4.5

Let $L = 0$ be a line, and let $G = 0$ be a curve of degree n. If L is not a factor of G, then L and G intersect at most n times, counting multiplicities, in the projective plane.

Proof

There is a transformation that maps two points of L to two points on the x-axis $y = 0$ (by Theorem 3.4). This transformation takes the line L and the curve G to the line $y = 0$ and a curve G' of degree n, as discussed after the proof of Theorem 3.4. G' does not have y as a factor, by the discussion before this theorem and the assumption G does not have L as a factor. Thus, $y = 0$ intersects $G' = 0$ at most n times, counting multiplicities, in the projective plane (by Theorem 4.4). Because transformations preserve intersection multiplicities (by Property 3.5), L and G intersect at most n times, counting multiplicities, in the projective plane. □

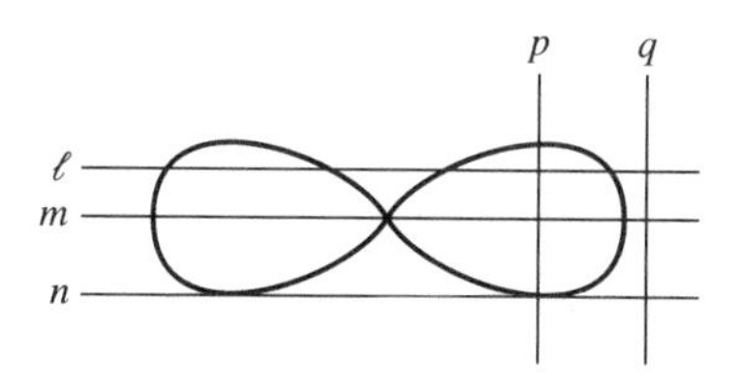

Figure 4.1

In short, a line intersects a curve of degree n which does not contain it at most n times, even when we count multiple intersections and intersections at infinity. For example, because (2) of the Introduction to Chapter I has degree 4, every line intersects the curve in Figure I.1 at most four times. Figure 4.1 shows the same curve, which appears to have four intersections with the lines l, m, and n (including one double intersection with m and two with n), two intersections with p, and none with q. Similarly, because the curve $y = x^3$ in Figure 1.3 has degree 3, it intersects any line at most three times. Theorem 4.1 shows that we can omit the words "at most" in Theorem 4.5 when $n = 1$: one line intersects another—that is, another curve of degree 1—exactly once, counting multiplicities.

Of course, we had to assume in Theorem 4.5 that L is not a factor of G. When L is a factor of G, they intersect infinitely many times at every point of L (by Theorem 3.6(vi)).

In order to introduce tangents to curves, we must analyze more carefully the number of times that a line and a curve intersect at a point. We start by looking at the origin.

Let $g(x, y) = 0$ be a curve that contains the origin. Then $g(x, y)$ has no constant term, and so we can write

$$g(x, y) = sx + ty + h(x, y),$$

where $h(x, y)$ is a polynomial in which every term has degree at least 2. We consider the intersection multiplicity of $g(x, y) = 0$ and the line $y = mx$ at the origin. If $g(x, mx)$ is nonzero, the intersection multiplicity is the smallest degree of any nonzero term in

$$g(x, mx) = sx + tmx + h(x, mx)$$

after collecting powers of x (by Theorems 1.9(ii) and 1.11). If $g(x, mx)$ is zero, the intersection multiplicity is ∞ (by Theorem 1.7, since Theorem 1.9(ii) shows that $y - mx$ is a factor of $g(x, y)$ in this case). Thus, the intersection multiplicity is at least 2 if and only if $s + tm$ equals 0, since every term of $h(x, mx)$ has degree at least 2. If $s + tm = 0$, then either s and t are both 0, or else $t \neq 0$ and the equations

$$sx + ty = -tmx + ty = t(y - mx)$$

show that $sx + ty = 0$ is the same line as $y = mx$. Conversely, $s + tm$ equals 0 if s and t are both 0. If $sx + ty = 0$ is the same line as $y = mx$, then t is nonzero and $y = (-s/t)x$ is the same line as $y = mx$, and so we have $m = -s/t$ and $s + tm = 0$. We have proved the following result for lines of the form $y = mx$:

Theorem 4.6
In the Euclidean plane, let l be a line through the origin, and let $g(x, y) = 0$ be a curve through the origin. Write

$$g(x, y) = sx + ty + h(x, y),$$

where $h(x, y)$ is a polynomial in which every term has degree at least 2. Then l and g intersect at least twice at the origin if and only if either s and t are both 0 or else $sx + ty = 0$ is the line l.

Proof
We have already proved this for lines of the form $y = mx$, and so the only line remaining is $x = 0$. Because the transformation switching x and y preserves intersection multiplicities (by Properties 3.1 and 3.5 and the discussion accompanying (8) of Section 3), and because we have proved the result when l is the line $y = 0$, it also holds when l is the line $x = 0$.

□

We can restate this theorem slightly by fixing the curve g, letting the line l vary, and using Properties 1.1 and 1.3. This gives the following result:

Theorem 4.7
In the Euclidean plane, let $g(x, y) = 0$ be a curve through the origin. Write

$$g(x, y) = sx + ty + h(x, y),$$

where $h(x, y)$ is a polynomial in which every term has degree at least 2.

(i) *If $s = 0 = t$, then every line through the origin intersects g at least twice at the origin.*
(ii) *If s and t are not both zero, then $sx + ty = 0$ is the unique line that intersects g more than once at the origin. Every other line through the origin intersects g exactly once there.* □

We can generalize Theorem 4.7 from the origin to any point P in the projective plane because there is a transformation that maps P to the origin and preserves intersection multiplicities (by Theorem 3.4 and Properties 3.1 and 3.5). Thus, Theorem 4.7 extends to the following result:

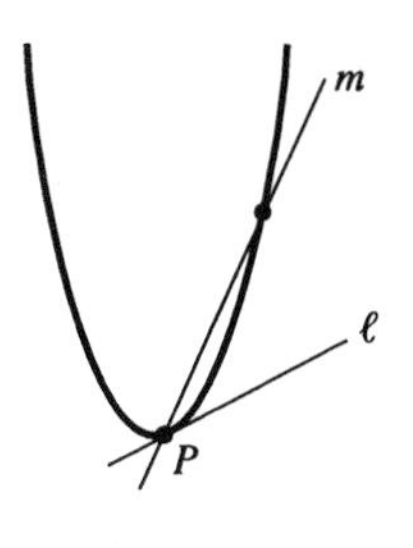

Figure 4.2

Theorem 4.8
Let P be a point on a curve $G(x, y, z) = 0$ in the projective plane. Then one of the following two conditions holds:

(i) *Every line through P intersects G at least twice at P.*
(ii) *There is a unique line that intersects G more than once at P. Every other line through P intersects G exactly once there.* □

Theorem 4.8 leads to the following definition:

Definition 4.9
Let P be a point on the curve $G(x, y, z) = 0$ in the projective plane. G is called *singular* at P if condition (i) of Theorem 4.8 holds, and *nonsingular* at P if condition (ii) holds. When condition (ii) holds, the unique line that intersects G more than once at P is called the *tangent* or *tangent line* to G at P. □

Intuition supports Definition 4.9. As we noted before Theorem 1.7, the intersection multiplicity of two curves at a point seems to measure how closely the curves approach each other there. Accordingly, Definition 4.9 characterizes the tangent l to a curve G at a point P as the line that best approximates the curve there. As in Figure 4.2, we can think of the multiple intersection of l and G at P as the coalescence of distinct intersections of G and a secant line m through P.

By Definition 4.9, a curve does not have a tangent at a singular point. This is a point where the curve has a complicated structure, such as the origin in Figure I.1 of the Introduction to Chapter I. In fact, Theorem 4.7 and Definition 4.9 show that (2) of the chapter introduction has a singular point at the origin because it has no terms of degree less than 2.

We have defined singularity and tangents of curves in terms of intersection multiplicities. Because transformations preserve intersection multiplicities (by Property 3.5), they preserve singularity and tangents of curves. Specifically, suppose that a transformation maps a curve F to a curve F' and maps a point P on F to a point P'. Then F is nonsingular at

P if and only if F' is nonsingular at P', and, if so, the transformation maps the tangent to F at P to the tangent to F' at P'.

Let $g(x, y)$ be a polynomial, and let (a, b) be a point in the Euclidean plane. We can write $g(x, y)$ as a sum of powers of $x - a$ and $y - b$ as follows. By substituting $x = x' + a$ and $y = y' + b$ in $g(x, y)$ and collecting terms, we can write

$$g(x' + a, y' + b) = \sum e_{ij} x'^i y'^j$$

for real numbers e_{ij}. Substituting $x' = x - a$ and $y' = y - b$ gives

$$g(x, y) = \sum e_{ij} (x - a)^i (y - b)^j, \tag{9}$$

which expands $g(x, y)$ in powers of $x - a$ and $y - b$. Readers familiar with multivariable calculus may recognize (9) as the Taylor expansion of $g(x, y)$ about (a, b).

We use (9) to translate Theorem 4.7 from the origin to any point in the Euclidean plane.

Theorem 4.10

Let (a, b) be a point on the curve $g(x, y) = 0$ in the Euclidean plane. We can write

$$g(x, y) = s(x - a) + t(y - b) + \sum e_{ij} (x - a)^i (y - b)^j, \tag{10}$$

where $i + j \geq 2$ for every term in the sum. Then g is nonsingular at (a, b) if and only if s and t are not both zero. Moreover, in this case, the tangent to g at (a, b) is the line

$$s(x - a) + t(y - b) = 0. \tag{11}$$

Proof

We have seen that we can write $g(x, y)$ in the form of (9). Because $g(a, b) = 0$, the constant term e_{00} in (9) is zero, and we can write $g(x, y)$ as in (10). Substituting $x = x' + a$ and $y = y' + b$ in (10) gives

$$g(x' + a, y' + b) = sx' + ty' + \sum e_{ij} x'^i y'^j,$$

where $i + j \geq 2$ for every term in the sum. By Theorem 4.7, s and t are not both zero if and only if there is a unique line that intersects $g(x' + a, y' + b) = 0$ more than once at the origin, and, if so, $sx' + ty' = 0$ is that line. Substituting $x' = x - a$ and $y = y' - b$ and applying Theorem 3.7(ii) shows that s and t are not both zero if and only if there is a unique line that intersects $g(x, y)$ more than once at (a, b), and, if so, (11) is that line. □

As in first-year calculus, we can use implicit differentiation with respect to x or y to find the tangent line to a curve $g(x, y) = 0$ in the

Euclidean plane at a point (a, b) on the curve. We claim that this gives the same tangent lines as Definition 4.9.

By Theorem 4.10, we can write $g(x, y)$ as in (10). Setting this expression equal to zero and differentiating implicitly with respect to x gives

$$s + t\,\frac{dy}{dx} + \sum \left[ie_{ij}(x-a)^{i-1}(y-b)^j + je_{ij}(x-a)^i(y-b)^{j-1}\frac{dy}{dx}\right] = 0, \quad (12)$$

where $i + j \geq 2$ for every term in the sum. When we evaluate this equation at $(x, y) = (a, b)$, every term in the sum is zero, since the fact that $i + j \geq 2$ implies that every term in the sum has a factor of $x - a$ or $y - b$. Thus, setting $x = a$ and $y = b$ in (12) gives

$$s + t\left(\left.\frac{dy}{dx}\right|_{(a,b)}\right) = 0.$$

If $t \neq 0$, we can rewrite this equation as

$$\left.\frac{dy}{dx}\right|_{(a,b)} = -\frac{s}{t}.$$

This shows that the tangent at (a, b), according to first-year calculus, is

$$y - b = -\frac{s}{t}(x - a),$$

which is equivalent to (11).

Similarly, if we write $g(x, y)$ as in (10), differentiate the equation $g(x, y) = 0$ implicitly with respect to y, and substitute $x = a$ and $y = b$, we obtain

$$s\left(\left.\frac{dx}{dy}\right|_{(a,b)}\right) + t = 0.$$

If $s \neq 0$, we can rewrite this equation as

$$\left.\frac{dx}{dy}\right|_{(a,b)} = -\frac{t}{s}.$$

According to first-year calculus, the tangent at (a, b) is

$$x - a = -\frac{t}{s}(y - b),$$

which is again equivalent to (11).

The last two paragraphs show that we can use implicit differentiation with respect to x or y to find the tangent to $g(x, y) = 0$ at (a, b) if and only if the numbers s and t in (10) are not both zero. This occurs if and only if g is nonsingular at (a, b) (by Theorem 4.10). Moreover, when this occurs, implicit differentiation with respect to x or y gives the same tangent line as Definition 4.9, by the last two paragraphs.

Intuition suggests two reasons why two curves on a point P would have a multiple intersection there. First, one of the curves could be singular at P; for example, the curve in Figure I.1 of the Introduction to Chapter I, which is singular at the origin, seems to intersect the x-axis twice there. Second, the two curves could approach each other so closely near P that they are tangent to the same line there, as in Figure 1.2 of Section 1. We end this section by formalizing these ideas when one of the curves is nonsingular at P.

Theorem 4.11

Let $F(x, y, z) = 0$ and $G(x, y, z) = 0$ be two curves on a point P in the projective plane, and assume that F is nonsingular at P. Then $I_P(F, G) \geq 2$ if and only if G is either singular at P or tangent to the same line there as F. Equivalently, $I_P(F, G) = 1$ if and only if G is nonsingular at P and tangent to a different line there than F.

Proof

Because F is nonsingular at P, it has a tangent there. There is a transformation that maps P to the origin and maps a second point on the tangent at P to a second point on the y-axis (by Theorem 3.4). We can replace F and G with their images under the transformation (by Property 3.5), and so we can assume that P is the origin and that F is tangent to the y-axis $x = 0$ at the origin. We can replace $F(x, y, z)$ and $G(x, y, z)$ with their restrictions $f(x, y) = F(x, y, 1)$ and $g(x, y) = G(x, y, 1)$ to the Euclidean plane (by Property 3.1).

Because $f(x, y) = 0$ is tangent to $x = 0$ at the origin, we can write

$$f(x, y) = sx + h(x, y), \tag{13}$$

where $s \neq 0$ and every term of h has degree at least 2 (by Theorem 4.7 and Definition 4.9). Since every term of f that is not divisible by x is divisible by y^2, we can write

$$f(x, y) = xp(x, y) + y^2q(y) \tag{14}$$

for polynomials p (in x and y) and q (in y alone). The constant term of $p(x, y)$ is the nonzero number s in (13), and so we have

$$p(0, 0) \neq 0. \tag{15}$$

Because $g = 0$ contains the origin, $g(x, y)$ has no constant term, and every term of g not divisible by x is divisible by y. Thus, we can write

$$g(x, y) = xu(x, y) + yv(y) \tag{16}$$

for polynomials u (in x and y) and v (in y alone).

By (14) and (16), we can rewrite $I_O(f, g)$ as

$$I_O(xp + y^2q, xu + yv), \tag{17}$$

where O is the origin $(0,0)$. Since $p(0,0) \neq 0$ (by inequality (15)), we can multiply the second polynomial in (17) by p without changing the intersection multiplicity (by Theorem 1.8). This gives

$$I_O(xp + y^2q, xpu + ypv).$$

We can subtract u times the first polynomial from the second (by Property 1.5), which gives

$$I_O(xp + y^2q, -y^2qu + ypv). \tag{18}$$

Factoring the second polynomial as $y(-yqu + pv)$ shows that the quantity in (18) equals

$$I_O(xp + y^2q, y) + I_O(xp + y^2q, -yqu + pv) \tag{19}$$

(by Property 1.6).

We can evaluate the first intersection multiplicity in (19) as follows. Since y^2q is a multiple of y, we have

$$\begin{aligned} I_O(xp + y^2q, y) &= I_O(xp, y) && \text{(by Property 1.5)} \\ &= I_O(x, y) && \text{(by Theorem 1.8 and (15))} \\ &= 1 && \text{(by Property 1.4).} \end{aligned}$$

Together with the previous paragraph, this shows that $I_O(f, g) \geq 2$ if and only if the second intersection multiplicity in (19) is at least 1.

We have

$$I_O(xp + y^2q, -yqu + pv) \geq 1$$

if and only if pv contains the origin (by Property 1.3), since the other terms xp, y^2q, and $-yqu$ contain the origin. By inequality (15), pv contains the origin if and only if v does. This is equivalent to the condition that $v(y)$ has no constant term. By (16), this happens exactly when $g(x, y)$ has no y term. This occurs when g is either singular at the origin or tangent there to $x = 0$ (by Theorem 4.7 and Definition 4.9). Since f is tangent to $x = 0$ at the origin, the theorem holds. □

Exercises

4.1. Consider the curve $x^2y = x + y$. Use Theorem 4.3 and the intersection properties to find the points of the projective plane where this curve intersects the following lines and to determine the multiplicity of each intersection. Determine the total number of intersections, counting multiplicities, and compare the result with Theorem 4.5. Illustrate your answers with a figure showing the curve, the line, and the points of intersection.

(a) $y = -x$. (b) $y = x$.
(c) $y = -x/2$. (d) $y = \frac{2}{3}$.
(e) $y = 0$. (f) $x = 2$.
(g) $x = 1$. (h) The line at infinity.

4.2. Follow the directions after the first sentence of Exercise 4.1 for the curve $x^2y + y = 2x^2$ and the following lines:
(a) $y = 0$. (b) $y = 1$.
(c) $y = 2$. (d) $y = 3$.
(e) $y = x/2$. (f) $y = x$.
(g) $y = 2x$. (h) $x = 0$.

4.3. Follow the directions after the first sentence of Exercise 4.1 for the curve $xy = x^2 - 1$ and the following lines:
(a) $y = x + 2$. (b) $y = x$.
(c) $y = 2x - 2$. (d) $y = 2x - 1$.
(e) $y = x/2$. (f) The line at infinity.

4.4. A curve $g(x, y)$ and a point (a, b) are given in each part of this exercise. For what real number k does (a, b) lie on the curve $g(x, y) = k$? By differentiating this equation implicitly with respect to x or y, as discussed after the proof of Theorem 4.10, determine whether the curve is nonsingular at (a, b) and, if so, find the equation of the tangent at (a, b).
(a) $x^3 - 3xy + 2y^3$, $(3, 1)$.
(b) $x^3 + xy^2 + 2y^3 - 2y$, $(2, -1)$.
(c) $x^3 + 6x^2 + 6xy + 4y^2 - 4y$, $(-2, 2)$.
(d) $x^3y + 5x^2 + y^3$, $(0, 2)$.
(e) $x^2 - 4xy + y^3 + 4y$, $(1, 0)$.
(f) $xy^2 + 5xy + 2x^2 - 3y$, $(1, -1)$.
(g) $x^2 - 3x^2y + y^3$, $(1, 2)$.
(h) $x^3 - 4x^2 + 4x - y^4 + 3y^2 + 2y$, $(2, 1)$.

4.5. Use Theorems 4.1 and 4.8 and Definition 4.9 to deduce that every line in the projective plane is nonsingular and equals its tangent at all of its points.

4.6. Let $F = 0$ be a curve of degree 3, and assume that F has no factors of degree 1.
(a) Prove that F has at most one singular point.
(*Hint*: If P is a singular point of F, and if Q is another point of F, one possible approach is to use Theorem 4.5 and Definition 4.9 to determine how many times line PQ intersects F at P and Q.)
(b) Prove that no line is tangent to F at more than one point.
(c) Prove that no line tangent to F contains a singular point of F.

4.7. (a) Let $F(x, y, z)$ be a homogeneous polynomial of degree 2. Prove that the curve $F = 0$ in the projective plane is singular at a point P and contains at least one other point if and only if we can write $F = LM$ for lines $L = 0$ and $M = 0$ that contain P. (See the Hint to Exercise 4.6(a).)
(b) Find a homogeneous polynomial $F(x, y, z)$ of degree 2 such that the curve $F = 0$ is singular at one point and contains no other point.

4.8. (a) Let $F(x, y, z)$ be a homogeneous polynomial of degree 4 that has no factors of degree 1. Assume that there is a point P such that every line through P intersects F at least three times at P. Prove that P is the only singular point of F.

(b) Prove that $F(x, y, z) = y^3 z - x^4$ has no factors of degree 1 and intersects every line through the origin at least three times there.

4.9. Let s and t be positive integers.

(a) Let $F(x, y, z)$ be a homogeneous polynomial of degree n that has no factors of degree 1. Assume that F contains two points P and Q such that every line through P intersects F at least s times at P and every line through Q intersects F at least t times at Q. Prove that $n \geq s + t$.

(b) Prove that $F(x, y, z) = y^s z^t - x^{s+t}$ has no factors of degree 1, intersects every line through the origin at least s times there, and intersects every line through $(0, 1, 0)$ at least t times there.

4.10. Let $F_1, \ldots, F_k$ be homogeneous polynomials, and let P be a point in the projective plane. Prove that the product $F_1 \cdots F_k$ is nonsingular at P if and only if exactly one of the curves $F_i = 0$ contains P and this curve is nonsingular at P.

4.11. In the projective plane, let $L = 0$ be a line, and let $G = 0$ be a curve of degree n. Prove that L is tangent to G at more than $n/2$ points if and only if G has L but not L^2 as a factor.

4.12. In the projective plane, let F be a curve of degree n, and let L be a line that is not contained in F.

(a) Prove that L and F cannot intersect exactly $n - 1$ times, counting multiplicities, in the projective plane.

(b) More generally, prove that L and F intersect $n - 2k$ times, counting multiplicities, in the projective plane, where k is an integer with $0 \leq k \leq n/2$. Use the fact, which follows from the Intermediate Value Theorem, that every polynomial $f(x)$ of odd degree in one variable has a root over the real numbers.

4.13. Let $m \geq 0$ and $n > 0$ be integers such that $m \leq n$ and $n - m$ is even. Let $s_1, \ldots, s_k$ be positive integers whose sum is m. Find a curve $F = 0$ of degree n such that F cannot be factored as a product of two polynomials of lower degree, and find a line $L = 0$ and distinct points $P_1, \ldots, P_k$ such that L and F intersect exactly s_i times at P_i for $i = 1, \ldots, k$ and have no other intersections.

4.14. Let $f(x)$ be a polynomial in x. Prove that the curve $y = f(x)$ has a singular point at infinity if and only if the degree of f is at least 3.

4.15. Let $f(x, y) = 0$ and $g(x, y) = 0$ be curves tangent to distinct lines $l = 0$ and $m = 0$ at a point (a, b) in the Euclidean plane. For any real number r, prove that $rf + g$ is tangent to $rl + m = 0$ at (a, b).

4.16. Let $f(x, y)$ and $g(x, y)$ be polynomials, and let (a, b) be a point of the Euclidean plane where the curves f and g intersect.

(a) If f is nonsingular at (a, b) and g is singular at (a, b), prove that $f + g$ is nonsingular at (a, b) and has the same tangent there as f.

(b) If f and g are both singular at (a, b), prove that $f + g$ is, as well.

(c) If f and g are nonsingular and tangent to the same line l at (a, b), prove that there is a unique real number s such that $f + sg$ is singular at (a, b) and that, for all other real numbers $r, f + rg$ is nonsingular and tangent to l at (a, b).

4.17. Let $f(x, y)$ be a polynomial of degree n, and let $F(x, y, z)$ be its homogenization. Let l be a line that intersects f a total of n times, counting multiplicities, in the Euclidean plane. Let $A_1, \ldots, A_n$ be the points of intersection of l and f in the Euclidean plane, with each point listed as many times as l and f intersect there; for example, if l and f intersect twice at a point, then the point appears twice in the list $A_1, \ldots, A_n$. Let $P = (v, w)$ be a point of l in the Euclidean plane.

(a) If l is a nonvertical line $y = mx + b$, prove that

$$f(x, mx + b) = F(1, m, 0)(x - r_1) \cdots (x - r_n),$$

where $r_1, \ldots, r_n$ are the x-coordinates of $A_1, \ldots, A_n$.

(b) If two points of the Euclidean plane lie on a nonvertical line of slope m in the Euclidean plane, prove that the distance between the points is $(m^2 + 1)^{1/2}$ times the absolute value of the difference between their x-coordinates.

(c) If l is a nonvertical line of slope m, use parts (a) and (b) to prove that the product of the distances from P to $A_1, \ldots, A_n$ is

$$\frac{(m^2 + 1)^{1/2} |f(v, w)|}{|F(1, m, 0)|}.$$

(d) State and prove the analogues of parts (a) and (c) when l is the vertical line $x = v$.

4.18. Let $f(x, y)$ be a polynomial of degree n. In the Euclidean plane, let a and b be two lines on a point P, and let c and d be lines parallel to a and b, respectively. (See Figure 4.3, which illustrates the case $n = 2$.) Let Q be the point of intersection of c and d. Assume that each of the lines a, b, c, d intersects f a total of n times, counting multiplicities, in the Euclidean plane. Let $A_1, \ldots, A_n$ be the points of the Euclidean plane where a and f

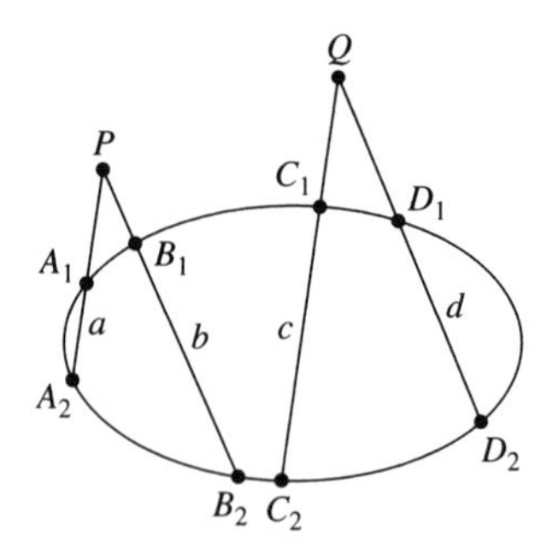

Figure 4.3

intersect, with each point listed as many times as a and f intersect there. Define points B_i, C_i, D_i for $i = 1, \ldots, n$ in the same way with respect to the lines b, c, d. Use Exercise 4.17 to prove that the product of the distances from P to the A_i divided by the product of the distances from P to the B_i equals the product of the distances from Q to the C_i divided by the product of the distances from Q to the D_i.

(In other words, we consider the ratio of the products of the distances from a point P to the points where two lines through P intersect f. Then the value of this ratio does not depend on the choice of P so long as the directions of the lines remain fixed and each line intersects f as many times as possible in the Euclidean plane. This result is due to Newton, who used it for $n = 2$ and $n = 3$ to study conics and cubics.)

4.19. Let A, S, T be three points on a line l in the Euclidean plane. The *division ratio* $\overline{AS}/\overline{AT}$ is $\pm$ the result of dividing the distance from A to S by the distance from A to T, where the minus sign is used when A lies between S and T (Figure 4.4) and the plus sign is used otherwise (Figures 4.5 and 4.6). If l is not vertical and A, S, T have x-coordinates a, s, t, prove that

$$\frac{\overline{AS}}{\overline{AT}} = \frac{s-a}{t-a}.$$

(This exercise is used in Exercises 4.20–4.22, 5.20, and 6.16–6.20.)

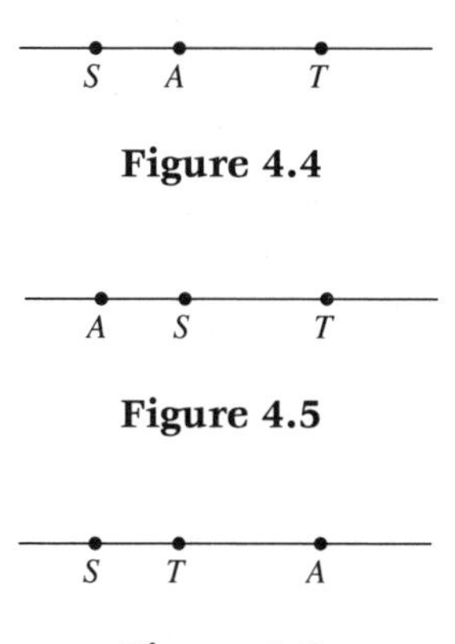

Figure 4.4

Figure 4.5

Figure 4.6

4.20. Define division ratios as in Exercise 4.19. Let $f(x, y) = 0$ be a curve of degree n. Let S and T be two points in the Euclidean plane that do not lie on f. Assume that f intersects line ST at n points $A_1, \ldots, A_n$ in the Euclidean plane, where each point is listed as many times as f intersects line ST there. Use Exercises 4.17(a) and (d) and 4.19 to prove that

$$\frac{\overline{A_1S}}{\overline{A_1T}} \cdots \frac{\overline{A_nS}}{\overline{A_nT}} = \frac{f(s, s')}{f(t, t')},$$

where (s, s') and (t, t') are the (x, y) coordinates of S and T.

4.21. Define division ratios as in Exercise 4.19. Use Exercise 4.20 to prove the following result:

Theorem

Let f be a curve of degree n. Let S, T, U be three points in the Euclidean plane that do not lie on f. Assume that f intersects line ST at n points $A_1, \ldots, A_n$, line TU at n points $B_1 \ldots, B_n$, and line US at n points $C_1, \ldots, C_n$, where the points A_i, B_i, C_i all lie in the Euclidean plane and are listed as many times as f intersects ST, TU, or US there. Then we have

$$\frac{\overline{A_1S}}{\overline{A_1T}} \cdots \frac{\overline{A_nS}}{\overline{A_nT}} \cdot \frac{\overline{B_1T}}{\overline{B_1U}} \cdots \frac{\overline{B_nT}}{\overline{B_nU}} \cdot \frac{\overline{C_1U}}{\overline{C_1S}} \cdots \frac{\overline{C_nU}}{\overline{C_nS}} = 1.$$

4.22. (a) Give a simple statement of the theorem in Exercise 4.21 when $n = 1$, and illustrate it with a figure. (This result, called *Menelaus' Theorem*, relates the ratios in which the three sides of a triangle STU are divided by their intersections with a line f.)

(b) In the Euclidean plane, let E, F, G, W be four points, no three of which are collinear. Assume that the lines EW and FG intersect at a point E', FW and GE intersect at at a point F', and GW and EF intersect at a point G'. Draw a figure to illustrate this arrangement of points and lines. Prove *Ceva's Theorem*, which states that

$$\frac{\overline{E'F}}{\overline{E'G}} \cdot \frac{\overline{F'G}}{\overline{F'E}} \cdot \frac{\overline{G'E}}{\overline{G'F}} = -1,$$

by applying Menelaus' Theorem from (a) to triangle $EE'F$ and line GW and to triangle $EE'G$ and line FW and combining the results.

4.23. Let $g(x, y)$ be a nonzero polynomial that contains the origin. Let d be the smallest degree of a nonzero term of g, and let $g_d(x, y)$ be the sum of the terms of degree d in g.

(a) Why can we factor

$$g_d(x, y) = (p_1x + q_1y)^{s_1} \cdots (p_kx + q_ky)^{s_k} r(x, y)$$

for distinct lines $p_ix + q_iy = 0$, where the s_i are positive integers and $r(x, y)$ is a polynomial that has no factors of degree 1?

(b) Let $l = 0$ be a line through the origin. Use Theorems 1.7, 1.9(ii), and 1.11 to prove that $I_O(l, g) > d$ if l is one of the lines $p_ix + q_iy = 0$ and that $I_O(l, g) = d$ otherwise.

(For example, if $g(x, y) = 0$ is the curve in (2) of the Introduction to Chapter I, we have $d = 2$ and

$$g_2(x, y) = -x^2 + y^2 = (-x + y)(x + y).$$

This exercise shows that every line through the origin except $y = x$ and $y = -x$ intersects g twice at the origin, and that these lines intersect g at least three times at the origin. Note the Figure I.1 suggests that $y = x$ and $y = -x$ are the lines that best approximate g at the origin. Exercises 1.2 and 1.3 provide further illustrations.)

4.24. Use Exercise 4.23 and Properties 3.1 and 3.5 to prove the following result:

Theorem
Let G be a curve and let P be a point in the projective plane. Then there is a nonnegative integer d such that all but a finite number of lines on P intersect G exactly d times there. All other lines on P intersect G more than d times there, and there are at most d such lines.

(We call P a *d-fold* point of G when the conditions of the theorem hold. Comparing this theorem with Definition 4.9 shows that $d = 1$ if and only if G is nonsingular at P.)

4.25. Define d-fold points as in Exercise 4.24. Consider the following result:

Theorem
Let F and G be two curves in the projective plane that intersect at a point P. Let P be a d-fold point of F and an e-fold point of G. If $1 \leq d \leq e$, then there is a curve H such that

$$I_P(F,G) = d + I_P(F,H)$$

and P is a k-fold point of H for $k \geq e-1$.

(a) Prove that there is a transformation that maps P to the origin O and maps F and G to curves F_1 and G_1 such that x^d has nonzero coefficient in $f(x,y) = F(x,y,1)$.
(b) Set $g(x,y) = G(x,y,1)$. Explain why we can write

$$f(x,y) = x^d p(x) + ys(x,y)$$

and

$$g(x,y) = x^e q(x) + yt(x,y)$$

for polynomials $p(x), q(x), s(x,y)$, and $t(x,y)$ such that $p(0) \neq 0$, every term of s has degree at least $d-1$, and every term of t has degree at least $e-1$.
(c) Deduce that

$$gp = fx^{e-d}q + y(tp - sx^{e-d}q),$$

and conclude that

$$I_O(f,g) = d + I_O(f, tp - sx^{e-d}q).$$

(d) Deduce that the theorem holds.

4.26. Define d-fold points as in Exercise 4.24. Deduce the following result from the theorem in Exercise 4.25.

Theorem
Let F and G be curves and let P be a point in the projective plane. If P is a d-fold point of F and an e-fold point of G, then we have $I_P(F,G) \geq de$.

(This exercise complements Theorem 4.11 by proving that two curves singular at a point intersect at least four times there. Exercise 14.6(a) does so in another way.)

4.27. Let $f(x, y) = 0$ and $g(x, y) = 0$ be curves in the Euclidean plane. In the terminology of Exercise 4.24, assume that the origin O is a d-fold point of f and an e-fold point of g. Let $f_d(x, y)$ be the sum of the terms of degree d in $f(x, y)$, and let $g_e(x, y)$ be the sum of the terms of degree e in $g(x, y)$.

(a) If $f_d(x, y)$ and $g_e(x, y)$ have no common factors of positive degree, prove that $I_O(f, g) = de$.

(b) If $f_d(x, y)$ and $g_e(x, y)$ have a common factor of positive degree, prove that $I_O(f, g) > de$.

(Theorem 4.11 is the case $d = 1 = e$ of this exercise. One possible approach to this exercise is to adapt the approaches of Exercises 4.25 and 4.26.)

4.28. Let d, e, and m be positive integers such that $m \geq de$. Find polynomials $f(x, y)$ and $g(x, y)$ such that, in the notation of Exercise 4.24, the origin is a d-fold point of f and an e-fold point of g and such that f and g intersect exactly m times at the origin. (Exercise 4.26 shows the need to assume that $m \geq de$.)

Conics

Introduction and History

Introduction

We developed the basic machinery for studying curves in Chapter I. We considered curves of degree 1, lines, in Section 4. We study curves of degree 2, conics and their degenerate forms, in this chapter. We consider curves of degree 3, cubics, in Chapter III.

We define a conic in Section 5 to be a nondegenerate curve of degree 2. We prove by completing squares that we can transform all conics into the same curve—for example, the unit circle or the parabola $y = x^2$. This is the algebraic equivalent of the geometric fact that all conics are sections of cones and, therefore, projections of circles. For example, Figure II.1 shows an ellipse K as a section of a cone and, consequently, as the projection of a circle C through a point O. Figures II.2 and II.3 show a parabola and a hyperbola as sections of a cone.

Because we can transform every conic into the parabola $y = x^2$, a statement holds for all conics if it is true for $y = x^2$ and is preserved by transformations. We use this idea in Section 5 to prove that a conic intersects any curve of degree n that does not contain it at most $2n$ times, counting multiplicities. This result holds for $y = x^2$ because its intersections with a curve $f(x, y) = 0$ of degree n correspond to the roots of the polynomial $f(x, x^2)$, which has degree at most $2n$ (although intersections at infinity must be considered as well).

We use a similar approach in Section 6 to prove that we can "peel off a conic" from the intersection of two curves of the same degree. Specifi-

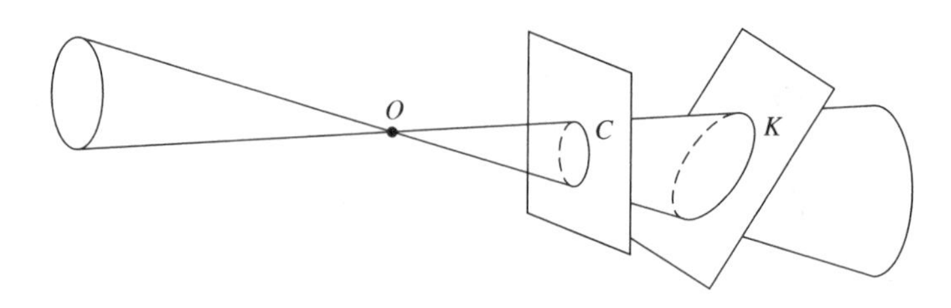

Figure II.1

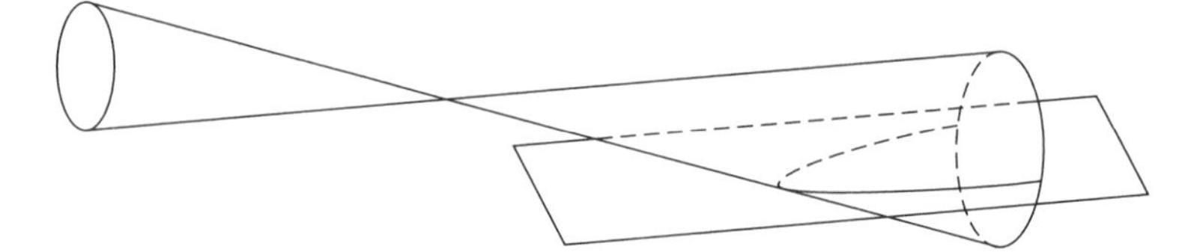

Figure II.2

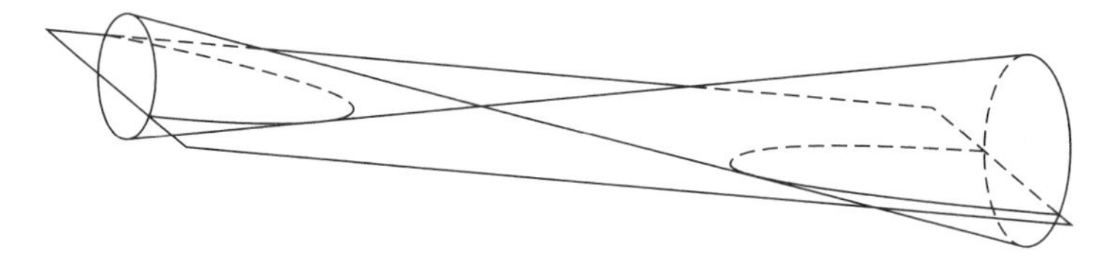

Figure II.3

cally, if two curves G and H of degree n intersect a conic at the same $2n$ points, then the remaining points of intersection of G and H are the points where each curve intersects a curve of degree $n-2$. This immediately gives Pascal's Theorem, which states that the three pairs of opposite sides of a hexagon inscribed in a conic intersect in collinear points. By considering multiple intersections, we obtain variations of Pascal's Theorem where sides of the hexagon are replaced by tangents to the conic. We also show that we can "peel off a line," and we use this result to prove Pappus' Theorem about a hexagon inscribed in two lines and, in Section 9, to derive the associative law of addition on a nonsingular, irreducible cubic.

We use homogeneous coordinates in Section 7 to show that we can dualize results about the projective plane by interchanging points and lines. Because this process interchanges the points of a conic with the tangents of a conic, Pascal's Theorem dualizes to Brianchon's Theorem, which states that the three pairs of opposite vertices of a hexagon circumscribed about a conic determine concurrent lines. We end Section 7 by using transformations between lines to construct the envelope of tangents of a conic.

History

Greek mathematicians such as Menaechmus, Aristaeus the Elder, and Euclid founded the study of conics in the fourth century B.C. Apollonius brought the subject to a high point in the third century B.C. By considering a conic as a section of a circular cone, he characterized the points of the conic by their distances from two lines. He deduced a wealth of geometric properties from this characterization, which is equivalent to the present-day equation of a conic. Apollonius, however, worked entirely in geometric terms, without algebraic notation.

Apollonius proved that a family of parallel chords of a conic have midpoints that lie on a line. Such a line is called a "diameter" of the conic, and Apollonius developed a number of connections between diameters and tangents. He also derived many properties of the foci of ellipses and hyperbolas. Apollonius founded the study of the "polar" of a point, which is a line determined by the point, with respect to a conic. Some of his results on polars are included in Exercise 13.6.

Euclid and Apllonius worked in Alexandria, the Egyptian city founded by Alexander the Great to be the capital of his empire and the intellectual center of many civilizations. Alexandria's distinguished tradition of geometry was revived in the third century A.D. by Pappus. We prove his theorem on hexagons inscribed in two lines as our Theorem 6.5. He also gave a geometric characterization of harmonic sets of points, which had been defined until then in terms of relative distances between points, as in Exercise 5.20 (b). The description of harmonic sets that we give before Exercise 5.18 is essentially that of Pappus.

In the first half of the 1600s, Girard Desargues reshaped the study of conics by introducing points at infinity and projections between planes. As in the discussion accompanying Figures II.1–II.3, the fact that all conics are sections of cones means that they are all projections of circles. Accordingly, if a property of circles is preserved by projections, then it holds for all conics. Desargues used this idea to redo and unify Apollonius' work on conics. He noted that diameters of conics are the polars of points at infinity, and he thereby derived many of Apollonius' results on diameters from properties of polars (as in Exercise 13.7.) Desargues' Involution Theorem, our Exercise 6.17, characterizes the pairs of points where the conics through four given points intersect a given line. In 1639, Blaise Pascal proved his famous theorem about hexagons inscribed in conics, our Theorem 6.2, by using Desargues' technique of projecting between planes to extend results from circles to conics.

At roughly the same time, Fermat began to use analytic geometry to study conics. He showed that equations of certain standard forms represent conics, and he claimed that any second degree equation can be reduced to one of these forms. In 1655, John Wallis proved conclusively that conics are exactly the nondegenerate curves of degree 2. By replacing geometric reasoning with algebra, he wrote the first treatment of conics that derived their properties directly from their equations.

In the first half of the 1800s, Gaspard Monge inspired a revival of interest in synthetic geometry that centered around projective geometry and conics. Charles Brianchon deduced his theorem on hexagons circumscribed about conics, our Theorem 7.6, by taking polars of the points in Pascal's Theorem on inscribed hexagons. He resolved long-standing problems about determining conics specified by five pieces of information, such as five points on the conic or four points and a tangent. Such problems date back at least to Pappus, and they fascinated Newton, who found complicated solutions based on analytic Euclidean geometry. Brianchon used synthetic projective geometry to obtain beautifully simple answers by applying Pascal's Theorem and its special cases, the results that follow from these by taking polars, and Desargues' Involution Theorem. Some of the simpler cases he analyzed are discussed after the proof of Theorem 6.2 and in Exercises 6.4–6.6 and 7.4–7.6.

Brianchon's use of polars is a special case of the duality principle, which states that we can interchange the roles of points and lines in the projective plane. Building on Brianchon's work, Jean Poncelet and Jacob Steiner developed duality as a general principle of projective geometry. Steiner and Michel Chasles gave geometric constructions of conics and, dually, their envelopes of tangents. Our Theorem 7.8 translates results of Steiner and Chasles into analytic form, using transformations between lines to construct envelopes of conics.

Julius Plücker clarified the logical basis of the duality principle when he justified the principle analytically in 1830. Following his approach, we show in Section 7 that we can simultaneously interchange the point (p, q, r) and the line $px + qy + rz = 0$ for all triples p, q, r of real numbers that are not all zero. We prove that this operation interchanges points of conics and tangents of conics.

Plücker developed duality into a powerful tool for studying algebraic curves of all degrees. The dual of the degree of a curve F is the *class* of F; this is the degree of the homogeneous polynomial in p, q, r whose solutions give the tangents $px + qy + rz = 0$ of F. The degree of F is the number of times, counting multiplicities, that F intersects any line it does not contain (if we consider points with complex coordinates, as in Theorem 11.1). Dually, the class of F is the number of tangents to F that pass through any point for which this number is finite. Plücker derived formulas that relate six quantities: first and second, the degree and class of F; third and fourth, the number of singular points of F of the two most basic types; fifth, the number of lines tangent to F at two different points; and, sixth,the number of inflection points of F. The duals of the third and fourth quantities are the fifth and sixth, and so the six quantities fall into three dual pairs. Plücker derived four formulas, which fall into two pairs, relating the six quantities. The only restrictions on F are that all of its singular points are of the two simplest types and the dual of this condition.

Plücker's role in the development of algebraic geometry was profound. Another of his key contributions was *abridged notation*, the technique of using a single letter to designate a polynomial instead of writing out every term. This technique is vital in studying curves because it makes algebraic combinations of polynomials easy to write. In particular, the families of curves $rF + G$ are important, where F and G are given curves of the same degree and r varies over all numbers. We use such families in Theorems 5.10, 6.1, 6.4, and 15.4 and Exercises 5.8, 5.11–5.15, 14.11, 14.12, 15.18, 15.22, and 15.23. Gabriel Lamé introduced abridged notation and the families $rF + G$ in 1818, a decade before Plücker, and Etienne Bobillier extended Lamé's work at the same time as Plücker. Nevertheless, it was Plücker who demonstrated the true power of abridged notation.

§5. Conics and Intersections

Conics are nondegenerate curves of degree 2 in the projective plane. We begin our study of conics in this section by proving that all conics can be transformed into one another. We discuss how to deduce theorems in the Euclidean plane about ellipses, parabolas, and hyperbolas from results about conics. Because any conic can be transformed into the parabola $y = x^2$, we can use Theorem 4.3 to deduce that a conic intersects any curve of degree n that does not contain it at most $2n$ times, counting multiplicities. It follows that any five points in the projective plane, no three of which are collinear, lie on a unique conic.

Our first goal is to classify the curves of degree 2 in the projective plane. These are the curves

$$ax^2 + bxy + cy^2 + dxz + eyz + fz^2 = 0, \tag{1}$$

where the coefficients a–f are not all zero. Setting $z = 1$ in (1) gives the curves

$$ax^2 + bxy + cy^2 + dx + ey + f = 0, \tag{2}$$

where a–f are not all zero. These are the curves of degree at most 2 in the Euclidean plane. They include two lines, one line doubled (with equation $(px + qy + r)^2 = 0$ for $p \neq 0$ or $q \neq 0$), one line, one point, and the empty set. We call these curves *degenerate*. Many precalculus and calculus books use rotations and translations to show that ellipses, parabolas, and hyperbolas are exactly the nondegenerate curves in the Euclidean plane given by (2). We prove the projective analogue of this result: any curve of degree 2 in the projective plane whose restriction to the Euclidean plane is nondegenerate can be transformed into $x^2 + y^2 = z^2$, the extension of the unit circle $x^2 + y^2 = 1$ to the projective plane. Transformations eliminate the distinctions among circles, ellipses, parabolas, and hyperbolas by interchanging points at infinity with points of the Euclidean plane and altering distances and angles in the Euclidean plane.

For any real numbers s and t, the equations

$$x' = x, \qquad y' = sx + y + tz, \qquad z' = z, \tag{3}$$

give a transformation because they are equivalent to the equations

$$x = x', \qquad y = -sx' + y' - tz', \qquad z = z'. \tag{4}$$

Likewise, the equations

$$x' = x + sy + tz, \qquad y' = y, \qquad z' = z, \tag{5}$$

give a transformation. We can also use transformations to interchange x, y, and z and to multiply them by nonzero numbers (by the discussions accompanying (8) and (9) of Section 3).

Theorem 5.1

Any curve of degree 2 in the projective plane can be transformed into one of the following curves:

(a) $x^2 = 0$, *a doubled line;*
(b) $x^2 + y^2 = 0$, *a point;*
(c) $x^2 - y^2 = 0$, *two lines;*
(d) $x^2 + y^2 + z^2 = 0$, *the empty set; and*
(e) $x^2 + y^2 - z^2 = 0$, *the unit circle.*

Proof

A curve of degree 2 has equation

$$ax^2 + bxy + cy^2 + dxz + eyz + fz^2 = 0, \tag{6}$$

where the coefficients a–f are not all zero. If the coefficients of x^2, y^2, and z^2 are all zero, the equations has the form

$$bxy + dxz + fyz = 0. \tag{7}$$

Because the coefficients are not all zero, we can assume that $b \neq 0$ (by using a transformation to interchange the variables, if necessary). Taking $s = -1$ and $t = 0$ in (3) and (4) gives a transformation that replaces y with $x' + y'$ and takes (7) to

$$bx(x+y) + dxz + f(x+y)z = 0,$$

where the coefficient of x^2 is now nonzero.

Thus, we can assume that the coefficients of x^2, y^2, and z^2 in (6) are not all zero. By interchanging the variables with a transformation, if necessary, we can assume that the coefficient a of x^2 is nonzero. We can divide (6) by a without changing the curve, as discussed after the proof of Theorem 3.6. By adjusting the values of b–f, we can assume that $a = 1$. We can eliminate the xy and xz terms by completing the square in x and rewriting (6) as

$$\left(x + \frac{b}{2}y + \frac{d}{2}z\right)^2 + cy^2 + eyz + fz^2 = 0$$

for revised values of c, e, and f. We transform this equation into

$$x^2 + cy^2 + eyz + fz^2 = 0 \tag{8}$$

by setting $x' = x + (b/2)y + (d/2)z$ (as in (5)). If c, e, and f are all zero, the curve is $x^2 = 0$, and it consists of two copies of the line $x = 0$, as in part (a) of the theorem's statement.

Thus, we can assume that c, e, f are not all zero in (8). If $c = 0 = f$, then e is nonzero, and taking $s = 0$ and $t = -1$ in (4) replaces y with $y' + z'$ and transforms (8) into

$$x^2 + e(y+z)z = 0,$$

where the coefficient of z^2 is nonzero. Thus, we can assume that c and f are not both zero in (8). By interchanging y and z with a transformation, if necessary, we can assume that c is nonzero. We can write $c = \pm s^2$ for $s = |c|^{1/2} > 0$. Replacing y with y/s (as in (9) of Section 3) transforms (8) into

$$x^2 \pm y^2 + eyz + fz^2 = 0 \tag{9}$$

for a revised value of e.

We can eliminate the yz term from (9) by completing the square in y, which gives

$$x^2 \pm \left(y \pm \frac{e}{2}z\right)^2 + fz^2 = 0$$

for a revised value of f. Setting $y' = y \pm (e/2)z$ (as in (3)) gives

$$x^2 \pm y^2 + fz^2 = 0. \tag{10}$$

If $f = 0$, we have $x^2 \pm y^2 = 0$. The curve $x^2 + y^2 = 0$ consists of one point $(0, 0, 1)$ as in (b) of the theorem's statement. The curve

$$0 = x^2 - y^2 = (x-y)(x+y)$$

consists of the two lines $y = x$ and $y = -x$ (as in (c)).

Thus, we can assume that f is nonzero in (10). We can write $f = \pm t^2$ for $t = |f|^{1/2} > 0$. Replacing z with z/t (as in the discussion accompanying (9) in Section 3) transforms (10) into

$$x^2 \pm y^2 \pm z^2 = 0.$$

The two $\pm$ signs are independent, which gives four possibilities. The graph of $x^2 + y^2 + z^2 = 0$ is the empty set (as in (d)), because $(0, 0, 0)$ is not a point in the projective plane. The curve

$$x^2 + y^2 - z^2 = 0 \tag{11}$$

is the unit circle $x^2 + y^2 = 1$ in the Euclidean plane. Interchanging y and z with a transformation takes $x^2 - y^2 + z^2 = 0$ into (11), as well. Interchanging x and z transforms $x^2 - y^2 - z^2 = 0$ into $z^2 - y^2 - x^2 = 0$, and multiplying this equation by -1 (as discussed after the proof of Theorem 3.6) also gives (11). □

We define a *conic* to be the set of points on a curve of degree 2 in the projective plane that does not consist of two lines, a doubled line, a point, or the empty set. It is clear from this definition that transformations preserve conics.

Theorem 5.1 shows that any conic can be transformed into a circle. Conversely, any curve that can be transformed into a circle has degree 2 (since transformations preserve degree) and does not consist of two lines, a line doubled, a point, or the empty set, and so it is a conic. Thus, *conics are exactly the curves in the projective plane that can be transformed into circles.*

As we observed after (2), ellipses, parabolas, and hyperbolas are exactly the nondegenerate restrictions to the Euclidean plane of curves of degree 2 in the projective plane. If we take two lines, a doubled line, a point, or the empty set in the projective plane, the restriction to the Euclidean plane is degenerate (as defined after (2)). On the other hand, if a curve in the projective plane can be transformed into a circle, its restriction to the Euclidean plane is nondegenerate (since, like a circle, such a curve contains infinitely many points, no three of which are collinear). Thus, Theorem 5.1 shows that *ellipses, parabolas, and hyperbolas are exactly the restrictions to the Euclidean plane of conics in the projective plane.*

An ellipse can be translated and rotated about the origin so that it has the equation

$$\frac{x^2}{a^2} + \frac{y^2}{b^2} = 1 \tag{12}$$

for positive numbers a and b (Figure 5.1). This extends to the curve

$$\frac{x^2}{a^2} + \frac{y^2}{b^2} = z^2$$

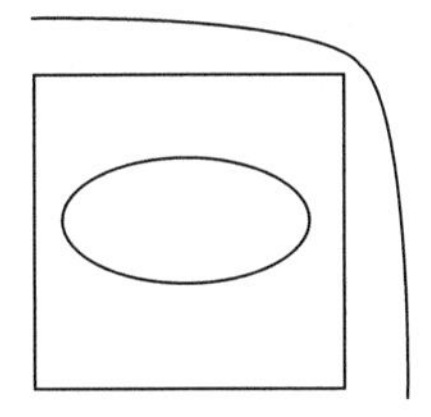

Figure 5.1

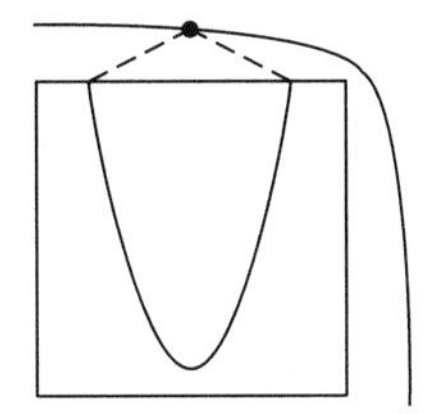

Figure 5.2

in the projective plane. If we set $z = 0$ in this equation, we see that x and y must also be zero. Since $(0, 0, 0)$ does not represent a point in homogeneous coordinates, an ellipse has no points at infinity, as its shape suggests. The ellipse in (12) can obviously be transformed into the unit circle by substituting ax for x and by for y.

A parabola can be translated and rotated about the origin so that it has the equation

$$y = ax^2 \tag{13}$$

for $a > 0$ (Figure 5.2). This extends to the curve

$$yz = ax^2$$

in the projective plane. Setting $z = 0$ in this equation gives $x = 0$, and so $(0, 1, 0)$ is the unique point at infinity on the extension of the parabola to the projective plane. Note that the lines of the Euclidean plane that contain this point at infinity are exactly the vertical lines $x = c$ (i.e., $x = cz$) for all real numbers c, the lines parallel to the axis of symmetry of the parabola. Figure 5.2 suggests that a parabola has the general shape of an ellipse when the point at infinity is added.

A hyperbola can be translated and rotated about the origin so that it has the equation

$$\frac{x^2}{a^2} - \frac{y^2}{b^2} = 1 \tag{14}$$

for positive numbers a and b (Figure 5.3). This extends to the curve

$$\frac{x^2}{a^2} - \frac{y^2}{b^2} = z^2$$

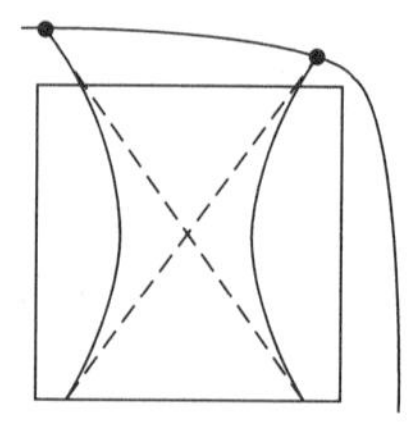

Figure 5.3

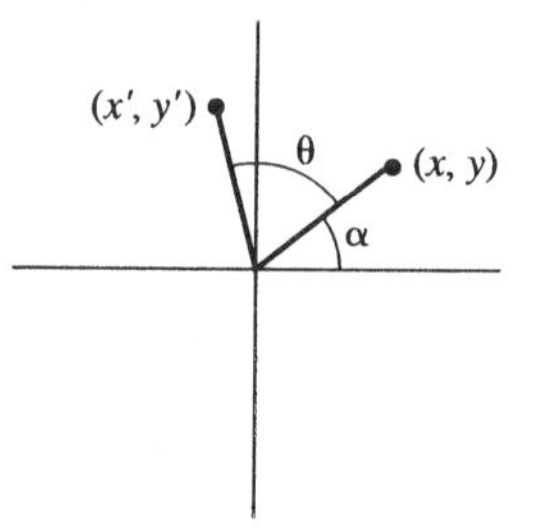

Figure 5.4

in the projective plane. Testing $(1, s, 0)$ and $(0, 1, 0)$ in this equation shows that the hyperbola contains two points at infinity $(1, \pm b/a, 0)$. These are the points at infinity on the two asymptotes $y = \pm(b/a)x$ of the hyperbola, and so the lines of the Euclidean plane that contain one of these points are the lines parallel to one of the asymptotes. The two points at infinity join the two branches of the hyperbola into a shape resembling an ellipse, as in the discussion accompanying Figures 3.1 and 3.2; the point at infinity on each asymptote seems to join the two ends of the hyperbola that approach the asymptote.

A rotation of the Euclidean plane about the origin extends to a transformation of the projective plane. In fact, if a point (x, y) of the Euclidean plane has polar coordinates (r, α), we have $x = r\cos\alpha$ and $y = r\sin\alpha$ (Figure 5.4). If we rotate the plane through angle θ about the origin, (x, y) maps to the point (x', y') with polar coordinates $(r, \alpha + \theta)$. The angle-addition formulas of trigonometry show that

$$\begin{aligned}
x' &= r\cos(\alpha + \theta) = r\cos\alpha\cos\theta - r\sin\alpha\sin\theta \\
&= x\cos\theta - y\sin\theta, \\
y' &= r\sin(\alpha + \theta) = r\cos\alpha\sin\theta + r\sin\alpha\cos\theta \\
&= x\sin\theta + y\cos\theta.
\end{aligned}$$

Thus, the linear change of variables

$$
\begin{aligned}
x' &= (\cos\theta)x - (\sin\theta)y, \\
y' &= (\sin\theta)x + (\cos\theta)y, \\
z' &= z,
\end{aligned}
\tag{15}
$$

extends the rotation to the projective plane. This change of variables is a transformation because we can reverse it by replacing θ with $-\theta$ and interchanging (x, y, z) and (x', y', z').

The equations in (15) show that a rotation of the Euclidean plane about the origin extends to a transformation of the projective plane that maps the points at infinity among themselves. Similarly, the equations in (7) of Section 3 show that a translation of the Euclidean plane extends to a transformation of the projective plane that maps the points at infinity among themselves. Thus, we can summarize the discussion from the proof of Theorem 5.1 and on as follows. *The restriction of a conic to the Euclidean plane is an ellipse, parabola, or hyperbola, depending on whether the conic has* 0, 1, *or* 2 *points at infinity. All ellipses, parabolas, and hyperbolas can be obtained in this way. The lines of the Euclidean plane through the unique point at infinity on a parabola are exactly the lines parallel to the axis of symmetry of the parabola. The lines of the Euclidean plane through either of the two points at infinity on a hyperbola are exactly the lines parallel to one of the asymptotes.* The discussions accompanying Figures II.1–II.3 and 5.1–5.3 help to explain the fact that all conics can be transformed into circles.

The general results about the intersections of a line and a curve in Section 4 specialize to the following theorem about the intersections of a line and a conic. Let $\tan A$ denote the tangent to a curve at a point A.

Theorem 5.2

Let A be any point on a conic K in the projective plane. Then K is nonsingular at A, and every line through A intersects K exactly twice, counting multiplicities. The tangent at A intersects K only at A, and it intersects twice there. Every other line l through A intersects K once at A and once at another point (Figure 5.5).

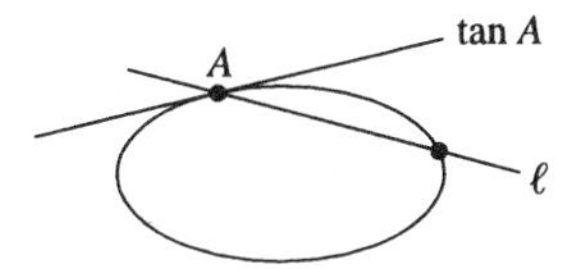

Figure 5.5

Proof

Let K have equation $G = 0$, where $G(x, y, z)$ is a homogeneous polynomial of degree 2. G does not have a polynomial of degree 1 as a factor; otherwise, G would factor as a product of two polynomials of degree 1, and K would consist of two lines or one line doubled, contradicting the definition of a conic. Thus, K intersects any line at most twice, counting multiplicities (by Theorem 4.5).

K contains infinitely many points other than A (since it can be transformed into the unit circle (by Theorem 5.1)). Let B be a point of K other than A. Because line AB intersects K at most twice, counting multiplicities, it intersects K once at A and once at B. It follows from Theorem 4.8 and Definition 4.9 that K is nonsingular at A and that the tangent at A intersects K only at A. The tangent at A intersects K exactly twice at A, since the intersection multiplicity at A is at most two (by the first paragraph of the proof), and it is at least two (by Definition 4.9).

Let l be any line through A other than $\tan A$. We claim that l and K intersect at a point other than A. To see this, we transform two points of l to two points on the x-axis $y = 0$ (by Theorem 3.4), and so we can assume that l is the line $y = 0$. Because the polynomial G giving K does not have y as a factor (by the first paragraph of the proof), Theorem 4.4 states that the number of times that l and K intersect, counting multiplicities, is 2 minus the degree of a polynomial $r(x)$ that has no roots. Since l and K intersect at A, $r(x)$ has degree at most 1. Thus, since $r(x)$ has no real roots, it must be constant, and so l and K intersect exactly twice, counting multiplicities. Because l and K intersect exactly once at A (by Theorem 4.8(ii) and the assumption that $l \neq \tan A$), they also intersect at another point. □

We could also have proved Theorem 5.2 by transforming K into the unit circle (by Theorem 5.1) and observing that the theorem obviously holds for the unit circle.

Theorem 5.2 shows that any line l intersects a conic in at most two points. When l is the line at infinity, this confirms that every conic restricts to an ellipse, a parabola, or a hyperbola in the Euclidean plane. Moreover, a conic intersects the line at infinity in only one point if and only if it is tangent to the line at infinity (by Theorem 5.2). Thus, *a conic is a parabola if and only if it is tangent to the line at infinity* (Figure 5.2).

We have seen that the two points at infinity on the hyperbola in (14) lie on the asymptotes $y = \pm(b/a)x$ (Figure 5.3). The asymptotes do not intersect the hyperbola in the Euclidean plane (since substituting $\pm(b/a)x$ for y makes the left side of (14) zero). Thus, each asymptote intersects the hyperbola at exactly one point of the projective plane, a point at infinity. It follows from Theorem 5.2 that *the asymptotes of a hyperbola are the tangents at the two points at infinity on the hyperbola.*

We use these ideas to obtain results about ellipses, parabolas, and

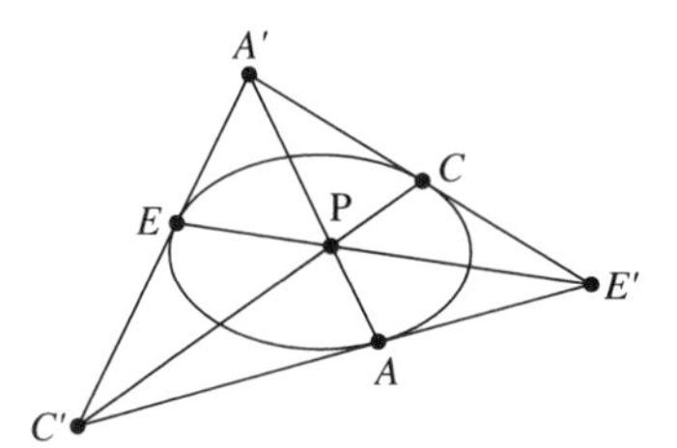

Figure 5.6

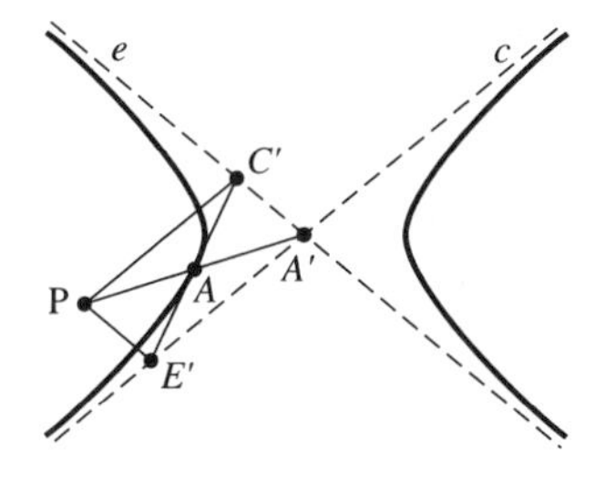

Figure 5.7

hyperbolas from theorems about conics by taking the line at infinity in various positions. For example, consider the following result, which we will prove in Theorem 7.7:

Theorem 5.3
In the projective plane, let A, C, E be three points on a conic. Set $A' = \tan C \cap \tan E$, $C' = \tan E \cap \tan A$, and $E' = \tan A \cap \tan C$. Then the lines AA', CC', EE' are concurrent at a point P (Figure 5.6). □

Suppose, for example, that we take CE to be the line at infinity. Because the conic now has two points C and E at infinity, it is a hyperbola, and the tangents at C and E are the asymptotes c and e, as discussed before Theorem 5.3. $A' = c \cap e$ is the point where the asymptotes intersect (Figure 5.7). $C' = e \cap \tan A$ and $E' = \tan A \cap c$ are the points where the tangent at A intersects the asymptotes. Because C is the point at infinity on c, CC' is the line through C' parallel to c. Likewise, since E is the point at infinity on e, EE' is the line through E' parallel to e. Thus, Theorem 5.3 gives the following result when CE is the line at infinity:

Theorem 5.4
In the Euclidean plane, let A be any point on a hyperbola with asymptotes c and e. Let A' be the point of intersection of the asymptotes, and let C' and E' be the points where the tangent at A intersects the asymptotes e and c, re-

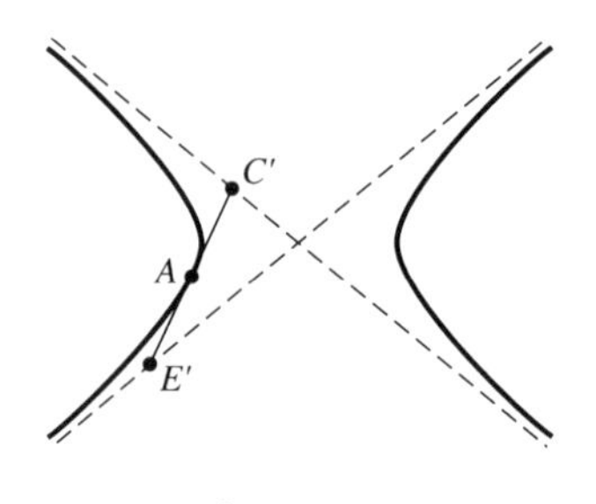

Figure 5.8

spectively. Then the line AA', the line through C' parallel to c, and the line through E' parallel to e lie on a common point P (Figure 5.7). □

We will prove after Theorem 7.5 that no three tangents of a conic are concurrent in the projective plane. In the notation of Theorem 5.4, this implies that $\tan A$ does not contain $A' = c \cap e$ (Figure 5.7), since the asymptotes c and e are tangent to the hyperbola at points at infinity. Thus, $C' = \tan A \cap e$ and $E' = \tan A \cap c$ are points on e and c other than A'. Together with the fact that P lies on the line through C' parallel to c and on the line through E' parallel to e, this shows that $PC'A'E'$ is a parallelogram. Because the diagonals of a parallelogram bisect each other, $A = PA' \cap C'E'$ is the midpoint of C' and E'. This gives the following simple restatement of Theorem 5.4:

Theorem 5.5
In the Euclidean plane, any point A on a hyperbola is the midpoint of the points C' and E' where the tangent at A intersects the asymptotes (Figure 5.8). □

As another example of the transfer of results about conics from the projective to the Euclidean plane, we take $\tan E$ in Theorem 5.3 to be the line at infinity. Then the conic restricts to a parabola in the Euclidean plane (as discussed before Theorem 5.3). $A' = \tan C \cap \tan E$ is the point at infinity on $\tan C$, and $C' = \tan E \cap \tan A$ is the point at infinity on $\tan A$. Thus, AA' is the line m through A parallel to $\tan C$, and CC' is the line n through C parallel to $\tan A$ (Figure 5.9). EE' is now the line through $E' = \tan A \cap \tan C$ parallel to the axis of symmetry of the parabola (by the discussion accompanying Figure 5.2, since E is the point at infinity on the parabola). Thus, we obtain the following result from Theorem 5.3 by taking $\tan E$ to be the line at infinity:

Theorem 5.6
In the Euclidean plane, let A and C be two points on a parabola. Let l be the line through $E' = \tan A \cap \tan C$ parallel to the axis of symmetry of the

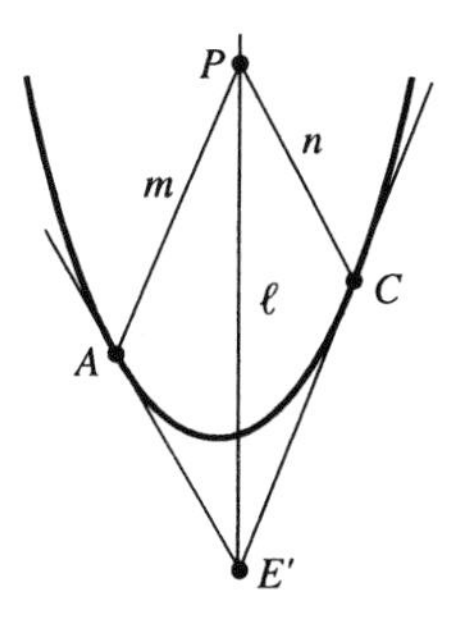

Figure 5.9

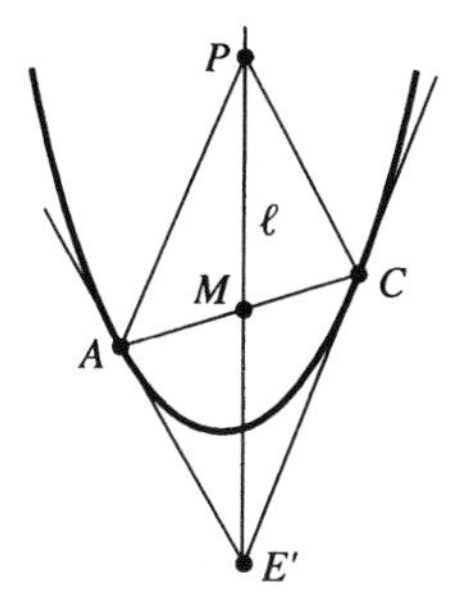

Figure 5.10

parabola. Let m be the line through A parallel to $\tan C$*, and let n be the line through C parallel to* $\tan A$*. Then l, m, and n lie on a common point P (Figure 5.9).* □

As we noted after Theorem 5.4, we will show after Theorem 7.5 that no three tangents of a conic are concurrent in the projective plane. It follows that no two tangents of a parabola are parallel in the Euclidean plane; otherwise, they would intersect at a point on the line at infinity, which is also tangent to the parabola. This observation ensures that the point $E' = \tan A \cap \tan C$ in Theorem 5.6 exists in the Euclidean plane. A, C, and E' are three distinct points in Theorem 5.6, since $\tan A$ and $\tan C$ intersect the parabolas only at A and C, respectively, by Theorem 5.2.

We can restate Theorem 5.6, like Theorem 5.4, in a particularly simple way. Since P lies on the line through A parallel to $\tan C$ and on the line through C parallel to $\tan A$, it follows from the previous paragraph that $APCE'$ is a parallelogram in Theorem 5.6 (Figure 5.9). Because the diagonals PE' and AC of the parallelogram bisect each other, PE' contains the midpoint M of A and C (Figure 5.10). Then $ME' = PE'$ is the line l through E' parallel to the axis of symmetry of the parabola. Thus, we can restate Theorem 5.6 as follows:

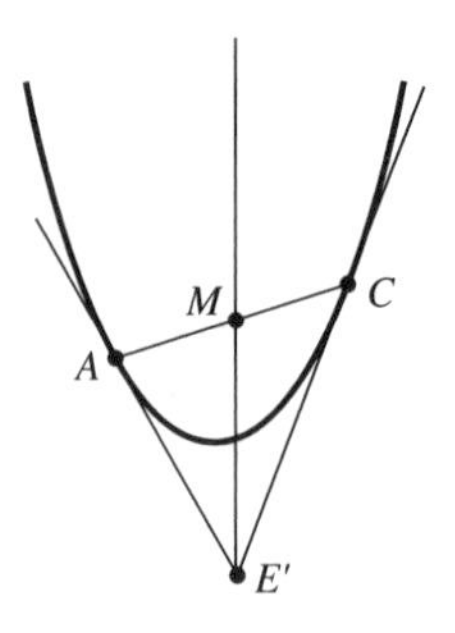

Figure 5.11

Theorem 5.7

In the Euclidean plane, let A and C be two points on a parabola. Let M be their midpoint, and let E' be the intersection of the tangents at A and C. Then the line ME' is parallel to the axis of symmetry of the parabola (*Figure* 5.11).

□

Our next goal is to prove that a conic intersects a curve of degree n which does not contain it at most $2n$ times in the projective plane, counting multiplicities. We start with the conic $y = x^2$. Substituting x^2 for y in a polynomial $g(x, y)$ of degree n gives a polynomial $g(x, x^2)$ of degree at most $2n$. Thus, $y = x^2$ intersects $g(x, y) = 0$ at most $2n$ times, counting multiplicities, in the Euclidean plane (by Theorem 4.3). We extend this result to the projective plane as follows:

Theorem 5.8

Let $G(x, y, z)$ be a homogeneous polynomial of degree n that does not have $yz - x^2$ as a factor. If we set $g(x, y) = G(x, y, 1)$, we can write

$$g(x, x^2) = (x - a_1)^{s_1} \cdots (x - a_v)^{s_v} r(x) \tag{16}$$

for distinct real numbers a_i, positive integers s_i, and a polynomial $r(x)$ that has no real roots. Then the number of times, counting multiplicities, that $yz = x^2$ and $G(x, y, z) = 0$ intersect in the projective plane is $2n$ minus the degree of $r(x)$.

Proof

If we could write

$$g(x, y) = (y - x^2)h(x, y)$$

for a polynomial $h(x, y)$, multiplying terms by appropriate powers of z would show that

$$G(x, y, z) = (yz - x^2)H(x, y, z)$$

for a homogeneous polynomial $H(x,y,z)$. Thus, since $yz - x^2$ is not a factor of $G(x,y,z)$, $y - x^2$ is not a factor of $g(x,y)$. Then Theorem 4.3 shows that we can factor $g(x,x^2)$ as in (16), and the total number of times, counting multiplicities, that $y = x^2$ and $g(x,y) = 0$ intersect in the Euclidean plane is $s_1 + \cdots + s_v$. This is the total number of times, counting multiplicities, that $yz = x^2$ and $G(x,y,z) = 0$ intersect in the Euclidean plane (by Theorem 3.7(iii)).

Equation (16) shows that $s_1 + \cdots + s_v$ is the degree d of $g(x,x^2)$ minus the degree of $r(x)$. Since G is homogeneous of degree n, we can write

$$G(x,y,z) = \sum e_{ij} x^i y^j z^{n-i-j} \tag{17}$$

for real numbers e_{ij}. It follows that

$$g(x,y) = G(x,y,1) = \sum e_{ij} x^i y^j$$

and

$$g(x,x^2) = \sum e_{ij} x^i x^{2j} = \sum e_{ij} x^{i+2j}.$$

Collecting terms shows that the degree d of $g(x,x^2)$ is the largest integer d such that the sum of all the e_{ij} with $i + 2j = d$ is nonzero.

How many times do $yz = x^2$ and $G(x,y,z) = 0$ intersect at infinity? Setting $z = 0$ in $yz = x^2$ gives $x = 0$, and so $(0,1,0)$ is the only point at infinity on the parabola $yz = x^2$. Interchanging y and z shows that the intersection multiplicity of $yz = x^2$ and $G(x,y,z) = 0$ at $(0,1,0)$ equals the intersection multiplicity of $zy = x^2$ and $G(x,z,y) = 0$ at $(0,0,1)$ (by Property 3.5 and the discussion accompanying (8) of Section 3). Setting $z = 1$ shows that this is the intersection multiplicity in the Euclidean plane of $y = x^2$ and $G(x,1,y) = 0$ at the origin (by Property 3.1). Since $yz - x^2$ is not a factor of $G(x,y,z)$, interchanging y and z shows that $zy - x^2$ is not a factor of $G(x,z,y)$, and it follows as in the first paragraph of the proof that $y - x^2$ is not a factor of $G(x,1,y)$. Thus the intersection multiplicity of $y = x^2$ and $G(x,1,y) = 0$ at the origin is the smallest degree of a nonzero term of $G(x,1,x^2)$ (by Theorem 1.11). Substituting 1 for y and x^2 for z in (17) shows that

$$G(x,1,x^2) = \sum e_{ij} x^i x^{2n-2i-2j} = \sum e_{ij} x^{2n-i-2j}.$$

Since $2n - i - 2j$ decreases as $i + 2j$ increases, the smallest degree of a nonzero term of $G(x,1,x^2)$ is $2n - d$, where, as in the previous paragraph, d is the largest integer such that the sum of all the e_{ij} with $i + 2j = d$ is nonzero. In short, $yz = x^2$ and $G(x,y,z) = 0$ intersect $2n - d$ times at infinity.

We have seen that the number of intersections, counting multiplicities, of $yz = x^2$ and $G(x,y,z) = 0$ in the Euclidean plane is d minus the degree of $r(x)$, and the number at infinity is $2n - d$. Summing these two numbers shows that the total number of intersections, counting multiplicities, in the projective plane is $2n$ minus the degree of $r(x)$. □

Theorem 5.1 and the discussion after its proof show that the parabola $yz = x^2$ can be transformed into the unit circle $x^2 + y^2 = z^2$. Since every transformation can be reversed, we can also transform the unit circle into the parabola. Specifically, substituting

$$x = x', \qquad y = \frac{y' - z'}{2}, \qquad z = \frac{y' + z'}{2}, \tag{18}$$

in $x^2 + y^2 = z^2$ gives $x'^2 = y'z'$. This change of variables is a transformation because the equations in (18) can be rewritten as

$$x' = x, \qquad y' = y + z, \qquad z' = -y + z.$$

Any conic can be transformed into $yz = x^2$, since it can be transformed first into the unit circle and then into $yz = x^2$ (by Theorem 5.1 and the previous paragraph). Transformations preserve intersection multiplicities and factorizations of polynomials (by Property 3.5 and the discussion before Theorem 4.5). Thus, Theorem 5.8 implies the following analogue of Theorem 4.5:

Theorem 5.9
Let $K = 0$ be a conic, and let $G = 0$ be a curve of degree n. If K is not a factor of G, then $K = 0$ and $G = 0$ intersect at most $2n$ times, counting multiplicities, in the projective plane. □

It follows from Theorem 5.9 that five points, no three of which are collinear, determine a unique conic.

Theorem 5.10
Five points in the projective plane, no three of which are collinear, lie on exactly one conic.

Proof
Let A–E be five points in the projective plane, no three of which are collinear. Let T, U, V, W be homogeneous polynomials of degree 1 such that $T = 0$, $U = 0$, $V = 0$, $W = 0$ are the lines AB, CD, AC, BD, respectively (Figure 5.12). The products TU and VW are homogeneous polynomials of degree 2 such that the curves $TU = 0$ and $VW = 0$ are two pairs of lines that both contain A, B, C, D.

Let E have homogeneous coordinates (f, g, h). Since no three of the points A–E are collinear, E does not lie on T or U, and so

$$T(f,g,h)U(f,g,h) \neq 0.$$

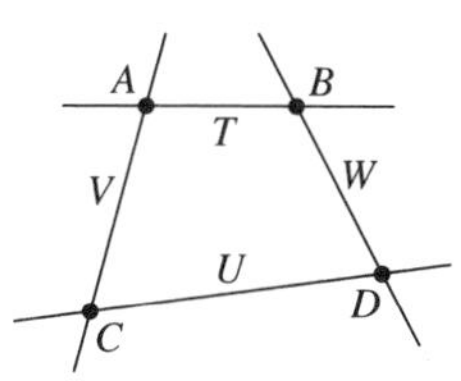

Figure 5.12

Thus, there is a real number r such that

$$rT(f,g,h)U(f,g,h) + V(f,g,h)W(f,g,h) = 0,$$

and so E lies on the curve $rTU + VW = 0$. A–D lie on this curve as well, since they lie on both $TU = 0$ and $VW = 0$.

A and B are the only two points of T on the curve $VW = 0$ (by Theorem 2.1, since no three of the points A–D are collinear), but every point of T lies on $-rTU = 0$. Thus, the polynomials VW and $-rTU$ are distinct. Then $rTU + VW$ is nonzero, and so it is a homogeneous polynomial of degree 2.

We have shown that $rTU + VW$ is a curve of degree 2 that contains A–E. Since no three of the points A–E are collinear, no two lines contain all five of these points. Accordingly, a curve of degree 2 containing A–E cannot consist of two lines, a doubled line, a point, or the empty set. Thus, the curve of degree 2 $rTU + VW = 0$ that contains A–E is a conic, by Theorem 5.1.

We must prove that A–E cannot lie on more than one conic. In fact, if $K = 0$ and $K' = 0$ are conics that both contain A–E, they intersect at least five times (by Theorem 3.6(iii)). Then the polynomials K and K' of degree 2 are each multiples of the other (by Theorem 5.9). It follows that $K = tK'$ for a nonzero constant t, and so $K = 0$ and $K' = 0$ are the same conic (as discussed after Theorem 3.6). Thus, A–E lie on a unique conic. □

A line and a conic intersect at most twice (by Theorem 5.2), and so *no three points on a conic are collinear*. Thus, Theorem 5.10 shows that a conic is determined by any five of its points. Theorem 5.2 shows the need for the hypothesis in Theorem 5.10 that no three of the points are collinear if five points are to lie on a conic.

Exercises

5.1. State the version of Theorem 5.3 that holds in the Euclidean plane when E is the only point at infinity named. Illustrate the result you state with a figure in the Euclidean plane. (Note that the conic restricts to a hyperbola in the Euclidean plane because it has at least one point at infinity and is not tangent to the line at infinity. The second point at infinity on the hyperbola is unnamed in the theorem.)

5.2. The following result is proved in Exercise 6.1:

Theorem
In the projective plane, let A, C, D, E, F be five points on a conic. Then the points $Q = \tan A \cap DE$, $R = AC \cap EF$, and $S = CD \cap FA$ are collinear (*Figure* 5.13).

State the version of this theorem that holds in the Euclidean plane in the following cases. Illustrate each result you state with a figure.
(a) A is the only point at infinity named.
(b) F is the only point at infinity named, and the conic is a parabola.
(c) F is the only point at infinity named, and the conic is a hyperbola.
(d) AQ is the line at infinity.
(e) AC is the line at infinity.
(f) AE is the line at infinity.

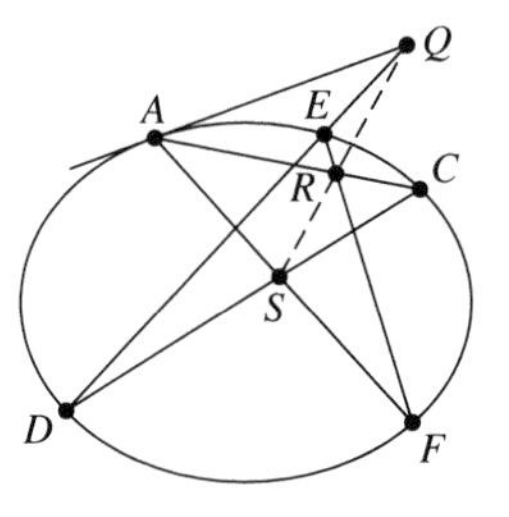

Figure 5.13

5.3. Follow the directions of Exercise 5.2 for the theorem below, which is proved in Exercise 6.2.

Theorem
In the projective plane, let A, C, E, F be four points on a conic. *Then the points $Q = \tan A \cap CE$, $R = AC \cap EF$, and $S = \tan C \cap FA$ are collinear* (*Figure* 5.14).

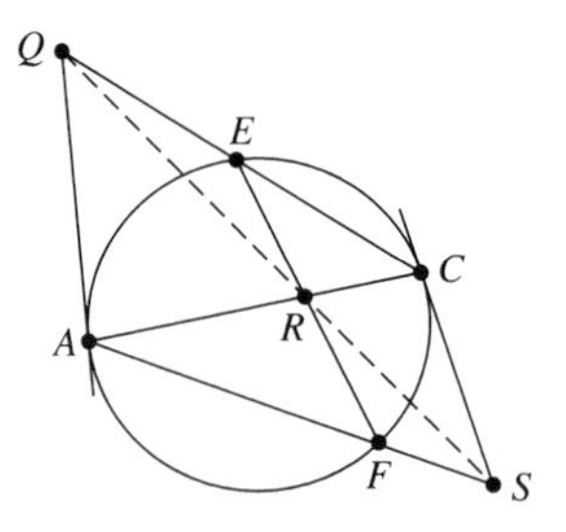

Figure 5.14

5.4. Follow the directions of Exercise 5.2 for the theorem below, which is proved in Exercise 6.3.

Theorem
In the projective plane, let A, C, E, F be four points on a conic. Then the points $Q = \tan A \cap \tan E$, $R = AC \cap EF$, and $S = CE \cap FA$ are collinear (Figure 5.15).

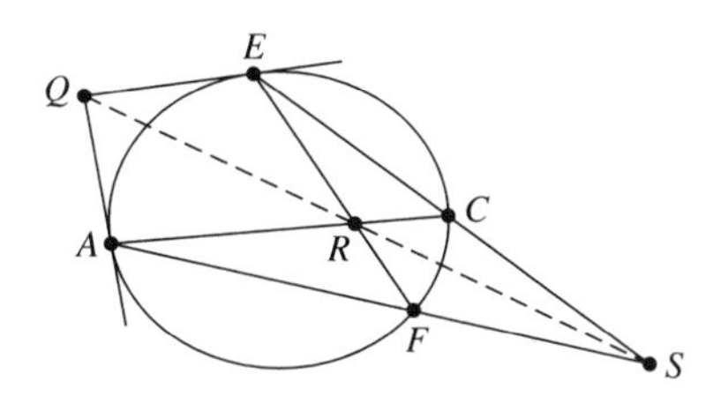

Figure 5.15

5.5. Use the theorem in Exercise 5.4 to prove the following result, and illustrate this result with a figure:

Theorem
In the projective plane, Let A, B, C, D be four points on a conic. Then the points $U = \tan A \cap \tan B$, $V = \tan C \cap \tan D$, and $W = AD \cap BC$ are collinear.

5.6. State the version of the theorem in Exercise 5.5 that holds in the Euclidean plane in the following cases. Illustrate each result you state with a figure.
(a) A is the only point at infinity named.
(b) AU is the line at infinity.
(c) AB is the line at infinity.
(d) AC is the line at infinity.
(e) AD is the line at infinity.

5.7. By using the quadratic formula to find the number of points at infinity on the curve in (1), derive conditions on the coefficients a–f that determine whether (2) gives an ellipse, parabola, or hyperbola.

5.8. Five points A–E, no three of which are collinear, are given in each part of this exercise. Find the equations $T = 0$, $U = 0$, $V = 0$, $W = 0$ of the lines AB, CD, AC, BD, respectively. Then find the real number r such that the curve $rTU + VW = 0$ contains E. Finally, write the equation $rTU + VW = 0$ in the form of (2). As the proof of Theorem 5.10 shows, this is the equation of the unique conic through the five points A–E.
(a) $A = (0,1)$, $B = (0,-1)$, $C = (1,2)$, $D = (1,-2)$, $E = (-2,0)$.
(b) A is the point at infinity on lines of slope 2, $B = (0,0)$, $C = (1,0)$, $D = (0,1)$, $E = (1,1)$.
(c) $A = (1,0)$, $B = (-1,0)$, $C = (2,1)$, $D = (-2,1)$, E is the point at infinity on vertical lines.
(d) $A = (0,0)$, $B = (2,2)$, $C = (1,-1)$, D is the point at infinity on vertical lines, E is the point at infinity on horizontal lines.

(e) $A = (1, 0)$, $B = (-1, 0)$, $C = (0, 1)$, $D = (0, -1)$, $E = (2, 2)$.
(f) $A = (3, 0)$, $B = (0, 3)$, $C = (-3, 0)$, $D = (0, -3)$, $E = (1, 1)$.
(g) $A = (0, 0)$, $B = (1, 0)$, $C = (0, 1)$, $D = (-1, -1)$, $E = (1, -1)$.

5.9. In each part of Exercise 5.8, draw the points A–E that lie in the Euclidean plane, and sketch the conic determined by A–E.

5.10. Prove that five points in the projective plane lie on a unique curve of degree 2 if and only if no four of the points are collinear.

5.11. Consider the following result:

Theorem
Let A, B, C, D be four points, no three of which are collinear. Let $T = 0$, $U = 0$, $V = 0$, $W = 0$ be the lines AB, CD, AC, BD, respectively. Then the curves of degree 2 containing A–D are $TU = 0$ and $rTU + VW = 0$ for all real numbers r, and every point except A–D lies on exactly one of these curves.

(a) Deduce the theorem from Theorems 5.1 and 5.10.
(b) Let $A = (1, 1)$, $B = (1, -1)$, $C = (-1, 1)$, $D = (-1, -1)$. Use the theorem to write the curves of degree 2 containing A–D in the form of (2). Which of these curves are not conics? Justify your answers.
(c) Illustrate the theorem by drawing the gamut of curves in (b) in a single figure, making it clear that each point in the Euclidean plane lies on exactly one of these curves.

5.12. Consider the following result:

Theorem
Let A–D be four points, no three of which are collinear, in the projective plane. Let a be a line through A that does not contain any of the points B–D. Then there is a unique conic that contains A–D and is tangent to a.

Let $T = 0$, $U = 0$, $V = 0$, $W = 0$ be the equations of the lines a, CD, AC, AD, respectively.

(a) For any nonzero number r, prove that $rTU + VW = 0$ is a curve of degree 2 that intersects $T = 0$ twice at A. Conclude that the curve is a conic that contains A, C, D and is tangent to a. (See Theorems 5.1 and 5.2.)
(b) Prove that there is a nonzero number r such that the curve $rTU + VW$ contains B.
(c) Deduce the theorem from parts (a) and (b) and Theorems 4.11 and 5.9.

5.13. Let A be the point at infinity on vertical lines, let $C = (1, 0)$ and $D = (-1, 0)$, and let a be the line at infinity. Let T, U, V, W be as in Exercise 5.12.

(a) Exercise 5.12 implies that the conics that contain A, C, D and are tangent to a are exactly the curves $rTU + VW = 0$ for nonzero numbers r. Write the equations of these conics in the form of (2).
(b) Draw a figure that shows the gamut of conics in (a) and the lines U, V, W. Make it clear in the figure that each point in the Euclidean plane except C and D lies on exactly one of the conics or lines, as Exercise 5.12 implies.

5.14. Consider the following result:

Theorem
Let A, B, C be three noncollinear points in the projective plane. Let a be a line on A that does not contain B or C, and let c be a line on C that does not contain A or B. Then there is a unique conic that contains A, B, C and is tangent to a and c.

Let $T = 0$, $U = 0$, $V = 0$ be the equations of the lines a, c, AC, respectively.

(a) For any nonzero number r, prove that $rTU + V^2 = 0$ is a curve of degree 2 that intersects $T = 0$ twice at A and $U = 0$ twice at C. Conclude from Theorems 5.1 and 5.2 that $rTU + V^2 = 0$ is a conic that is tangent to a at A and tangent to c at C.

(b) Prove that there is a nonzero number r such that $rTU + V^2 = 0$ contains B.

(c) Deduce the theorem from parts (a) and (b) and Theorems 4.11 and 5.9.

5.15. Let $A = (1, 0)$ and $C = (-1, 0)$, and let a and c be the vertical lines through A and C. Let T, U, V be as in Exercise 5.14.

(a) Exercise 5.14 implies that the conics that contain A and C and are tangent to a and c are exactly the curves $rTU + V^2 = 0$ for nonzero numbers r. Write the equations of these conics in the form of (2).

(b) Draw a figure that shows the gamut of conics in (a) and the lines a, c, and AC. Make it clear that each point except A and C lies on exactly one of the conics or lines, as Exercise 5.14 implies.

5.16. In the projective plane, let A, B, C be three points on a conic K, and let A', B', C' be three points on a conic K'. Prove that there is a transformation that maps K to K' and A, B, C to A', B', C', respectively.

(*Hint*: One possible approach is to set $D = \tan A \cap \tan B$ and $D' = \tan A' \cap \tan B'$ and deduce that there is a transformation mapping A, B, C, D to A', B', C', D'. Why do Theorems 4.11 and 5.9 imply that the transformation maps K to K'?)

5.17. Prove that no three tangents to a conic are concurrent, by using Exercise 5.16 to reduce to the case of tangents at three particular points of a particular conic. (The discussion after Theorem 7.5 provides another proof of this result.)

Exercises 5.18–5.22, *which we use in Exercises* 6.17–6.20 *and* 13.6–13.12, *are based on the following terminology.* Four points, P, Q, R, S, no three of which are collinear, determine a *harmonic set* $A, B; C, D$, where $A = PQ \cap RS$, $B = PR \cap QS$, $C = PS \cap AB$, and $D = QR \cap AB$ (Figure 5.16).

5.18. Let A, B, C be three collinear points. This exercise shows that there is a unique point D such that $A, B; C, D$ is a harmonic set. We call D the *harmonic conjugate* of C with respect to A and B.

(a) Let P and S be two points collinear with C that do not lie on a line AB. Describe how to construct points Q, R, D such that P, Q, R, S determine the harmonic set $A, B; C, D$.

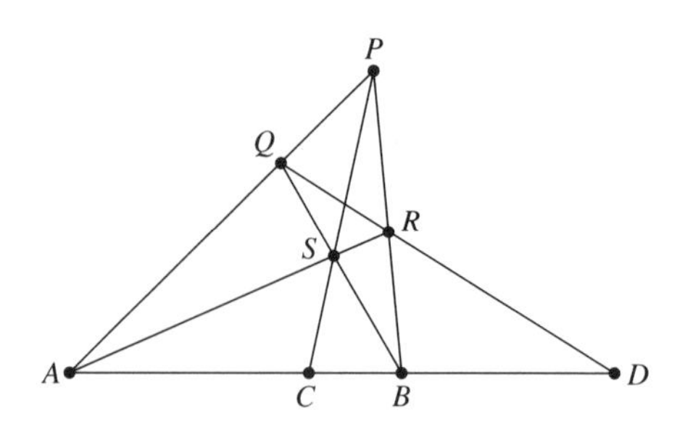

Figure 5.16

(b) Let P–S, P'–S', D, D' be points such that P, Q, R, S determine the harmonic set $A, B; C, D$ and P', Q', R', S' determine the harmonic set $A, B; C, D'$. Prove as follows that $D = D'$ (Figure 5.17): show that there is a transformation that maps P, Q, R, S to P', Q', R', S', deduce from Exercise 3.11(b) that this transformation fixes every point of line AB, and conclude that $D = D'$.

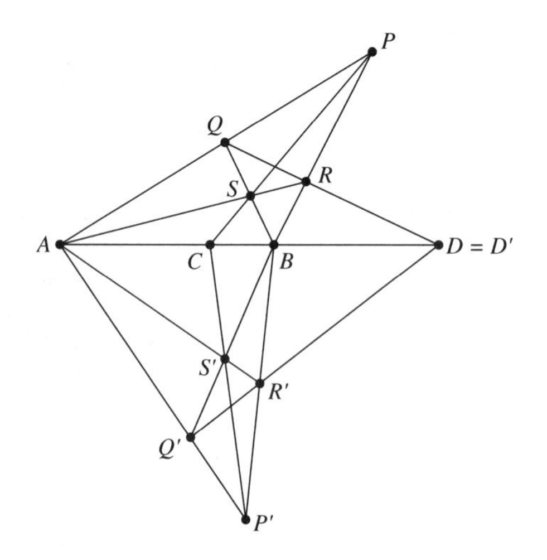

Figure 5.17

5.19. Let $A, B; C, D$ be a harmonic set, as defined before Exercise 5.18, and assume that A and B lie in the Euclidean plane. Prove that C lies at infinity if and only if D is the midpoint of A and B.

(*Hint*: One possible approach is to use Exercise 5.18 to choose P and S so that line AB is their perpendicular bisector.)

5.20. Let A, B, C, D be four collinear points in the Euclidean plane. Define division rations as in Exercise 4.19.

(a) Prove the $\overline{CA}/\overline{CB} \neq \overline{DA}/\overline{DB}$ by arguing geometrically or by using Exercise 4.19 to argue algebraically.

(b) Prove that $A, B; C, D$ is a harmonic set if and only if

$$\overline{CA}/\overline{CB} = -\overline{DA}/\overline{DB}. \tag{19}$$

(Equation (19) shows that C and D divide A and B internally and externally in the same ratio. If $A, B; C, D$ is a harmonic set, Exercise 5.18 implies that it can be determined by points P, Q, R, S in the Euclidean plane. Applying Menelaus' Theorem from Exercise 4.22(a) to triangle PAB and line QR, applying Ceva's Theorem from Exercise 4.22(b) to the four points P, A, B, S, and combining the results gives (19). Part (b) follows from this, part (a), Exercise 5.18, and possibly Exercise 5.19.)

5.21. Use Exercise 5.18 to prove the theorem in Exercise 2.10.

5.22. Let $A, B; C, D$ be a harmonic set.

(a) Prove that no two of the points A, B, C, D are equal. (One or more of the Exercises 3.10, 3.14, 5.19, and 5.20 may help.)

(b) Prove that $C, D; A, B$ is a harmonic set. Illustrate this fact with a figure that shows points P–S that determine a harmonic set $A, B; C, D$ and also shows points P'–S' that determine the harmonic set $C, D; A, B$. (See part (a) and Exercise 3.15).

5.23. Let K be a nondegenerate curve of degree 2 in the Euclidean plane. This exercise reviews the proof of the assertion after (2) that K is an ellipse, parabola, or hyperbola. Use the equations in (15) to show that K can be rotated so that it is given by (2) with $b = 0$. Deduce by completing squares that K can be translated and rotated so that it is given by (12), (13), or (14) for positive numbers a and b.

5.24. Let K be a curve of degree 2 that consists of a single point P, and let F be any curve nonsingular at P. Prove that $I_P(K, F) = 2$ by using Theorem 5.1 and (15) to reduce to the case where K is $x^2 + y^2 = 0$ and F is tangent to the y-axis at the origin and by using the proof of Theorem 4.11 to write the restriction of F to the Euclidean plane as in (14) of Section 4. (We use this exercise in Exercises 10.8, 15.20, 15.22, and 15.23.)

§6. Pascal's Theorem

This section is devoted to Pascal's Theorem and its variants. Pascal's Theorem states that the three pairs of opposite sides of a hexagon inscribed in a conic intersect in three collinear points. We vary the theorem in two ways. First, we replace sides of the hexagon with tangents to the conic. Second, we inscribe the hexagon in two lines instead of a conic, which gives Pappus' Theorem 2.3.

The following result is the key to proving Pascal's Theorem. If a conic K intersects each of two curves G and H of degree n in the same $2n$ points, counting multiplicities, we prove that there is a curve W of degree $n-2$ such that the intersections of G and H are the intersections of either curve with K together with its intersections with W. As indicated after Theorem 3.6, we say that $G=0$ and $H=0$ are *distinct curves* when G and H are homogeneous polynomials that are not scalar multiples of each other.

Theorem 6.1

Let $G=0$ and $H=0$ be distinct curves of degree n. Assume that there is a conic $K=0$ such that $I_P(G,K)=I_P(H,K)$ for every point P in the projective plane and such that K intersects G or H a total of $2n$ times, counting multiplicities. Then there is a curve $W=0$ of degree $n-2$ such that

$$I_P(G,H)=I_P(G,K)+I_P(G,W)=I_P(H,K)+I_P(H,W)$$

for every point P in the projective plane.

Proof

We can transform $K=0$ into the parabola $yz=x^2$ (as discussed before Theorem 5.9), and transformations preserve intersection multiplicities and degrees of homogeneous polynomials (by Property 3.5 and the discussion after the proof of Theorem 3.4). Thus, we can assume that K is $yz-x^2$. Since $yz=x^2$ intersects $G=0$ and $H=0$ $2n$ times, $yz-x^2$ is not a factor of G or H (by Theorem 3.6(iii) or (vi)).

Let $g(x,y)=G(x,y,1)$ and $h(x,y)=H(x,y,1)$ be the restrictions of G and H to the Euclidean plane. Theorem 5.8 shows that

$$g(x,x^2)=r(x-a_1)^{s_1}\cdots(x-a_v)^{s_v} \tag{1}$$

for a real number $r\neq 0$, because the assumption that K intersects G a total of $2n$ times, counting multiplicities, implies that the polynomial $r(x)$ in Theorem 5.8 has degree 0 and is thus a constant r. Each exponent s_i is the number of times that K and G intersect at the point (a_i,a_i^2), and these are the only points of the Euclidean plane where K and G intersect, by Theorem 4.3.

Because H intersects K the same number of times at every point as G does, Theorem 4.3 implies that

$$h(x,x^2)=t(x-a_1)^{s_1}\cdots(x-a_v)^{s_v}$$

for a real number $t\neq 0$. As discussed after Theorem 3.6, we can multiply H, and hence h, by $-r/t$, which gives

$$h(x,x^2)=-r(x-a_1)^{s_1}\cdots(x-a_v)^{s_v}.$$

Adding this equation to (1) shows that

$$g(x,x^2)+h(x,x^2)=0.$$

Then $y-x^2$ is a factor of $g(x,y)+h(x,y)$ (by Theorem 1.9(ii)), and we can write

$$g(x,y) + h(x,y) = (y - x^2)w(x,y) \tag{2}$$

for a polynomial $w(x,y)$.

Because $G = 0$ and $H = 0$ are distinct curves, G and H are not scalar multiples of each other, and so $G + H$ is nonzero. Thus, since G and H are homogeneous polynomials of degree n, so is $G + H$. Multiplying every term of each polynomial in (2) by an appropriate power of z shows that

$$G(x,y,z) + H(x,y,z) = (yz - x^2)W(x,y,z) \tag{3}$$

for a homogenous polynomial $W(x,y,z)$ of degree $n-2$. For any point P of the projective plane, it follows that

$$\begin{aligned} I_P(G,H) &= I_P(G, G+H) && \text{(by Theorem 3.6(iv))} \\ &= I_P(G, KW) && \text{(by (3))} \\ &= I_P(G,K) + I_P(G,W) && \text{(by Theorem 3.6(v)).} \end{aligned}$$

Interchanging G and H in the last sentence shows that

$$I_P(H,G) = I_P(H,K) + I_P(H,W),$$

and the left-hand side equals $I_P(G,H)$ (by Theorem 3.6(ii)). □

A conic intersects a curve of degree n that does not contain it at most $2n$ times, counting multiplicities, by Theorem 5.9. Thus, the hypotheses of Theorem 6.1 state that G and H are curves of the same degree n that intersect the conic K as many times as possible without containing it and that have the same intersections with K, taking into account multiplicities. The conclusion of Theorem 6.1 shows that, if we list the points where G and H intersect and remove the points where either curve intersects K, then we are left with the points where either curve intersects a curve W of degree $n-2$, provided that we repeat each point of intersection as many times as its multiplicity. We think of Theorem 6.1 as "peeling off a conic" from the intersection of two curves of the same degree.

We can now prove the main result of this section, Pascal's Theorem.

Theorem 6.2 (Pascal's Theorem)
Let A–F be six points on a conic K in the projective plane. Then the points $Q = AB \cap DE$, $R = BC \cap EF$, and $S = CD \cap FA$ are collinear (*Figure* 6.1).

Proof
Let $L = 0, M = 0, N = 0, T = 0, U = 0, V = 0$ be the lines

$$AB, \quad CD, \quad EF, \quad BC, \quad DE, \quad FA, \tag{4}$$

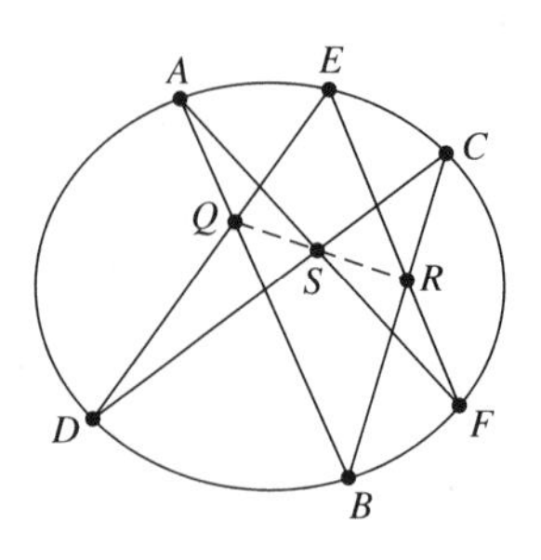

Figure 6.1

respectively. Set

$$G = LMN \qquad \text{and} \qquad H = TUV. \tag{5}$$

G and H are homogeneous polynomials of degree 3, since they are each the product of three homogeneous polynomials of degree 1. The curve $G = 0$ consists of the three lines AB, CD, EF, and the curve $H = 0$ consists of the three lines BC, DE, FA. We prove the theorem by peeling off the conic K from the intersection of G and H.

Theorem 5.2 implies that K intersects line AB once at A and once at B, line CD once at C and once at D, and line EF once at E and once at F. Thus K intersects G once at each of the six points A–F (by (5) and Theorem 3.6(v)). Likewise, Theorem 5.2 implies that K intersects line BC once at B and once at C, line DE once at D and once at E, and line FA once at F and once at A. Thus K also intersects H once at each of the points A–F (by (5) and Theorem 3.6(v)). In short, the hypotheses of Theorem 6.1 hold with $n = 3$: G and H are curves of degree 3 that intersect the conic K in the same $6 = 2 \cdot 3$ points A–F.

No three of the points A–F on K are collinear (by Theorem 5.2). Thus, the six lines in (4) are distinct, and any two intersect exactly once, counting multiplicities (by Theorem 4.1). If we intersect each of the three lines AB, CD, EF forming G with each of the three lines BC, DE, FA forming H, we obtain the nine points $AB \cap BC = B$, $AB \cap DE = Q$, $AB \cap FA = A$, $CD \cap BC = C$, $CD \cap DE = D$, $CD \cap FA = S$, $EF \cap BC = R$, $EF \cap DE = E$, and $EF \cap FA = F$. Thus, G and H intersect at the nine points A–F, Q, R, S (by (5) and Theorem 3.6(v)).

If we remove the six points A–F where G and H intersect K from the nine points A–F, Q, R, S where G and H intersect each other, we are left with the three points Q, R, S. We can apply Theorem 6.1 (by the second paragraph of the proof), and we deduce that Q, R, S are the points where G and H intersect a curve of degree $3 - 2 = 1$. This curve is a line that contains Q, R, S (by Theorem 3.6(iii)), as desired. □

Five points A–E, no three of which are collinear, lie on a unique conic

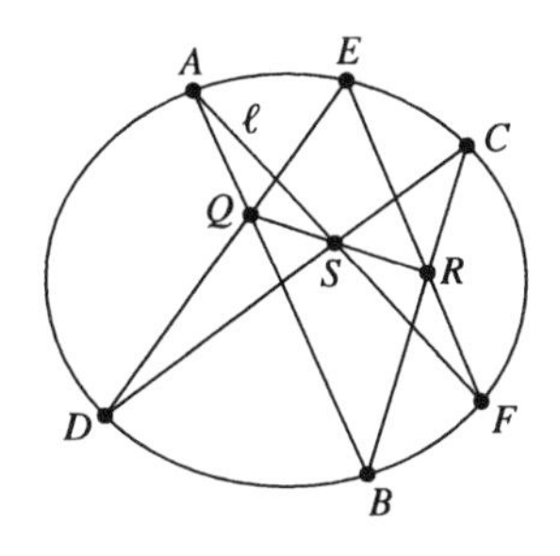

Figure 6.2

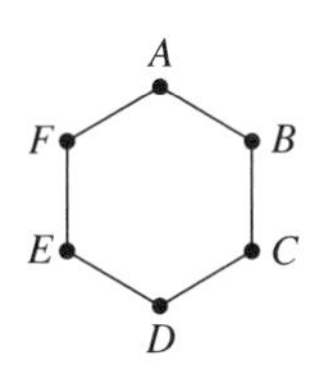

Figure 6.3

K (by Theorem 5.10). Pascal's Theorem implies that we can use a straightedge and the five given points A–E to construct any number of points of K. In fact, let l be any line through A other than $\tan A$, AB, AC, AD, AE (Figure 6.2). K intersects l in a point F other than A (by Theorem 5.2). By Pascal's Theorem, we can use a straightedge to construct F as follows: we construct $Q = AB \cap DE$, $S = l \cap CD$, $R = QS \cap BC$, and $F = ER \cap l$. These points exist (by Theorems 2.1, 2.2, and 5.2).

If A–F are six points such that the lines in (4) are distinct, we think of *hexagon ABCDEF* as the figure formed by the six points A–F and the six lines in (4) (Figure 6.3). We call the points A–F the *vertices* of the hexagon, and we call the six lines in (4) the *sides* of the hexagon. As Figure 6.3 suggests, we call AB and DE, BC and EF, and CD and FA the three pairs of *opposite sides* of the hexagon. These are the three pairs of lines that intersect in the points Q, R, S in Pascal's Theorem 6.2 (Figure 6.1). Accordingly, we can restate Pascal's Theorem as follows: *If a hexagon is inscribed in a conic, the three pairs of opposite sides intersect in collinear points.* The curves G and H in (5) used to prove Pascal's Theorem are the two triples of lines AB, CD, EF and BC, DE, FA formed by taking every other side of hexagon $ABCDEF$ (Figure 6.3).

Let A–F be six points on a conic. If we arrange the points in different orders to form hexagons, we obtain different lines through the three intersections of opposite sides of each hexagon. For instance, hexagon $ABCDEF$ shows that the points Q, R, S in Figure 6.1 are collinear, and hexagon $ADCFBE$ shows that the points $T = AD \cap FB$, $U = DC \cap BE$, and $V = CF \cap EA$ are collinear (Figure 6.4).

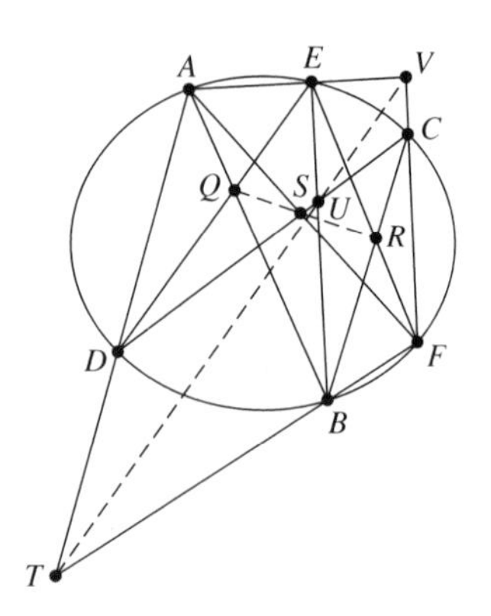

Figure 6.4

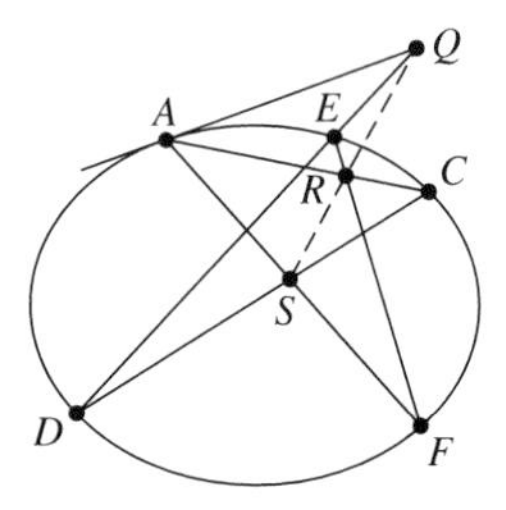

Figure 6.5

If the point B in Pascal's Theorem moves around the conic K until it approaches the point A, the lines AB and BC approach $\tan A$ and AC. Thus, the conclusion of Pascal's Theorem that the points $Q = AB \cap DE$, $R = BC \cap EF$, and $S = CD \cap FA$ are collinear suggests that the points.

$$Q = \tan A \cap DE, \qquad R = AC \cap EF, \qquad S = CD \cap FA, \tag{6}$$

are collinear for any five points A, C, D, E, F on a conic (Figure 6.5).

If the point D moves around the conic to approach C, the lines DE and CD in (6) approach CE and $\tan C$. Thus, the collinearity of the points in (6) suggests that the points

$$Q = \tan A \cap CE, \qquad R = AC \cap EF, \qquad S = \tan C \cap FA, \tag{7}$$

are collinear for any four points A, C, E, F on a conic (Figure 6.6).

If the point F moves around the conic to approach E, the lines EF and FA in (7) approach $\tan E$ and EA. Thus, the collinearity of the points in (7) suggests that the points

$$Q = \tan A \cap CE, \qquad R = AC \cap \tan E, \qquad S = \tan C \cap EA, \tag{8}$$

are collinear for any three points A, C, E on a conic (Figure 6.7).

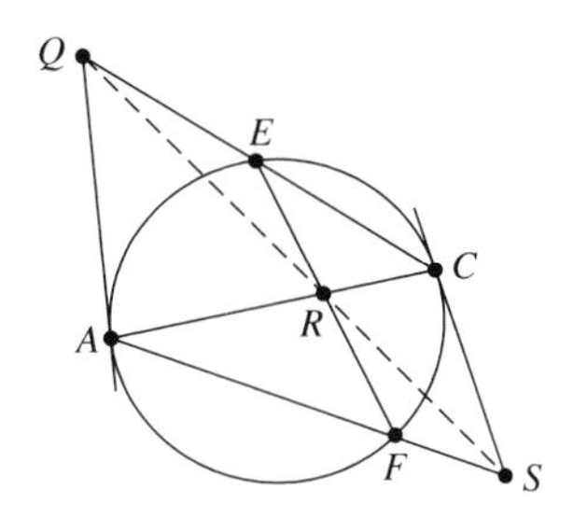

Figure 6.6

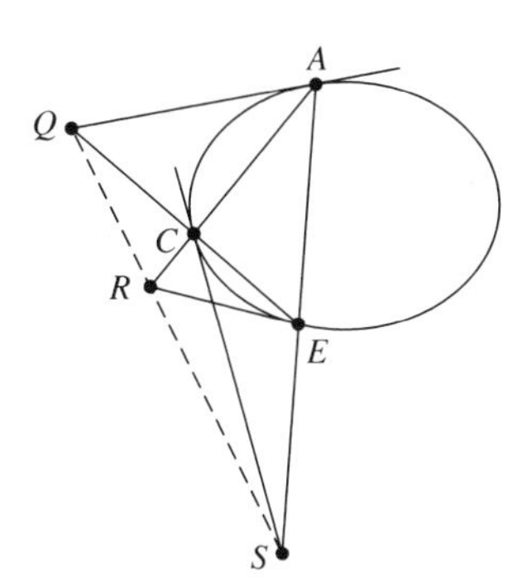

Figure 6.7

Pascal's Theorem refers to a hexagon inscribed in a conic. In the three preceding paragraphs, we have replaced the hexagon with an n-gon inscribed in a conic and the tangents at $6-n$ of its vertices. In (6), we considered an inscribed pentagon and the tangent at one of its vertices. In (7), we considered an inscribed quadrilateral and the tangents at two of its vertices. In (8), we considered an inscribed triangle and the tangents at its three vertices.

As a point Y on a conic K approaches a point X on K, the line XY approaches $\tan X$. Accordingly, we think of "line XX" as the tangent at X. We can then think of the points in (6)–(8) as the intersections of opposite sides of "hexagon $ABCDEF$" when consecutive vertices are equal. For example, if we set $B=A, D=C$, and $F=E$, hexagon $ABCDEF$ becomes "hexagon $AACCEE$" (Figure 6.8). Opposite sides of this hexagon intersect in the points

$$Q = AA \cap CE, \qquad R = AC \cap EE, \qquad S = CC \cap EA, \tag{9}$$

listed in (8).

Intersection multiplicities make it possible to prove these variations in essentially the same way as Pascal's Theorem. To illustrate this, we

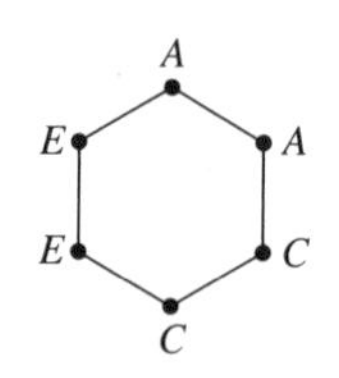

Figure 6.8

prove that the points in (8)—or, equivalently, (9)—are collinear. We proved Pascal's Theorem by considering the sides

$$AB, \quad CD, \quad EF, \quad BC, \quad DE, \quad FA,$$

of hexagon $ABCDEF$, as listed in (4). Replacing B with A, D with C, and F with E, we now consider the sides

$$\tan A, \quad \tan C, \quad \tan E, \quad AC, \quad CE, \quad EA,$$

of "hexagon $AACCEE$."

Theorem 6.3

Let A, C, E be three points on a conic K in the projective plane. Then the points $Q = \tan A \cap CE, R = \tan E \cap AC$, and $S = \tan C \cap EA$ are collinear (Figure 6.7).

Proof

Let $L = 0, M = 0, N = 0, T = 0, U = 0, V = 0$ be the lines

$$\tan A, \quad \tan C, \quad \tan E, \quad AC, \quad CE, \quad EA, \tag{10}$$

respectively. Set

$$G = LMN \qquad \text{and} \qquad H = TUV. \tag{11}$$

G and H are homogeneous polynomials of degree 3, since they are each the product of three homogeneous polynomials of degree 1. The curves $G = 0$ and $H = 0$ consist of alternate sides of "hexagon $AACCEE$" (Figure 6.8): G consists of the three lines $\tan A, \tan C, \tan E$, and H consists of the three lines AC, CE, EA.

Theorem 5.2 implies that the conic K intersects $\tan A$ twice at A, $\tan C$ twice at C, and $\tan E$ twice at E. Thus, K intersects G twice at each of the points A, C, E (by (11) and Theorem 3.6(v)). Theorem 5.2 also implies that K intersects line AC once at A and once at C, line CE once at C and once at E, and line EA once at E and once at A. Thus, K intersects H twice at each of the points A, C, E (by (11) and Theorem 3.6(v)). In short, the hypotheses of Theorem 6.1 hold with $n = 3$: G and H are curves of degree 3 that intersect the conic K the same $6 = 2 \cdot 3$ times—twice at A, twice at C, and twice at E.

Theorem 5.2 shows that no three points of K are collinear and that the tangent at any point X of K intersects K only at X. Thus, the six lines in (10) are distinct, and any two of them intersect exactly once, counting multiplicities (by Theorem 4.1). Accordingly, if we intersect each of the three lines $\tan A, \tan C, \tan E$ forming G with each of the three lines AC, CE, EA forming H, we obtain the points $\tan A \cap AC = A$, $\tan A \cap CE = Q$, $\tan A \cap EA = A$, $\tan C \cap AC = C$, $\tan C \cap CE = C$, $\tan C \cap EA = S$, $\tan E \cap AC = R$, $\tan E \cap CE = E$, and $\tan E \cap EA = E$. Thus, G and H intersect nine times: twice at each of the points A, C, E, and once at each of the points Q, R, S (by (11) and Theorem 3.6(v)).

If we remove the intersections of G or H with K from the intersections of G and H, taking into account multiplicities, we are left with the points Q, R, S. We can apply Theorem 6.1 (by the second paragraph of the proof), and we deduce that Q, R, S are the points where G or H intersect a curve of degree $3 - 2 = 1$. This curve is a line containing Q, R, S (by Theorem 3.6(iii)). $\square$

We can also think of Pappus' Theorem 2.3 as a variation of Pascal's Theorem (Figure 6.9). In Pappus' Theorem, hexagon $AB'CA'BC'$ is inscribed in two lines e and f in the following sense: alternate vertices A, C, B of the hexagon are points of e other than $e \cap f$, and the remaining alternate vertices B', A', C' are points of f other than $e \cap f$ (Figure 6.10). The three pairs of opposite sides of hexagon $AB'CA'BC'$ intersect in three

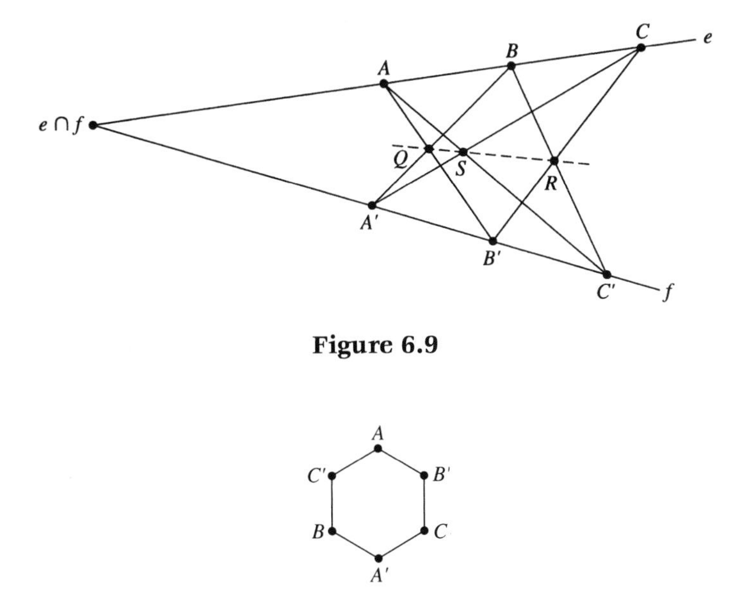

Figure 6.9

Figure 6.10

points $AB' \cap A'B = Q, B'C \cap BC' = R$, and $CA' \cap C'A = S$, and the conclusion of Pappus' Theorem is that these points are collinear. Accordingly, we can restate Pappus' Theorem as follows: *If a hexagon is inscribed in two lines, the three pairs of opposite sides intersect in collinear points.* In short, the conic in which a hexagon is inscribed in Pascal's Theorem is replaced in Pappus' Theorem with two lines, a degenerate conic.

The proof of Pascal's Theorem 6.2 was based on Theorem 6.1, which lets us "peel off a conic" from the intersection of two curves of the same degree. The conic is replaced with two lines in Pappus' Theorem. The following analogue of Theorem 6.1 lets us "peel off a line" from the intersection of two curves of the same degree:

Theorem 6.4

Let $G = 0$ and $H = 0$ be distinct curves of degree n. Assume that there is a line $L = 0$ such that $I_P(G, L) = I_P(H, L)$ for every point P in the projective plane and such that L intersects G or H a total of n times, counting multiplicities. Then there is a curve W of degree $n - 1$ such that

$$I_P(G, H) = I_P(G, L) + I_P(G, W) = I_P(H, L) + I_P(H, W)$$

for every point P in the projective plane.

Proof

There is a transformation that maps two points of the line $L = 0$ to two points of the line $y = 0$ (by Theorem 3.4), and transformations preserve intersection multiplicities and degrees of homogeneous polynomials (by Property 3.5 and the discussion after the proof of Theorem 3.4). Thus, we can assume that $L = 0$ is the line $y = 0$. Because $y = 0$ intersects $G = 0$ and $H = 0$ $2n$ times, y is not a factor of G or H (by Theorem 3.6(iii) or (vi)).

Let $g(x, y) = G(x, y, 1)$ and $h(x, y) = H(x, y, 1)$ be the restrictions of G and H to the Euclidean plane. Theorem 4.4 shows that

$$g(x, 0) = r(x - a_1)^{s_1} \cdots (x - a_v)^{s_v} \tag{12}$$

for a real number $r \neq 0$, because the assumption that $y = 0$ intersects $G = 0$ a total of n times, counting multiplicities, implies that the polynomial $r(x)$ in Theorem 4.4 has degree 0 and is thus a constant r. Each exponent s_i is the number of times that $y = 0$ and $G = 0$ intersect at the point $(a_i, 0)$, and these are the only points of the Euclidean plane where $y = 0$ and $G = 0$ intersect (by Theorem 4.3).

Because $y = 0$ intersects $G = 0$ and $H = 0$ the same number of times at every point, Theorem 4.3 implies that

$$h(x, 0) = t(x - a_1)^{s_1} \cdots (x - a_v)^{s_v}$$

for a real number $t \neq 0$. As discussed after Theorem 3.6, we can multiply

H, and hence h, by $-r/t$, which gives

$$h(x,0) = -r(x-a_1)^{s_1}\cdots(x-a_v)^{s_v}.$$

Adding this equation to (12) shows that

$$g(x,0) + h(x,0) = 0.$$

Then y is a factor of $g(x,y) + h(x,y)$ (by Theorem 1.9(ii)), and we can write

$$g(x,y) + h(x,y) = yw(x,y) \tag{13}$$

for a polynomial $w(x,y)$.

Because $G = 0$ and $H = 0$ are distinct curves, G and H are not scalar multiples of each other, and so $G + H$ is nonzero. Thus, since G and H are homogeneous polynomials of degree n, so is $G + H$. Multiplying every term of each polynomial in (13) by an appropriate power of z shows that

$$G(x,y,z) + H(x,y,z) = yW(x,y,z) \tag{14}$$

for a homogeneous polynomial $W(x,y,z)$ of degree $n-1$. For any point P in the projective plane, it follows that

$$\begin{aligned} I_P(G,H) &= I_P(G, G+H) && \text{(by Theorem 3.6(iv))} \\ &= I_P(G, yW) && \text{(by (14))} \\ &= I_P(G,y) + I_P(G,W) && \text{(by Theorem 3.6(v))}. \end{aligned}$$

Interchanging G and H shows that

$$I_P(H,G) = I_P(H,y) + I_P(H,w),$$

and the left-hand side equals $I_P(G,H)$ (by Theorem 3.6(ii)). □

A line intersects a curve of degree n that does not contain it at most n times, counting multiplicities (by Theorem 4.5). Thus, the hypotheses of Theorem 6.4 state that G and H are curves that have the same degree n, intersect the line L as many times as possible without containing it, and intersect L in the same points, counting multiplicities. The conclusion of Theorem 6.4 is that, if we list the points where G and H intersect and remove the points where either curve intersects L, then we are left with the points where G or H intersects a curve W of degree $n-1$, provided that we take into account the multiplicities of intersections. We think of Theorem 6.4 as "peeling off a line" from the intersection of two curves of the same degree.

We use this result in Section 9 to prove the associative law for multiplication of points on a cubic. We use it now to prove Pappus' Theorem 2.3 in a manner analogous to Pascal's Theorem. If a hexagon is inscribed

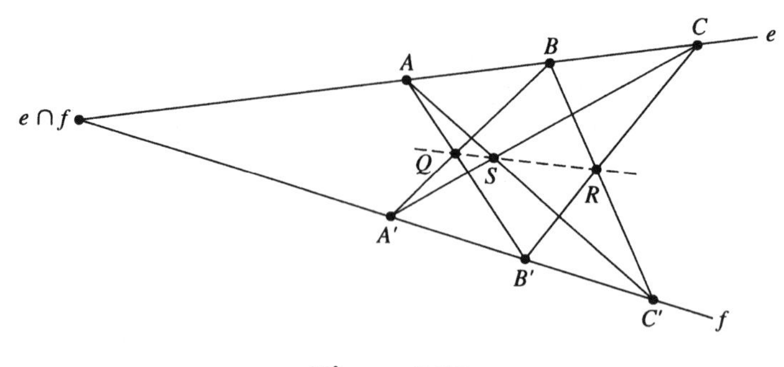

Figure 6.11

in two lines e and f, we peel off e and f from the intersection of the two cubics formed by the three pairs of opposite sides of the hexagon. It follows that the three pairs of opposite sides of the hexagon intersect in three collinear points.

Theorem 6.5 (Pappus' Theorem).
Let e and f be two lines in the projective plane. Let A, B, C be three points of e other than $e \cap f$, and let A', B', C' be three points of f other than $e \cap f$. Then the points $Q = AB' \cap A'B, R = BC' \cap B'C$, and $S = CA' \cap C'A$ are collinear (Figure 6.11).

Proof
Let $L = 0, M = 0, N = 0, T = 0, U = 0, V = 0$ be the lines

$$AB', \quad CA', \quad BC', \quad B'C, \quad A'B, \quad C'A, \tag{15}$$

respectively. Set

$$G = LMN \quad \text{and} \quad H = TUV. \tag{16}$$

G and H are homogeneous polynomials of degree 3, since they are each the product of three homogeneous polynomials of degree 1. The curves $G = 0$ and $H = 0$ consist of alternate sides of hexagon $AB'CA'BC'$ (Figure 6.10): G consists of the three lines AB', CA', BC', and H consists of the three lines $B'C, A'B, C'A$.

Since none of the points A', B', C' equals $e \cap f$, Theorem 4.1 implies that e intersects line AB' once at A, line CA' once at C, and line BC' once at B. Thus, e intersects G once at each of the points A, B, C (by (16) and Theorem 3.6(v)). Likewise, e intersects line $B'C$ once at C, $A'B$ once at B, and $C'A$ once at A, and so e intersects H once at each of the points A, B, C (by (16) and Theorem 3.6(v)). In short, the hypotheses of Theorem 6.4 hold with $n = 3$: G and H are curves of degree 3 that intersect e in the same three points A, B, C.

The six lines in (15) are distinct because each one is determined by the points where it intersects e and f (by Theorems 2.1 and 2.2). Thus, any two of the lines in (15) intersect exactly once, counting multiplicities (by Theorem 4.1). If we intersect each of the three lines AB', CA', BC' forming G with each of the three lines $B'C$, $A'B$, $C'A$ forming H, we obtain the points $AB' \cap B'C = B'$, $AB' \cap A'B = Q$, $AB' \cap C'A = A$, $CA' \cap B'C = C$, $CA' \cap A'B = A'$, $CA' \cap C'A = S$, $BC' \cap B'C = R$, $BC' \cap A'B = B$, and $BC' \cap C'A = C'$. Thus, G and H intersect at the nine points

$$A, B, C, A', B', C', Q, R, S, \tag{17}$$

(by (16) and Theorem 3.6(v)).

If we remove the three points A, B, C where G and H intersect e from the nine points in (17) where G and H intersect, we are left with the six points

$$A', B', C', Q, R, S. \tag{18}$$

We can apply Theorem 6.4 (by the second paragraph of the proof) and deduce that there is a curve W of degree $3 - 1 = 2$ that intersects both G and H at the six points in (18).

In particular, W contains the six points in (18) (by Theorem 3.6(iii)). Since f also contains A', B', C', it intersects W at least once at each of these three points (by Theorem 3.6(iii)). Then f intersects the curve W of degree 2 at least three times. Thus, if $F = 0$ is the equation of f in homogeneous coordinates, F is a factor of W (by Theorem 4.5). We write $W = FD$, where D is a homogeneous polynomial of degree 1, and so $D = 0$ is a line.

The lines AB' and $A'B$ intersect f at distinct points A' and B' (by Theorem 2.1), and so their intersection $Q = AB' \cap A'B$ does not lie on f. Likewise, neither R nor S lies on f. On the other hand, the six points in (17) lie on $W = FD = 0$, and so they each lie on either $F = 0$ or $D = 0$. Since Q, R, S do not lie on f, they lie on the line $D = 0$ and are therefore collinear. □

Exercises

6.1. Prove the theorem in Exercise 5.2 by adapting the proof of Pascal's Theorem 6.2. (This shows that the points in (6) are collinear. These points are the intersections of the three pairs of opposite sides of "hexagon" $AACDEF$.)

6.2. Prove the theorem in Exercise 5.3 by adapting the proof of Pascal's Theorem 6.2. (This shows that the points in (7) are collinear. These points are the intersections of the three pairs of opposite sides of "hexagon" $AACCEF$.)

6.3. Prove the theorem in Exercise 5.4 by adapting the proof of Pascal's Theorem 6.2. (This shows that the three pairs of opposite sides of "hexagon" $AACEEF$ intersect in collinear points. This result, like the theorem in Exercise 5.3, concerns a quadrilateral inscribed in a conic and the tangents at two of the vertices. The tangents are opposite sides of the "hexagon" in Exercise 5.4 but not in Exercise 5.3.)

6.4. Let A, C, D, E, F be five points on a conic. Describe how to use a straightedge to construct the tangent at A by applying the theorem in Exercise 5.2.

6.5. Let four points A, C, E, F on a conic and the tangent at A be given.
   (a) Use the theorem in Exercise 5.4 to describe how to construct the tangent at E with a straightedge.
   (b) If l is a line through E other than $AE, CE, EF, \tan E$, use the theorem in Exercise 5.2 to describe how to use a straightedge to construct the point other than E where l intersects the conic.

6.6. Let three points A, C, E on a conic and the tangents at A and C be given.
   (a) Use Theorem 6.3 to describe how to construct the tangent at E with a straightedge.
   (b) If l is any line through A other than $AC, AE, \tan A$, use the theorem in Exercise 5.3 to describe how to use a straightedge to construct the point other than A where l intersects the conic.

6.7. Consider the following converse of Pascal's Theorem 6.2:

Theorem
Let A–F be six points, no three of which are collinear, in the projective plane. If the points $Q = AB \cap DE, R = BC \cap EF$, and $S = CD \cap FA$ are collinear, then the six points A–F lie on a conic (Figure 6.1).

Prove this theorem by using Theorem 6.4, taking L to be the line through Q, R, S, and taking G and H as in (5).

6.8. Consider the following converse of Theorem 6.3:

Theorem
Let A, C, E be three noncollinear points in the projective plane. Let a be a line through A other than AC and EA, let c be a line through C other than AC and CE, and let e be a line through E other than EA and CE. If the points $Q = a \cap CE, R = e \cap AC$, and $S = c \cap EA$ are collinear, then there is a conic that is tangent to a at A, tangent to c at C, and tangent to e at E (Figure 6.7).

Prove this theorem by using Theorem 6.4, taking L to be the line through Q, R, S.

6.9. Use Theorem 6.4 to prove the following converse of the theorem in Exercise 5.2:

Theorem
Let A, C, D, E, F be five points, no three of which are collinear, in the projective plane. Let a be a line through A other than AC, AD, AE, AF. If the points $Q = a \cap DE, R = AC \cap EF$, and $S = CD \cap FA$ are collinear, then the points A, C, D, E, F lie on a conic tangent to a.

6.10. The following theorems arise from Pascal's Theorem 6.2 by replacing the lines AB and CD with a second conic K'. Prove these theorems by adapting the proof of Theorem 6.2.

(a) **Theorem.** *Let K and K' be two conics through four points A–D in the projective plane. Let E and F be two points of K that do not lie on K' and are such that E does not lie on the tangent to K' at D and F does not lie on the tangent to K' at A. Then DE intersects K' at a point Q other than D, FA intersects K' at a point S other than A, and EF intersects BC at a point R collinear with Q and S (Figure 6.12).*

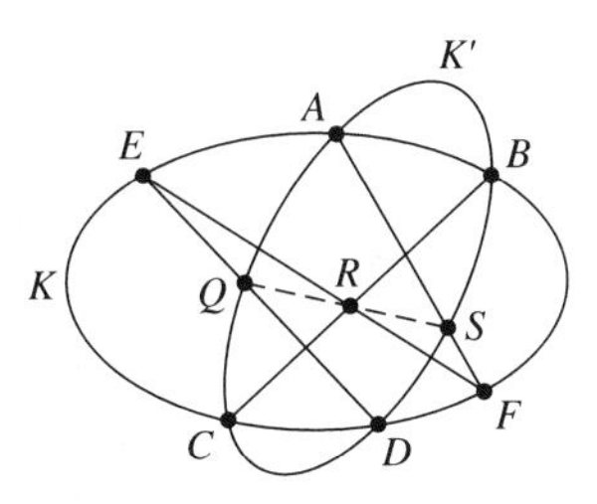

Figure 6.12

(b) **Theorem.** *Let K and K' be two conics through four points A–D in the projective plane. Let E be a point of K that does not equal B or C or lie on the tangents to K' at A and D. Then the tangent to K' at A intersects K at a point F other than A, DE intersects K' at a point Q other than D, and EF intersects BC at a point R collinear with Q and A (Figure 6.13).*

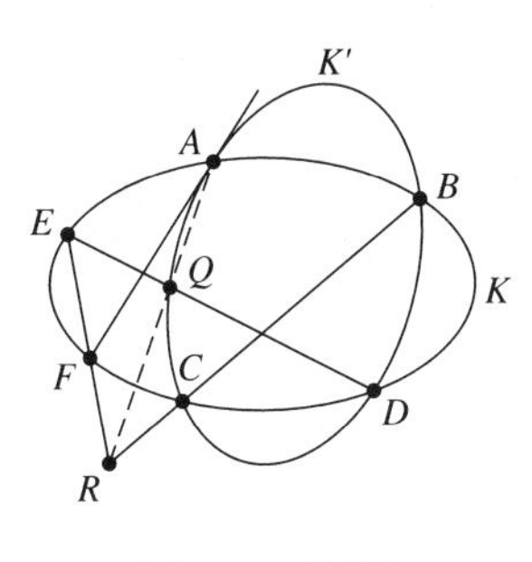

Figure 6.13

(c) **Theorem.** *Let K and K' be two conics through four points A–D in the projective plane. Assume that the tangents to K' at A and D do not intersect at a point of K. Then the tangent to K' at D intersects K at a point E other than D, the tangent to K' at A intersects K at a point F other than A, and EF intersects BC at a point R collinear with A and D (Figure 6.14).*

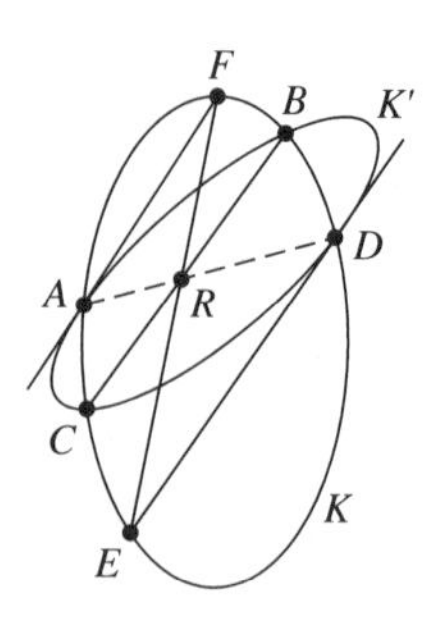

Figure 6.14

6.11. The following result arises by replacing the lines $\tan A$ and $\tan C$ in Theorem 6.3 with a conic K'. Illustrate the result with a figure, and prove it by adapting the proof of Theorem 6.3.

Theorem
Let A and C be two points in the projective plane, let l be a line on A, and let m be a line on C. Let K and K' be two conics that are both tangent to l at A and tangent to m at C. Let E be a point on K other than A and C. Then line CE intersects K' at a point Q other than C, the tangent to K at E intersects line AC at a point R, line EA intersects K' at a point S other than A, and the points Q, R, S are collinear.

6.12. The following result arises by replacing the lines AC and CE in Theorem 6.3 with a conic K'. Illustrate the result with a figure, and prove it by adapting the proof of Theorem 6.3.

Theorem
Let K and K' be two conics that both contain three points A, C, E in the projective plane and are tangent to the same line at C. Assume that the tangents to K at A and E do not intersect at a point of K'. Let S be the point where the common tangent to K and K' at C intersects line AE. Then the tangent to K at A intersects K' at a point Q other than A, the tangent to K at E intersects K' at a point R other than E, and the points Q, R, S are collinear.

6.13. (a) Prove the following result:

Theorem
Let A–H be eight points on a conic in the projective plane. Then the points $P = AB \cap DE$, $Q = BC \cap EF$, $R = CD \cap FG$, $S = DE \cap GH$, $T = EF \cap HA$, $U = FG \cap AB$, $V = GH \cap BC$, and $W = HA \cap CD$ lie either on a conic or on two lines.

(This theorem arises by replacing the hexagon $ABCDEF$ in Pascal's Theorem with the octagon $ABCDEFGH$ in Figure 6.15. We intersect each side of the octagon with the two sides adjacent to the opposite side; for example, we intersect AB with the two sides DE and FG adjacent to the side EF opposite AB.)

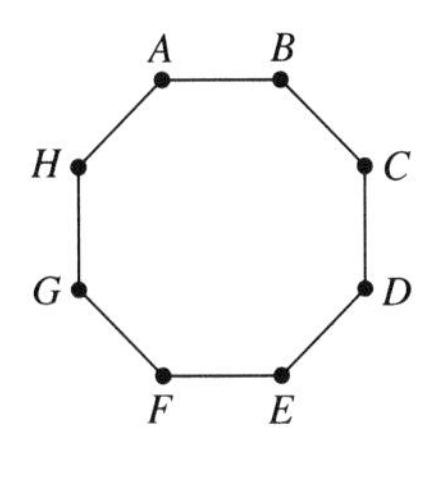

Figure 6.15

(b) Illustrate the theorem when P–W lie on a conic.
(c) Illustrate the theorem when P–W lie on two lines.
(*Hint*: One possible approach is to choose points A–C and E–G on the conic and then choose D and H on the conic so that $S = DE \cap GH$ is collinear with $Q = BC \cap EF$ and $U = FG \cap AB$. Since no three points on a conic are collinear, P–W must lie on two lines.)

6.14. (a) Prove the following result, which arises from Pappus' Theorem in the same way as the theorem in Exercise 6.13 arises from Pascal's Theorem:

Theorem
Let e and f be two lines in the projective plane. Let A, C, E, G be four points of e other than $e \cap f$, and let B, D, F, H be four points of f other than $e \cap f$. Then the points $P = AB \cap DE$, $Q = BC \cap EF$, $R = CD \cap FG$, $S = DE \cap GH$, $T = EF \cap HA$, $U = FG \cap AB$, $V = GH \cap BC$, and $W = HA \cap CD$ lie either on a conic or on two lines.

(b) Illustrate the theorem when P–W lie on a conic.
(c) Illustrate the theorem when P–W lie on two lines. (See the hint to Exercise 6.13(c).)

6.15. Use Desargues' Theorem (Exercise 3.20) and Exercise 5.17 to deduce Theorem 5.3 from Theorem 6.3.

6.16. State the version of the theorem in Exercise 4.21 that holds in the following cases when $n = 2$ and f is a conic K. Illustrate each version with a figure. (These results are known as *Carnot's Theorem*.)
(a) K is not tangent to any of the lines ST, TU, or US.
(b) K is tangent to line ST but not to TU or US.
(c) K is tangent to lines ST and TU but not US.
(d) K is tangent to all of the lines ST, TU, and US.

6.17. Define harmonic conjugates as in Exercise 5.18. Let E, F, G, H be four points, no three of which are collinear, in the projective plane. Let l be a line that does not contain any of the points $E, F, G, H, EF \cap GH$. Assume that there is a curve of degree 2 that contains E–H and intersects l twice at a point P. Prove that the harmonic conjugate of P with respect to $EF \cap l$ and $GH \cap l$ is the unique point other than P at which l intersects twice a curve of degree 2 containing E–H.

(This is a version of *Desargues' Involution Theorem*. One possible approach is to apply Exercises 4.21 and 5.20 after using Theorem 3.4 to ensure that no relevant points lie at infinity.)

6.18. Let E, F, G, H be four points, no three of which are collinear, in the projective plane. Let l be a line that does not contain any of these points. Prove that there are either zero or exactly two points at which l intersects twice a curve of degree 2 containing E–H. (See Exercises 3.14, 5.22(a), and 6.17.)

6.19. Use Exercise 6.18 and Theorem 5.1 to prove the following result:

Theorem
Let E, F, G, H be four points, no three of which are collinear, in the projective plane. Let l be a line that does not contain any of these points:
(i) *If l contains none of the points*

$$EF \cap GH, \qquad EG \cap FH, \qquad EH \cap FG, \tag{19}$$

then either zero or two conics contain E–H and are tangent to l.
(ii) *If l contains exactly one of the points in* (19), *then there is exactly one conic that contains E–H and is tangent to l.*
(iii) *If l contains two of the points in* (19), *then no conic contains E–H and is tangent to l.*

(Exercise 3.14 shows that l cannot contain all three points in (19).)

6.20. Illustrate the theorem in Exercise 6.19 by drawing four figures, one for each of the two possibilities in (i), one for (ii), and one for (iii).

§7. Envelopes

The *envelope* of a conic is the set of tangent lines. We study envelopes in this section, and our main tool is a map that interchanges the points and lines of the projective plane. We prove that this map interchanges conics and envelopes, and so results about conics imply results about envelopes. We end the section by showing how to construct the envelopes of a conic by joining the points of a line with their images under a transformation.

Our study of envelopes is based on the map

$$(a, b, c) \quad \rightarrow \quad ax + by + cz = 0 \tag{1}$$

that sends each point (a, b, c) of the projective plane to the line $ax + by + cz = 0$ whose coefficients are the homogeneous coordinates of the point. The homogeneous coordinates a, b, c of the point are not all zero, and so $ax + by + cz = 0$ is, in fact, a line. As t varies over all nonzero numbers, (ta, tb, tc) varies over all triples of homogeneous coordinates that represent one point; the corresponding equations $tax + tby + tcz = 0$ all represent the same line, and so (1) gives a well-defined map of points to lines in the projective plane. There does not seem to be a generally recognized name for the map in (1); we call it the *basic polarity*.

What is the image of a line under the basic polarity? The line has equation $px + qy + rz = 0$ for real numbers p, q, r that are not all zero. A point (a, b, c) lies on this line if and only if the equation

$$pa + qb + rc = 0 \tag{2}$$

holds. The basic polarity maps the point (a, b, c) to the line $ax + by + cz = 0$. Note that we can rewrite (2) as

$$ap + bq + cr = 0, \tag{3}$$

and this equation holds if and only if the line $ax + by + cz = 0$ contains the point (p, q, r). Thus, the basic polarity matches up the points (a, b, c) of the line $px + qy + rz = 0$ with the lines $ax + by + cz = 0$ that contain the point (p, q, r). Accordingly, the basic polarity determines a map

$$px + qy + rz = 0 \quad \to \quad (p, q, r) \tag{4}$$

of lines to points in the sense that it matches up the points of the line $px + qy + rz = 0$ with the lines through the point (p, q, r).

Note that the maps in (1) and (4) are inverses: a point maps to a line in (1) if and only if the line maps to the point in (4). Thus, the basic polarity interchanges points and lines in pairs. As we have seen, the equivalence of (2) and (3) shows that the basic polarity preserves *incidence*, the property of points lying on lines. In other words, if the basic polarity interchanges a point P with a line m and interchanges a line l with a point Q, then P lies on l if and only if m contains Q.

Given a theorem about points and lines in the projective plane, the *dual* is the statement obtained by applying the basic polarity to the points and lines in the original theorem. As we have seen, this means that we interchange points and lines while preserving incidence. The dual of a theorem holds automatically, without further work; once we have proved that a certain relationship holds among points and lines, applying the basic polarity gives another true statement.

For example, suppose that we start with Pappus' Theorem 6.5.

Pappus' Theorem

Let e and f be two lines in the projective plane. Let A, B, C be three points of e other than $e \cap f$, and let A', B', C' be three points of f other than $e \cap f$. Then the points $Q = AB' \cap A'B$, $R = BC' \cap B'C$, and $S = CA' \cap C'A$ are collinear (*Figure* 6.11).

If we apply the basic polarity to the points and lines in Pappus' Theorem, we interchange points and lines while preserving incidence. In particular, we interchange the terms "line XY" and "point $x \cap y$": XY

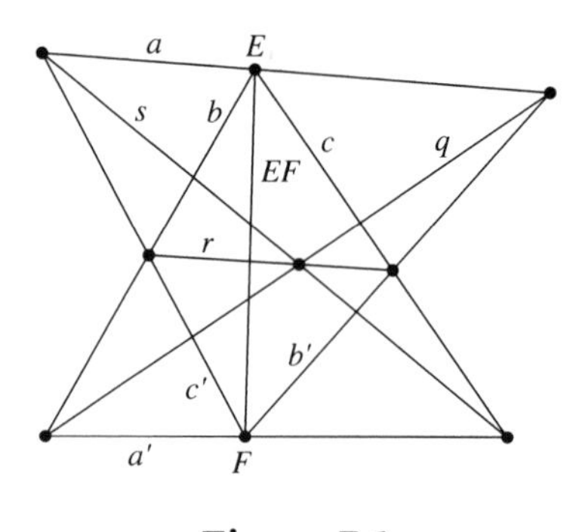

Figure 7.1

is the unique line through two points X and Y, and $x \cap y$ is the unique point on two lines x and y. Thus, dualizing Pappus' Theorem gives the following result. It requires no proof beyond the observation that it is the dual of Pappus' Theorem, which we have already proved.

Theorem 7.1
Let E and F be two points in the projective plane. Let a, b, c be three lines on E other than EF, and let a', b', c' be three lines on F other than EF. Then the lines $q = (a \cap b')(a' \cap b)$, $r = (b \cap c')(b' \cap c)$, and $s = (c \cap a')(c' \cap a)$ are concurrent (*Figure* 7.1) □

The basic polarity interchanges points and lines in pairs, as (1) and (4) show. Thus, dual theorems occur in pairs; we obtain each theorem in a pair by interchanging the points and lines of the other. For instance, dualizing Theorem 7.1 gives Pappus' Theorem.

In order to dualize our results on conics, we must determine the effect of the basic polarity on conics. Because all conics can be transformed into each other, we consider the relationship between transformations and the basic polarity.

We call the dual of a transformation a *line transformation*. Dualizing Definition 3.3 shows that a line transformation is a map of lines

$$px + qy + rz = 0 \quad \rightarrow \quad p'x + q'y + r'z = 0$$

given by equations

$$p' = ap + bq + cr,$$
$$q' = dp + eq + fr,$$
$$r' = gp + hq + ir,$$

that are equivalent to equations of the same form expressing p, q, r in terms of p', q', r'.

The discussion accompanying (12) and (13) of Section 3 show that a transformation

$$(x, y, z) \to (x', y', z')$$

given by (5) and (6) of Section 3 induces a map of lines

$$px + qy + rz = 0 \quad \to \quad p'x' + q'y' + r'z' = 0$$

given by the equations

$$\begin{aligned} p' &= Ap + Dq + Gr, \\ q' &= Bp + Eq + Hr, \\ r' &= Cp + Fq + Ir. \end{aligned}$$

Because transformations are reversible, we can likewise express p, q, r in terms of p', q', r'. Thus, any transformation induces a line transformation. Applying the basic polarity shows that any line transformation induces a transformation. The processes of the last two sentences reverse each other because both preserve incidence: the image of a line is determined by the images of the points it contains, and the image of a point is determined by the images of the lines that contain it. Thus, a line transformation induces a transformation, which in turn induces the original line transformation. We have proved the following result:

Theorem 7.2
Any transformation induces a line transformation, and every line transformation arises in this way. □

If we think of a transformation as a map of points together with the map of lines that it induces, Theorem 7.2 states that the dual of a transformation is a transformation.

We recall that the envelope of a conic is the set of tangent lines. We can now prove that the basic polarity maps any conic to the envelope of a conic. We prove this first for the parabola $4y = x^2$. The result for all conics follows from Theorem 7.2 and the fact that all conics are equivalent under transformations.

Theorem 7.3
Let K be a conic of the projective plane. Then the basic polarity matches up the points of K with the tangent lines of a conic K^*.

Proof
We start by considering the tangents to the parabola $yz = x^2$. The intersection of this parabola with the Euclidean plane has equation $y = x^2$, and it consists of the points (a, a^2) for all real numbers a. Calculus gives

the formula $dy/dx = 2x$, and so the tangent to $y = x^2$ at (a, a^2) is the line

$$y - a^2 = 2a(x - a).$$

We can rewrite this equation as $-2ax + y + a^2 = 0$, which becomes

$$-2ax + y + a^2 z = 0 \tag{5}$$

in homogeneous coordinates (x, y, z). Taking $p = -2a$, $q = 1$, and $r = a^2$ shows that the line in (5) has the form

$$px + qy + rz = 0, \tag{6}$$

where

$$4qr = p^2. \tag{7}$$

Conversely, consider any line (6) whose coefficients satisfy (7) with $q \neq 0$. Dividing (6) by q gives

$$\frac{p}{q}\, x + y + \frac{r}{q}\, z = 0.$$

This has the form of (5) for $a = -p/2q$, since the coefficient of x is $p/q = -2a$ and the coefficient of z is

$$\frac{r}{q} = \frac{4qr}{4q^2} = \frac{p^2}{4q^2} = \left(\frac{-p}{2q}\right)^2 = a^2.$$

Together with the previous paragraph, this shows that the tangents to $yz = x^2$ at points of the Euclidean plane are exactly the lines in (6) whose coefficients satisfy (7) with $q \neq 0$.

As we saw in the discussions accompanying Figure 5.2 and following the proof of Theorem 5.2, the parabola $yz = x^2$ has one point at infinity, and it is tangent there to the line at infinity $z = 0$. On the other hand, setting $q = 0$ in (9) gives $p = 0$, and so (6) becomes the line at infinity $rz = 0$ for $r \neq 0$. Thus, the tangent to $yz = x^2$ at its one point at infinity is the one line (6) given by (7) with $q = 0$.

The last sentences of the two previous paragraphs show that the tangent lines to $yz = x^2$ are exactly the lines in (6) as p, q, r vary over all triples of real numbers that satisfy (7) and are not all zero. These lines are the images under the basic polarity of the points (p, q, r) on the parabola $4qr = p^2$. In short, the basic polarity matches up the points of the parabola $4yz = x^2$ with the tangents of the parabola $yz = x^2$. This establishes the theorem when K is $4yz = x^2$ and K^* is $yz = x^2$.

Now let K be any conic. We can transform K first into $yz = x^2$ (as discussed before Theorem 5.9) and then into $4yz = x^2$ (by substituting $x/2$ for x, as in (9) of Section 3). The net effect is a transformation

$$(s, t, u) \to (s', t', u') \tag{8}$$

that takes K to $4yz = x^2$.

The basic polarity takes each point (s,t,u) of K to the line $sx+ty+uz=0$. We have seen that the basic polarity matches up the points (s',t',u') of $4yz=x^2$ with the tangent lines $s'x+t'y+u'z=0$ of $yz=x^2$. Thus, if we apply the basic polarity to the transformation in (8), we obtain a line transformation

$$sx+ty+uz=0 \quad\rightarrow\quad s'x+t'y+u'z=0$$

that matches up the images of the points of K under the basic polarity with the tangent lines of $yz=x^2$.

By Theorem 7.2, this line transformation is induced by a transformation. The transformation matches up the points of a conic K^* with the points of $yz=x^2$. The corresponding line transformation matches up the tangents of K^* with the tangents of $yz=x^2$ (since transformations preserve tangents, as noted after Definition 4.9). Together with the last sentence of the previous paragraph, this shows that the images of the points of K under the basic polarity are the tangents of K^*. □

The basic polarity interchanges points and lines in pairs. By the last theorem, the basic polarity interchanges the points of any conic K with the tangents of a conic K^*. Likewise, the basic polarity interchanges the points of K^* with the tangents of a conic K^{**}. The next result shows that $K^{**}=K$, and so the basic polarity interchanges the points of each of the conics K and K^* with the tangents of the other.

Theorem 7.4

Let K be a conic in the projective plane. Then the basic polarity interchanges the points of K with the tangents of a conic K^*, and it interchanges the points of K^* with the tangents of K. In fact, if the basic polarity interchanges a point X of K with the tangent to K^* at a point X^*, then it also interchanges X^* with the tangent to K at X.

Proof

The basic polarity interchanges points and lines in pairs while preserving incidence. It interchanges the points of K with the tangents of a conic K^* (by Theorem 7.3). Let X be any point of K, and let its image under the basic polarity be tan X^*, the tangent to K^* at a point X^* (Figure 7.2). Tan X, the tangent to K at X, is the unique line that intersects K only at X (by Theorem 5.2). Thus the basic polarity interchanges tan X with the unique point on tan X^* that lies on no other tangents of K^*. Since X^* lies on tan X^* and on no other tangents of K^*, the basic polarity interchanges tan X and X^*, as the last sentence of the theorem states. Moreover, as X varies over all points of K, tan X^* varies over all tangents of K^*, and so X^* varies over all points of K^*. Thus, the basic polarity interchanges the points of K^* with the tangents of K. □

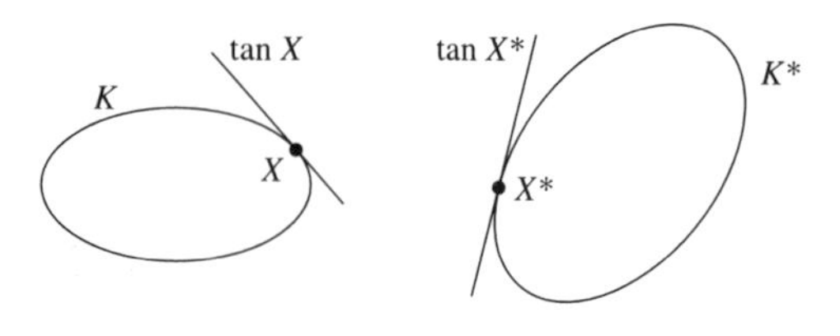

Figure 7.2

The conics K and K^* in Theorem 7.4 have a symmetric relationship; the basic polarity interchanges the points of each one with the tangents of the other. Since K can be any conic, so can K^*. Thus, we can dualize results about conics as follows: the points and tangents of a conic K dualize, respectively, to the tangents and points of a conic K^*. Specifically, if a point X of K dualizes to the tangent to K^* at a point X^*, then the tangent to K at X dualizes to the point X^*.

We can now obtain a number of results about the envelope—the set of tangents—of a conic by dualizing results about the points of a conic. For example, Theorem 5.10 states that five points in the projective plane, no three of which are collinear, lie on exactly one conic. Dualizing this theorem gives the following result:

Theorem 7.5
Five lines in the projective plane, no three of which are concurrent, are tangent to exactly one conic. □

As we observed after the proof of Theorem 5.10, Theorem 5.2 implies that no three points on a conic are collinear. Dualizing this result shows that *no three tangents of a conic are concurrent*. This shows why we need to assume in Theorem 7.5 that no three of the given lines are concurrent.

Let A be any point on a conic K. Theorem 5.2 states that any line through A except tan A intersects K at exactly two points, A and one other. Dualizing this result shows that *every point on* tan A *except A lies on exactly two tangents of K,* tan A *and one other*. This strengthens the result that no three tangents of K are concurrent.

Pascal's Theorem 6.2 states that the points $Q = AB \cap DE$, $R = BC \cap EF$, and $S = CD \cap FA$ are collinear for any six points A–F on a conic. Dualizing Pascal's Theorem gives the following result, known as *Brianchon's Theorem*:

Theorem 7.6 (Brianchon's Theorem)
Let a–f be six tangents of a conic in the projective plane. Then the lines $q = (a \cap b)(d \cap e)$, $r = (b \cap c)(e \cap f)$, and $s = (c \cap d)(f \cap a)$ are concurrent (*Figure* 7.3). □

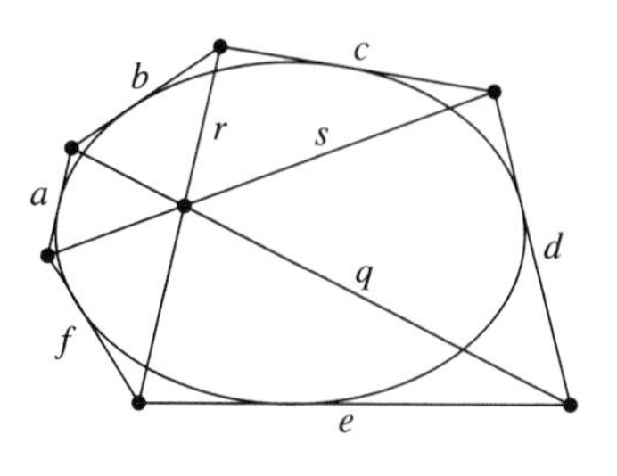

Figure 7.3

Pascal's Theorem states that the three pairs of opposite sides of a hexagon $ABCDEF$ inscribed in a conic intersect in three collinear points. The dual result, Brianchon's Theorem, refers to a hexagon *abcdef* whose sides are tangents of a conic, a hexagon circumscribed about a conic (Figure 7.3). We call

$$\{a \cap b, d \cap e\}, \quad \{b \cap c, e \cap f\}, \quad \{c \cap d, f \cap a\},$$

the three pairs of opposite vertices of hexagon *abcdef*. They determine the lines q, r, s in Brianchon's Theorem. Thus, Brianchon's Theorem states that *the three pairs of opposite vertices of a hexagon circumscribed about a conic determine concurrent lines.*The fact that no three tangents of a conic are concurrent, as noted before Theorem 7.6, implies that no two vertices of a circumscribed hexagon are equal.

As a final example, we dualize Theorem 6.3, which states that $Q = \tan A \cap CE$, $R = \tan E \cap AC$, and $S = \tan C \cap EA$ are collinear for any three points A, C, E on a conic (Figure 6.7). The three tangents tan A, tan C, tan E and the three points A, C, E dualize to three points A, C, E on a conic and to tan A, tan C, tan E, respectively (by the last sentence of the proof of Theorem 7.4). Thus, Theorem 6.3 dualizes to the following result:

Theorem 7.7

Let A, C, E be three points on a conic in the projective plane. Then the three lines $q = A(\tan C \cap \tan E)$, $r = E(\tan A \cap \tan C)$, and $s = C(\tan E \cap \tan A)$ are concurrent (Figure 7.4). □

We can simplify the statement of Theorem 7.7 by setting $A' = \tan C \cap \tan E$, $C' = \tan E \cap \tan A$, and $E' = \tan A \cap \tan C$. Theorem 7.7 states that the three lines AA', CC', EE' are concurrent for any three points A, C, E on a conic (Figure 7.4). This proves Theorem 5.3.

We end this section by developing a simple way to construct the envelope of a conic. We show that we obtain all the tangents of a conic if we join each point of a line l to its image under a transformation that maps l to another line m and does not fix $l \cap m$.

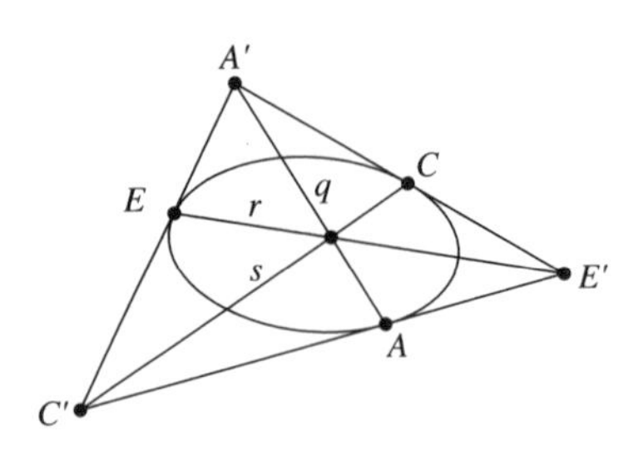

Figure 7.4

The proof of Theorem 7.3 shows that the tangents of the parabola $yz = x^2$ are the lines

$$px + qy + rz = 0, \tag{9}$$

where $4qr = p^2$. The transformation that interchanges x and z acts on the line in (9) by interchanging p and r. Thus, the tangents of $xy = z^2$ are the lines given by (9) for

$$4pq = r^2. \tag{10}$$

Setting $r = 0$ in (10) gives $p = 0$ or $q = 0$. Thus, the coordinate axes $y = 0$ and $x = 0$ are tangents of $xy = z^2$; in fact, they are the asymptotes of the hyperbola $xy = 1$ (Figure 3.1), in agreement with the discussion before Theorem 5.3. If r is nonzero in (10), then so is p, and we can divide (9) by p and relabel q and r. Thus, we can assume that $p = 1$ and $4q = r^2$. In short, the tangents to $xy = z^2$ are the coordinate axes and the lines

$$x + \frac{r^2}{4} y + rz = 0, \tag{11}$$

for all nonzero real numbers r. Setting $z = 1$ and either $y = 0$ or $x = 0$ in (11) shows that this line intersects the x-axis at $(-r, 0)$ and the y-axis at $(0, -4/r)$.

On the other hand, consider the equations

$$x' = y, \qquad y' = 4z, \qquad z' = x. \tag{12}$$

These equations give a transformation because they can obviously be solved for x, y, z in terms of x', y', z'. Since this transformation maps $(t, 0, 1)$ to $(0, 4, t)$, it takes the point $(t, 0)$ on the x-axis to the point $(0, 4/t)$ on the y-axis for all nonzero numbers t. Setting $t = -r$ in the last two sentences of the previous paragraph shows that we obtain all the tangents to the hyperbola $xy = 1$ except the asymptotes by joining each point $(t, 0)$ on the x-axis for $t \neq 0$ to its image $(0, 4/t)$ on the y-axis under the transformation in (12). The two remaining points on the x-axis are

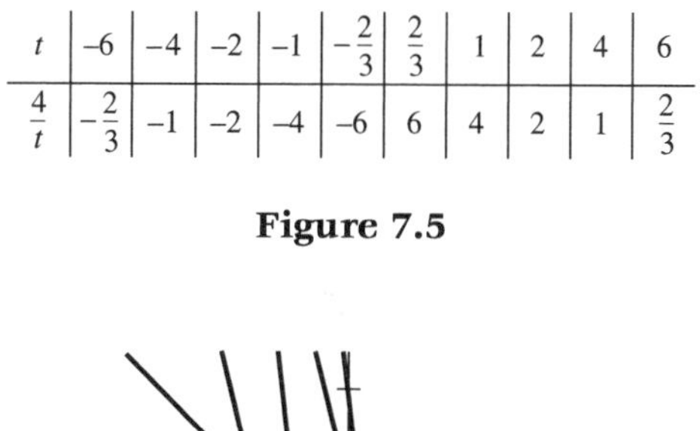

t	-6	-4	-2	-1	$-\frac{2}{3}$	$\frac{2}{3}$	1	2	4	6
$\frac{4}{t}$	$-\frac{2}{3}$	-1	-2	-4	-6	6	4	2	1	$\frac{2}{3}$

Figure 7.5

Figure 7.6

the origin and the point at infinity. The transformation in (12) maps the origin $(0, 0, 1)$ to the point $(0, 4, 0)$ at infinity on vertical lines, and these two points determine the y-axis $x = 0$. The transformation maps the point $(1, 0, 0)$ at infinity on the x-axis $y = 0$ to the origin $(0, 0, 1)$, and these two points determine the x-axis $y = 0$.

In short, the transformation in (12) maps points on the x-axis to points on the y-axis, and the lines that join corresponding points form the envelope—the set of tangents—of the hyperbola $xy = z^2$. The transformation maps the point $(t, 0)$ on the x-axis to the point $(0, 4/t)$ on the y-axis for any $t \neq 0$. Figure 7.5 gives various values of t and $4/t$, and Figure 7.6 shows the lines through the corresponding points $(t, 0)$ and $(0, 4/t)$. As t varies, these lines are the tangents to the hyperbola $xy = 1$ (sketched in Figure 3.1) at points of the Euclidean plane.

We can generalize this result by replacing the transformation in (12) that maps the x-axis to the y-axis with any transformation that maps a line l to another line m and does not fix the point $l \cap m$. This generalization follows from the previous example and Theorem 3.4, which shows that four points, no three of which are collinear, can be transformed into any other such points.

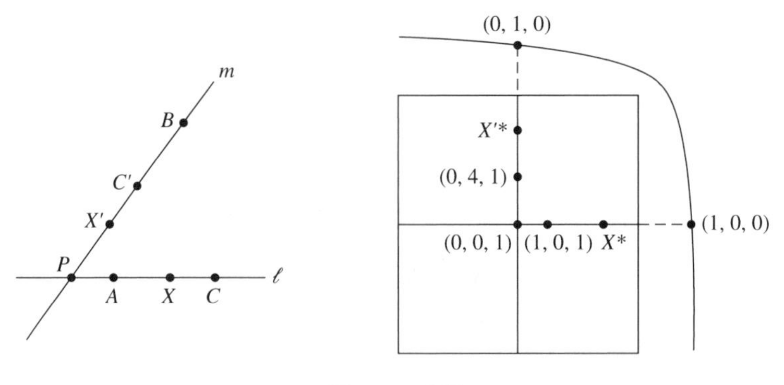

Figure 7.7

Theorem 7.8

Let $X \to X'$ be a transformation that maps a line l to a line $m \neq l$ and does not fix $l \cap m$. As X varies over all points of l, the lines XX' form the envelope of a conic.

Proof

The given transformation $X \to X'$ matches up the points of l and m, and it does not map the point $P = l \cap m$ to itself. Thus, there is a point A on l other than P that the transformation maps to P, and the image of P under the transformation is a point B on m other than P (Figure 7.7).

Let C be a point of l other than A and P. The transformation maps C to a point C' on m other than P and B. Since none of the points A, B, C, C' equals P, neither A nor C lies on $m = BC'$, and neither B nor C' lies on $l = AC$. Thus, no three of the four points A, B, C, C' are collinear.

The x-axis $y = 0$ contains the point $(1, 0, 0)$ at infinity and the point $(1, 0, 1)$ one unit from the origin. The y-axis $x = 0$ contains the point $(0, 1, 0)$ at infinity and the point $(0, 4, 1)$ four units from the origin. Neither $(1, 0, 0)$ nor $(1, 0, 1)$ lies on the y-axis—the line through $(0, 1, 0)$ and $(0, 4, 1)$—and neither $(0, 1, 0)$ nor $(0, 4, 1)$ lies on the x-axis—the line through $(1, 0, 0)$ and $(1, 0, 1)$. Thus, no three of the four points $(1, 0, 0)$, $(1, 0, 1)$, $(0, 1, 0)$, $(0, 4, 1)$ are collinear.

By Theorem 3.4 and the last two paragraphs, there is a transformation $X \to X^*$ that maps $A \to (1, 0, 0)$, $B \to (0, 1, 0)$, $C \to (1, 0, 1)$, and $C' \to (0, 4, 1)$. This transformation maps $l = AC$ to the line through $(1, 0, 0)$ and $(1, 0, 1)$—the x-axis—and it maps $m = BC'$ to the line through $(0, 1, 0)$ and $(0, 4, 1)$—the y-axis. Thus, the transformation maps $P = l \cap m$ to the intersection of the x- and y-axes—the origin $(0, 0, 1)$.

We take the given transformation $X \to X'$, precede it with the reverse of the transformation $X \to X^*$ just defined, and follow it with the transformation $X \to X^*$. This gives the sequence of transformations

$$X^* \to X \to X' \to X'^*. \tag{13}$$

As X^* varies over the x-axis, X varies over l, X' varies over m, and X'^* varies over the y-axis. The sequence of transformations in (13) maps

$$(1,0,0) \to A \to P \to (0,0,1), \tag{14}$$

$$(0,0,1) \to P \to B \to (0,1,0), \tag{15}$$

$$(1,0,1) \to C \to C' \to (0,4,1). \tag{16}$$

The sequence of transformation in (13) is itself a transformation, and so there are constants a–i such that this transformation maps

$$(x,y,z) \to (ax+by+cz, dx+ey+fz, gx+hy+iz).$$

This map takes $(1,0,0)$ to $(0,0,1)$ (by (14)), and so we have $a=0$ and $d=0$. The map also takes $(0,0,1)$ to $(0,1,0)$ (by (15)), and so we have $c=0$ and $i=0$. Thus, the transformation in (13) maps

$$(x,y,z) \to (by, ey+fz, gx+hy).$$

This map takes $(1,0,1)$ to $(0,4,1)$ (by (16)), and so we have $f=4g$. In short, the transformation in (13) maps

$$(x,y,z) \to (by, ey+4gz, gx+hy).$$

Setting $y=0$ shows that $(x,0,z)$ maps to $(0,4gz,gx)$. It follows that g is nonzero, and so $(0,4gz,gx)$ represents the same point as $(0,4z,x)$ for x and z not both zero. Thus the transformation in (13) maps

$$(x,0,z) \to (0,4z,x) \tag{17}$$

for any point $(x,0,z)$ on the x-axis. Comparing (17) with (12) shows that the transformation in (13) maps each point on the x-axis to the same point on the y-axis as the transformation in (12). The discussion after (12) shows that we get the envelope of $xy=z^2$ by joining each point on the x-axis to its image on the y-axis under the transformation in (12). Thus, the same result holds for the transformation in (13).

We now know that the tangents of $xy=z^2$ are the lines $X^*X'^*$ for all points X^* on the x-axis (since the transformation in (13) maps X^* to X'^*). Because the reverse of the transformation $X \to X^*$ is itself a transformation, it preserves conics and tangents. Thus, the lines XX' are the tangents of a conic as X varies over the points of l. □

Exercises

7.1. A theorem is stated in each of the following exercises. Use Theorem 7.4 to state the dual of each theorem in terms of conics and tangents, as in Theorems 7.5–7.7. Illustrate the results you state by drawing one figure for each of the parts (a)–(p) and the four possibilities in (q) (as in Exercise 6.20).

(a) Exercise 5.2. (b) Exercise 5.3.
(c) Exercise 5.4. (d) Exercise 5.5.
(e) Exercise 5.12. (f) Exercise 5.14.
(g) Exercise 6.7. (h) Exercise 6.8.
(i) Exercise 6.9. (j) Exercise 6.10(a).
(k) Exercise 6.10(b). (l) Exercise 6.10(c).
(m) Exercise 6.11. (n) Exercise 6.12.
(o) Exercise 6.13. (p) Exercise 6.14.
(q) Exercise 6.19.

7.2. State the duals of the following theorems, which are proved in Exercise 13.8. Draw figures to illustrate the stated theorems and their duals.

(a) **Theorem.** *Let A, B, C, D, E be five points on a conic. Set $F = AB \cap CD$, $G = AD \cap BC$, and $H = \tan A \cap \tan B$. Then* tan E *contains F if and only if E lies on line GH.*

(b) **Theorem.** *Let A, B, C, D be four points on a conic. Set $P = \tan A \cap \tan B$ and $Q = \tan C \cap \tan D$. Then P lies on CD if and only if Q lies on AB.*

7.3. Use single-variable calculus and the discussion after the proof of Theorem 4.10 to show that the tangents to $xy = 1$ in the Euclidean plane are the lines determined by the pairs of points $(t, 0)$ and $(0, 4/t)$ for all real numbers $t \neq 0$, as observed after (12). Do not transform $xy = 1$ into another curve.

7.4. Let five tangents a, c, d, e, f of a conic K be given.
(a) Use Exercise 7.1(a) to describe how to use a straightedge to construct the point at which a is tangent to K.
(b) Let P be any point on a that does not lie on c, d, e, f, or K. Use Brianchon's Theorem 7.6 to describe how to use a straightedge to construct the line through P other than a that is tangent to K. (Such a line exists, by the discussion after Theorem 7.5.)

7.5. Let a point A on a conic K, the tangent a at A, and three other tangents c, e, and f be given.
(a) Use Exercise 7.1(b) to describe how to use a straightedge to construct the point at which c is tangent to K.
(b) Let P be any point on c that does not lie on a, e, f, or K. Use Exercise 7.1(a) to described how to use a straightedge to construct the line through P other than a that is tangent to K. (Such a line exists, by the discussion after Theorem 7.5.)

7.6. Suppose that we are given two points A and C on a conic K, the tangents a and c at A and C, and a third tangent e.
(a) Use Theorem 7.7 to describe how to use a straightedge to construct the point at which e is tangent to K.
(b) For any point P on a other than A, $a \cap c$, and $a \cap e$, use Exercise 7.1(b) to describe how to use a straightedge to construct the line through P other than a that is tangent to K. (Such a line exists, by the discussion after Theorem 7.5.)

7.7. Let $X \to X'$ be a transformation that maps a line l to a line $m \neq l$ and does not fix the point $l \cap m$. By Theorem 7.8, the lines XX' form the envelope of a conic K as X varies over all points of l. Let A be the point that the transformation maps to $l \cap m$, and let B be the image of $l \cap m$ under the transformation. Prove that K is tangent to l at A and tangent to m at B.

7.8. Let p and q be nonzero real numbers.

(a) Prove that there is a transformation that maps the points $(t, 0)$ on the x-axis to the point $(0, pt + q)$ on the y-axis for all real numbers t and that maps the point at infinity on the x-axis to the point at infinity on the y-axis. Conclude that there is a parabola K whose tangents in the Euclidean plane are the lines joining the points $(t, 0)$ and $(0, pt + q)$ for all real numbers t.

(b) Prove that K is tangent to the x-axis at $(-q/p, 0)$ and to the y-axis at $(0, q)$.

(c) Prove that K is tangent to the line at infinity at the point on lines of slope $-p$. (This shows that $-p$ is the slope of the axis of symmetry of K. One possible approach is to determine what real numbers are the slopes of tangents of K.)

7.9. Nonzero real numbers p and q are given in each part of this exercise. By Exercise 7.8(a), there is a parabola K whose tangents in the Euclidean plane are the lines through the points $(t, 0)$ and $(0, pt + q)$ for all real numbers t. Construct a chart analogous to Figure 7.5 that gives a number of corresponding values of t and $pt + q$. Then draw a figure analogous to Figure 7.6 showing the lines through the points $(t, 0)$ and $(0, pt + q)$ for the values in the chart. Sketch K itself on the same figure, and mark the points in Exercise 7.8(b) where K is tangent to the x- and y-axes.

(a) $p = 1$ and $q = 3$.
(b) $p = -1$ and $q = 4$.
(c) $p = 2$ and $q = -3$.
(d) $p = -\frac{1}{2}$ and $q = -2$.

7.10. Let p, q, r be real numbers such that $p \neq 0$ and $r \neq 0$.

(a) Prove that the equations $x' = x + py$, $y' = qy + rz$, and $z' = y$ give a transformation by solving these equations for x, y, z in terms of x', y', z'.

(b) Prove that the transformation in (a) maps the y-axis $x = 0$ to the line $x = p$ and does not fix the point at infinity where these lines intersect.

(c) Conclude from parts (a) and (b) and Theorem 7.8 that there is an ellipse or a hyperbola K whose tangents (including the asymptotes of a hyperbola) are the lines $x = 0$ and $x = p$ and the lines through the points $(0, t)$ and $(p, (qt + r)/t)$ in the Euclidean plane for all real numbers $t \neq 0$.

(d) Prove that K is tangent to the y-axis at the origin and tangent to $x = p$ at the point (p, q).

7.11. An expression of the form $(qt + r)/t$ is given in each part of this exercise for real numbers q and r such that $r \neq 0$. Take $p = 4$, and let K be the ellipse or hyperbola determined in Exercise 7.10(c). Construct a chart analogous to Figure 7.5 that gives a number of corresponding values of t and $(qt + r)/t$. Then draw a figure analogous to Figure 7.6 showing the lines through the points $(0, t)$ and $(p, (qt + r)/t)$ for the values in the chart.

Sketch K itself on the same figure, and mark the points in Exercise 7.10(d) where K is tangent to the y-axis and $x = p$.

(a) $6/t$. (b) $-6/t$. (c) $\dfrac{2t+4}{t}$.

(d) $\dfrac{2t-4}{t}$. (e) $\dfrac{-t+3}{t}$. (f) $\dfrac{t-3}{t}$.

7.12. Let K be a parabola that is tangent to the x-axis at a point A and tangent to the y-axis at a point B. Prove that there are nonzero numbers p and q such that the construction in Exercise 7.8 gives a parabola K' that is also tangent to the x-axis at A and tangent to the y-axis at B. Conclude from Theorems 4.11 and 5.9 that $K = K'$. (This shows that the construction in Exercise 7.8 gives all parabolas tangent to the x- and y-axes. Since every parabola has two perpendicular tangents, it follows that the construction gives every parabola in an appropriate coordinate system.)

7.13. Let K be an ellipse or a hyperbola that is tangent to the y-axis at the origin O and tangent to the line $x = p$ for a real number $p \neq 0$ at a point B in the Euclidean plane. Let n be a line in the Euclidean plane that is tangent to K and not vertical. Prove that there are real numbers q and r such that $r \neq 0$, and the construction in Exercise 7.10 gives an ellipse or hyperbola K' that is tangent to $x = 0$ at the origin, tangent to $x = p$ at B, and tangent to n in the projective plane. Conclude from Theorems 4.11 and 5.9 that $K = K'$.

(This shows that the construction in Exercise 7.10 gives every ellipse or hyperbola tangent to the y-axis at the origin and tangent to another vertical line. Since every ellipse and hyperbola has two parallel tangents, it follows that the construction in Exercise 7.10 gives every ellipse or hyperbola in an appropriate coordinate system.)

7.14. In the projective plane, let A and B be two points on a conic K, and let l and m be the tangents at A and B, respectively (Figure 7.7). Set $P = l \cap m$. Let n be a tangent of K other than l and m, and set $C = l \cap n$ and $C' = m \cap n$. Deduce from Exercise 3.10, Theorem 5.2, and the discussion after Theorem 7.5 that there is a transformation that maps $A \to P$, $P \to B$, and $C \to C'$. Conclude from Theorems 4.11 and 5.9 that this transformation gives rise to K via the construction in Theorem 7.8.

(This shows that *any conic can be constructed as in Theorem 7.8 with respect to any two of its tangents.*)

7.15. Let p, q, r be real numbers such that $p \neq 0$ and $r \neq 0$.

(a) For any nonzero number t, prove that the line through the points $(0, t)$ and $(p, (qt + r)/t)$ has slope m, where

$$t^2 + (mp - q)t - r = 0.$$

(b) Let K be the conic in Exercise 7.10(c). If $r > 0$, prove that K has two tangents of every slope and is therefore an ellipse. If $r < 0$, prove that K does not have tangents of every slope and is therefore a hyperbola.

(c) Show that a tangent in the projective plane of a hyperbola is an asymptote if and only if there is no other tangent parallel to it in the Euclidean plane. If $r < 0$, prove that $\pm|r|^{1/2}$ are the two values of t that give asymptotes of the hyperbola constructed in Exercise 7.10.

7.16. Let A, B, C, A', B', C' be six points such that no three of these points are collinear and no three of the six lines $a = BC$, $b = CA$, $c = AB$, $a' = B'C'$, $b' = C'A'$, and $c' = A'B'$ are concurrent. Noncorresponding sides of triangles ABC and $A'B'C'$ intersect at the six points

$$a \cap b', \quad a \cap c', \quad b \cap c', \quad b \cap a', \quad c \cap a', \quad c \cap b', \tag{18}$$

and noncorresponding vertices of the triangles determine the six lines

$$AB', \quad AC', \quad BC', \quad BA', \quad CA', \quad CB'. \tag{19}$$

Assume that no three of the points in (18) are collinear and that no three of the lines in (19) are concurrent. Prove that the six points in (18) lie on a conic if and only if the six lines in (19) are tangent to a conic. Illustrate this result with a figure.

(*Hint*: Pascal's Theorem 6.2 and its converse in Exercise 6.7 give a criterion for the points in (18) to lie on a conic. Duality gives a criterion for the lines in (19) to be tangents of a conic. These criteria are related by Desargues' Theorem, from Exercise 3.20.)

7.17. Let $F = 0$ be a curve of degree 4 that contains four singular points and at least one other point. Prove that F has a factor of degree 1 or 2.

(*Hint*: One possible approach is to show that there is curve of degree 2—a conic or two lines—through the four singular points and a fifth point of F. Conclude from Theorems 4.5, 4.11, and 5.9 that F has a factor of degree 1 or 2.)

7.18. This exercise shows that we cannot omit the assumption in Exercise 7.17 that F contains at least one point besides the four singular points. Consider the polynomial of degree 4

$$g(x, y) = (x^2 - 1)^2 + (y^2 - 1)^2.$$

(a) Prove that the curve $g = 0$ consists of exactly four points in the projective plane.
(b) Prove that g is singular at each of the points in (a).
(c) Prove that g has no factors of degree 1 or 2. (See part (a) and Theorem 5.1.)

7.19. This exercise shows that we cannot reduce the number of singular points in Exercise 7.17. Consider the polynomial

$$H(x, y, z) = x^2y^2 + x^2z^2 + y^2z^2 + 3x^2yz + 3xy^2z + 3xyz^2.$$

(a) Prove that H is singular at the three points $(0, 0, 1)$, $(0, 1, 0)$, $(1, 0, 0)$.
(b) Prove that H contains infinitely many points but has no singular points besides those in (a).
(c) Prove that H has no factors of degree 1 or 2.

Cubics

Introduction and History

Introduction

This chapter is devoted to classifying irreducible cubics. These are curves of degree 3 given by polynomials that do not have factors of degree 1 or 2. We prove that every irreducible cubic can be transformed into the form

$$y^2 = x^3 + fx^2 + gx + h. \tag{1}$$

for real numbers f, g, h.

The proof has two main steps. First, we prove in Section 8 that, if an irreducible cubic C has a flex—i.e., a generalized inflection point—or a singular point, then C can be transformed into (1). Second, we prove in Section 12 that there is a flex on every irreducible cubic that is nonsingular (i.e., has no singular points), and so the previous sentence applies to all irreducible cubics.

In Section 9, we interrupt our work on the classification of cubics to discuss one of their most important properties. We use collinearity of points to define addition on a nonsingular, irreducible cubic C that has a flex O. This definition makes C an abelian group, which means that the sum of two points of C is again a point of C, addition is commutative and associative, O is an identity element, and every point of C has an additive inverse. A central problem in number theory is to determine the set C^* of points of C that have rational coordinates, when C is given by (1) for rational numbers f, g, h. The key to this problem is to observe that C^* is itself a group whose structure can be analyzed.

Sections 10 and 11 lay the groundwork for us to complete the classification of cubics in Section 12. We introduce the complex numbers in Section 10 and prove the Fundamental Theorem of Algebra, which states that every polynomial in one variable factors completely over the complex numbers. We introduce points with complex coordinates in Section 11. The Fundamental Theorem of Algebra ensures that curves have "as many intersections as possible" over the complex numbers. This result is Bezout's Theorem, which we prove in Section 11.

We complete the classification of irreducible cubics in Section 12 by proving that every nonsingular, irreducible cubic C has a flex. The Hessian H of C is a cubic formed from the second partial derivatives of C. The points of intersection of C and H are the flexes of C. We use Bezout's theorem from Section 11 to prove that C and H intersect exactly nine times, counting multiplicities, over the complex numbers. Because nine is odd, and because the intersections of C and H over the complex numbers are interchanged in pairs by conjugating their coordinates, it follows that C and H intersect at least once over the real numbers. Thus, every nonsingular, irreducible cubic C has a flex over the real numbers, as desired.

History

Newton's classification of cubics in the late 1600s was the first great success of analytic geometry apart from its role in calculus. Newton claimed that the equation of every cubic in the Euclidean plane could be simplified to one of the forms

$$xy^2 + ey = ax^3 + bx^2 + cx + d, \tag{2}$$

$$xy = ax^3 + bx^2 + cx + d, \tag{3}$$

$$y^2 = ax^3 + bx^2 + cx + d, \tag{4}$$

$$y = ax^3 + bx^2 + cx + d, \tag{5}$$

by an appropriate choice of the coordinate axes, which were not required to be perpendicular in Newton's time. James Stirling published a proof of Newton's claim in 1717, possibly in collaboration with Newton. The key to Stirling's proof is to consider the family of chords of the cubic parallel to an asymptote, find the locus of the midpoints of the chords, and choose coordinate axes to simplify the equation of the locus.

Newton multiplied (2) by x and completed the square of the left-hand side to obtain the equation

$$(xy + \tfrac{1}{2}e)^2 = ax^4 + bx^3 + cx^2 + dx + \tfrac{1}{4}e^2. \tag{6}$$

By considering the roots of the right-hand sides of (2)–(6), he divided cubics into 72 species. Stirling identified four more species, and Jean-Paul de Gua de Malves found another two in 1740, giving a total of 78 species of cubics.

Newton also made the remarkable assertion that all cubics could be obtained from those in (4) by projecting between planes. This chapter centers around proving Newton's assertion for irreducible cubics, with the change that projections are replaced by their algebraic equivalent, transformations.

The first proofs of Newton's assertion appeared in 1731, due independently to Alexis Clairaut and Francois Nicole. Clairaut considered the graph of the equation

$$zy^2 = ax^3 + bx^2z + cxz^2 + dz^3 \tag{7}$$

in three-dimensional-Euclidean space. Equation (7) is homogeneous and yields (4) when we set $z = 1$. It follows that (7) describes a cubical cone having the origin as vertex; that is, the graph consists of the lines joining the origin to the cubic given by (4) in the plane $z = 1$. Clairaut showed that every cubic in (2)–(5) is the intersection of a plane and a cubical cone given by (7), which proves Newton's assertion.

Among the important attributes of cubics are flexes, which are generalizations of inflection points. Clairaut asserted in 1731 that an irreducible cubic has from one through three inflection points over the real numbers (Exercises 12.8 and 12.18). For irreducible cubics having three inflection points, de Gua proved in 1740 that the inflection points are collinear (Exercises 8.6 and 9.2(c)). Plücker argued in 1834 that a nonsingular cubic has nine flexes over the complex numbers that lie by threes on twelve lines (Exercise 12.24). His argument was completed in 1844 by Ludwig Hesse, who characterized flexes with a determinant of second partial derivatives that is now called a Hessian (Theorem 12.4).

Suppose that the cubic C in (1) has rational coefficients. If we take the tangent line through a point of C with rational coordinates, or if we take the secant line through two points of C with rational coordinates, the line intersects C at another point that has rational coordinates, as we discuss in Section 9. Applying this tangent–secant construction repeatedly can produce any number of points of C with rational coordinates from just one. This addresses a central problem of number theory, finding the rational solutions of equations. An ad hoc algebraic version of the tangent–secant construction was introduced by Diophantus, who lived in Alexandria during the third century A.D. Fermat systematized the construction algebraically, and Newton interpreted it geometrically in terms of tangents and secants.

Complex numbers were introduced in the 1500s to solve cubic equations in one variable. They were reintroduced in the 1700s to facilitate integration by partial fractions. Mathematicians gradually developed proficiency in working with complex numbers, and their confidence increased when Carl Friedrich Gauss gave four proofs of the Fundamental Theorem of Algebra in the first decade of the 1800s. Jean d'Alembert had given an incomplete proof of the Fundamental Theorem in 1746. The main gap in his proof was filled in 1806 when Jean Argand proved a result generally called "d'Alembert's Lemma" (Claim 5 of Section 10). In fact, no truly complete proof of the Fundamental Theorem could be given until the 1870s, when Georg Cantor and Richard Dedekind developed the real numbers formally and Karl Weierstrass derived the basic properties of continuous functions. The proof in Section 10 of the Fundamental Theorem is based on the paper of Charles Fefferman cited in the References, which modernizes and simplifies the work of d'Alembert and Argand.

The idea of a complex curve—an algebraic curve whose coefficients and variables are complex numbers—emerged over centuries. Analytic geometers from Newton onward considered "imaginary points" on curves without clearly specfying the nature of these points. In the 1820s, Jean Poncelet and Michel Chasles argued for using imaginary points systematically in synthetic projective geometry. In 1830, Plücker clarified the nature of imaginary points when the homogeneous coordinates he introduced made it possible to consider points with complex coordinates. Nevertheless, complex curves were not generally considered natural objects of study until Georg Riemann proposed in 1851 a way to consider them topologically: the "Riemann surface" of a polynomial equation $f(w, z) = 0$ consists of sheets that lie over the complex z-plane and correspond to the values of w determined by the equation. In the 1860s, Alfred Clebsch and Paul Gordan recast Riemann's ideas from complex analytic to geometric form, and the modern view of complex curves was established.

"Elliptic integrals" are, speaking roughly, integrals that involve the square root of a polynomial of degree 3 or 4. Unlike integrals that involve the square root of a polynomial of degree 2, elliptic integrals cannot generally be evaluated in closed form. Examples of elliptic integrals arose from scientific and geometric considerations in the last half of the 1600s and the first half of the 1700s. The first examples involved arc lengths of ellipses and led to the name "elliptic integrals." In the mid-1700s, Leonhard Euler revolutionized the study of elliptic integrals by establishing the identity

$$\int_a^{x_1} g(t)^{-1/2}\,dt + \int_a^{x_2} g(t)^{-1/2}\,dt = \int_a^{x_3} g(t)^{-1/2}\,dt \tag{8}$$

for any polynomial $g(t)$ of degree 3 of 4, where x_3 is a rational function of $x_1, x_2, a, g(x_1)^{1/2}, g(x_2)^{1/2}, g(a)^{1/2}$.

Certain cubics are now called "elliptic curves" because of their connection with elliptic integrals. This connection was discovered by Gauss,

Niels Abel, and Carl Jacobi in the 1820s. Their results were clarified and extended by Riemann in the 1850s, Weierstrass in 1863, and Henri Poincaré in 1901. We summarize a small part of this work below.

Let

$$g(t) = 4t^3 + ct + d$$

be a polynomial of degree 3 without repeated roots. The Weierstrass P-function

$$x = P(u) \tag{9}$$

parametrizes the nonsingular, irreducible complex cubic

$$y^2 = g(x) \tag{10}$$

in the following sense: Equation (9) and the equation

$$y = P'(u) \tag{11}$$

match up the complex numbers u on and inside a parallelogram in the complex plane with the points (x, y) of the complex cubic (10), except that any two complex numbers u in corresponding positions on opposite sides of the parallelogram map to the same point (x, y). The function $P(u)$ can be written in the form

$$P(u) = \frac{1}{u^2} + a_2 u^2 + a_4 u^4 + \cdots$$

for complex numbers a_2, a_4, ...

Equations (9)–(11) imply that

$$\frac{dx}{du} = P'(u) = y = g(x)^{1/2},$$

and taking reciprocals gives

$$\frac{du}{dx} = g(x)^{-1/2}.$$

This implies that

$$u = \int g(x)^{-1/2}\, dx, \tag{12}$$

which means that u is a multivalued indefinite integral of the two-valued function $g(x)^{-1/2}$ of the complex variable x. In short, we obtain the Weierstrass P-function in (9) by inverting the elliptic integral in (12) and considering x as a function of u.

The idea of parametrizing the complex cubic in (10) by inverting the elliptic integral in (12) arose by drawing analogies with the following familiar facts: the unit circle

$$y^2 = 1 - x^2 \tag{13}$$

is parametrized by setting

$$x = \sin(u) \tag{14}$$

and

$$y = \sin'(u) = \cos(u), \tag{15}$$

where the relation given by (14) is the inverse of the relation

$$u = \arcsin(x) = \int_0^x (1 - t^2)^{-1/2}\, dt. \tag{16}$$

Drawing parallels between the cubic $g(t)$ and the quadratic $1 - t^2$ and between the Weierstrass P-function $x = P(u)$ and the sine function $x = \sin(u)$ creates analogies between (9) and (14), (10) and (13), (11) and (15), and (12) and (16).

In Section 9, we use secants and tangents to define additon of points on a nonsingular cubic given by (10). This method of adding points on the cubic corresponds via the Weierstrass P-function to addition of complex numbers. Specifically, for any complex numbers u_1 and u_2, the point of the complex cubic (10) that corresponds via (9) and (11) to the complex number $u_1 + u_2$ is the sum of the points on the complex cubic that correspond to u_1 and u_2. This is the geometric form of Euler's relation (8). It corresponds to the angle-addition formula for sines via the analogies in the previous paragraph.

The discussion accompanying (9)–(11) shows that the points of a nonsingular complex cubic correspond to the points of a parallelogram whose two pairs of opposite sides are glued together. Gluing together one pair of opposite sides of a parallelogram gives a cylinder. Gluing together the opposite sides of the cylinder gives a torus—the surface of a doughnut. Thus, a nonsingular complex cubic is topologically equivalent to a torus; that is, it can be continously bent into the surface of a doughnut.

A nonsingular curve $f(x, y) = 0$ over the real numbers can be divided into pieces that are each parametrized by x or y. For example, the unit circle $x^2 + y^2 = 1$ can be divided into the upper and lower half-circles

$$y = (1 - x^2)^{1/2} \qquad \text{and} \qquad y = -(1 - x^2)^{1/2},$$

which are each parametrized by x. The analogous result holds over the complex numbers. In this sense, we can think of complex curves as "one-dimensional over the complex numbers." On the other hand, the complex numbers are themselves two-dimensional over the reals. That explains why a nonsingular complex cubic is topologically equivalent to a two-dimensional surface, the torus. In this text, we always think of complex cuves as "one-dimensional over the complex numbers." We work with complex curves algebraically just like real curves.

We discuss the history of Bezout's Theorem in Chapter IV as part of the history of the study of intersection multiplicities.

§8. Flexes and Singular Points

We classify cubics by using changes of variables to transform their equations into particularly simple form. When we classified curves of degree 2 in Theorem 5.1, we did not need additional information about the curves in order to simplify their equations. Cubics are too complicated to analyze so directly. In this section, we classify irreducible cubics that have a notable point, either a flex—which is a generalized inflection point—or a singular point. The fact that a cubic has such a point gives us enough information about the equation to simplify it algebraically.

We prove in Section 12 that every irreducible cubic has a flex or a singular point. Thus, we actually classify all irreducible cubics in this section, but we cannot justify this statement until Section 12.

Formally, a *cubic* is a curve of degree 3 in the projective plane. Thus, a cubic is a curve

$$\begin{aligned} &ax^3 + bx^2y + cxy^2 + dy^3 + ex^2z + fxyz + gy^2z \\ &\quad + hxz^2 + iyz^2 + jz^3 = 0 \end{aligned} \tag{1}$$

in homogeneous coordinates, where a–j are real numbers that are not all zero. The restriction of the cubic to the Euclidean plane is the curve

$$\begin{aligned} &ax^3 + bx^2y + cxy^2 + dy^3 + ex^2 + fxy + gy^2 \\ &\quad + hx + iy + j = 0 \end{aligned} \tag{2}$$

of degree at most 3.

Let $p(x,y)$ be a polynomial of positive degree, and let $F(x,y,z)$ be a homogeneous polynomial of positive degree. We call p or F *reducible* if it factors as a product of polynomials of lower degrees, and we call it *irreducible* if there is no such factorization. We also refer to the curves $p = 0$ and $F = 0$ and their algebraic equivalents as reducible or irreducible.

If a cubic is reducible, it consists of a line and a curve of degree 2. Because we have already studied lines and curves of degree 2, we concentrate on irreducible cubics. When the cubic in (1) is irreducible, it does not have z as a factor; then at least one of the coefficients a–d is nonzero, and the restriction of the cubic to the Euclidean plane in (2) has degree exactly 3.

We need a generalization of inflection points that applies to points at infinity as well as points in the Euclidean plane and that is preserved by transformations. A *flex* of a curve G is a point P of G such that G is nonsingular at P and G intersects the tangent at P at least three times at P. That is, G has a flex at P if it has a tangent l at P and $I_P(l, G) \geq 3$. Transformations preserve flexes because they preserve tangents and intersection multiplicities.

The tangent l to a curve G at any nonsingular point P intersects the curve at least twice there (by Definition 4.9). The stronger condition that

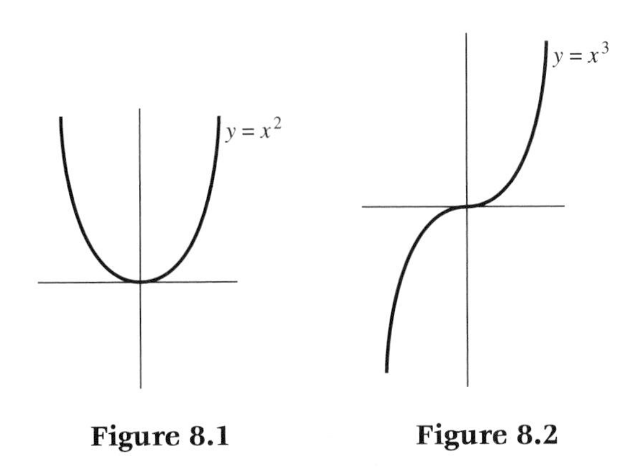

Figure 8.1 **Figure 8.2**

$I_P(l, G) \geq 3$ characterizes flexes. For example, the curves $y = x^2$ and $y = x^3$ are both tangent to the x-axis $y = 0$ at the origin (by Theorem 4.7(ii) and Definition 4.9). The x-axis intersects $y = x^2$ twice at the origin and $y = x^3$ three times (by the third paragraph after Example 1.12). Then $y = x^2$ does not have a flex at the origin, but $y = x^3$ does. As Figures 8.1 and 8.2 illustrate, this suggests that flexes are generalized inflection points. We explore this idea further in Exercises 12.13 and 12.15.

A cubic is tangent to the x-axis $y = 0$ at the origin if and only if it has equation

$$ay + bx^2 + cxy + dy^2 = ex^3 + fx^2y + gxy^2 + hy^3 \tag{3}$$

for constants a–h with $a \neq 0$ (by Theorem 4.7 and Definition 4.9). We have collected terms of degree 1 and 2 on the left of (3) and terms of degree 3 on the right. We divide (3) by a, as discussed after the proof of Theorem 3.6. By adjusting the other coefficients, we can assume that $a = 1$ in (3).

We assume that the cubic in (3) is irreducible. Then b and e are not both zero, since y is not a factor of the cubic. The number of times that the cubic intersects the x-axis $y = 0$ at the origin is the exponent of the least power of x remaining when we substitute $y = 0$ in (3) (by Theorem 1.11). This exponent is 2 if $b \neq 0$, and it is 3 if $b = 0$ and $e \neq 0$. Thus, an irreducible cubic is tangent to the x-axis at the origin and has a flex there if and only if it has equation

$$y + cxy + dy^2 = ex^3 + fx^2y + gxy^2 + hy^3$$

for $e \neq 0$.

In homogeneous coordinates, we have shown that an irreducible cubic is tangent to $y = 0$ at $(0, 0, 1)$ and has a flex there if and only if it has equation

$$yz^2 + cxyz + dy^2z = ex^3 + fx^2y + gxy^2 + hy^3$$

for $e \neq 0$. Interchanging y and z with a transformation shows that an irreducible cubic is tangent to $z = 0$ at $(0, 1, 0)$ and has a flex there if and only if the cubic has equation

$$y^2z + cxyz + dyz^2 = ex^3 + fx^2z + gxz^2 + hz^3 \tag{4}$$

for $e \neq 0$. Since $z = 0$ is the line at infinity and $(0, 1, 0)$ is the point at infinity on vertical lines (as discussed after (8) of Section 2), setting $z = 1$ in (4) gives part (i) of the next result once we prove that (4) is irreducible for $e \neq 0$.

Theorem 8.1

(i) *A cubic is tangent to the line at infinity at the point at infinity on vertical lines, has a flex there, and is irreducible if and only if it has equation*

$$y^2 + cxy + dy = ex^3 + fx^2 + gx + h \tag{5}$$

for real numbers e–h with $e \neq 0$.

(ii) *A cubic C is irreducible and has a flex at a point P if and only if there is a transformation that takes C to*

$$y^2 = x^3 + fx^2 + gx + h \tag{6}$$

for real numbers f, g, h and takes P to the point at infinity on vertical lines.

Proof

We start by proving that we can transform (5) to (6) when $e \neq 0$. Completing the square in y on the left-hand side of (5) gives

$$\left(y + \frac{c}{2}x + \frac{d}{2}\right)^2 = ex^3 + \left(f + \frac{c^2}{4}\right)x^2 + \left(g + \frac{cd}{2}\right)x + \left(h + \frac{d^2}{4}\right). \tag{7}$$

The transformation

$$x' = x, \qquad y' = y + \frac{c}{2}x + \frac{d}{2}z, \qquad z' = z,$$

takes (7)—or, more accurately, its homogenized form—to

$$y^2 = ex^3 + fx^2 + gx + h \tag{8}$$

for revised values of f, g, h. Because the value of e has not been changed, it is still nonzero. Thus, the transformation

$$x' = e^{1/3}x, \qquad y' = y, \qquad z' = z,$$

takes (8) to (6), as desired, for revised values of f and g..

We claim next that the homogeneous polynomial

$$y^2z - x^3 - fx^2z - gxz^2 - hz^3$$

corresponding to (6) is irreducible. Because this polynomial does not have z as a factor, any factorization of it into homogeneous polynomials of lower degree would give a factorization of

$$y^2 - x^3 - fx^2 - gx - h \tag{9}$$

into nonconstant polynomials in x and y. However, no such factorization exists: if it did, the absence of any term having y to the first power would imply that the polynomial in (9) factors as

$$(y - q(x))(y + q(x))$$

for a polynomial $q(x)$, but $q(x)^2$ cannot have a leading term x^3 of odd degree.

Transformations preserve irreducibility (as discussed before Theorem 4.5), equation (5) with $e \neq 0$ can be transformed into (6) (by the first paragraph of the proof), and (6) is irreducible (by the previous paragraph). Thus, (5) is irreducible. Together with the discussion before the theorem, this proves part (i).

Let C be a cubic that is irreducible and has a flex at a point P. There is a transformation that takes P to the point at infinity on vertical lines and takes a second point on the tangent at P to a second point at infinity (by Theorem 3.4). This transforms C into (5) for $e \neq 0$ (by part (i)), which can be transformed into (6) while fixing $(0, 1, 0)$ (by the first paragraph of the proof). Conversely, any curve that can be transformed into (6) is irreducible and has a flex because these properties are preserved by transformations and (6) is the special case of (5) with $c = d = 0$ and $e = 1$. □

Let $q(x)$ be a nonzero polynomial in one variable x, and let r be a real number. By Theorem 4.3, $x - r$ has the same exponent whenever we factor $q(x)$ as far as possible: this exponent is the intersection multiplicity of $y = q(x)$ and $y = 0$ at $(r, 0)$. We call $x - r$ a *repeated factor* of $q(x)$ when this exponent is greater than 1. We use this terminology to determine when the cubic in (6) has a singular point.

Theorem 8.2

Let C be the cubic $y^2 - q(x)$ for

$$q(x) = x^3 + fx^2 + gx + h.$$

(i) *Then C is nonsingular at all of its points in the Euclidean plane that do not lie on the y-axis, and the tangents at these points are not vertical.*

(ii) *A point $(r, 0)$ on the x-axis in the Euclidean plane lies on C if and only if $x - r$ is a factor of $q(x)$. If $x - r$ is not a repeated factor of $q(x)$, then C is nonsingular at $(r, 0)$ and has a vertical tangent there. If $x - r$ is a repeated factor of $q(x)$, then C is singular at $(r, 0)$.*

(iii) *The one point of C at infinity is the point at infinity on vertical lines, and C is nonsingular there and tangent to the line at infinity.*

Proof

Let (a, b) be a point of the Euclidean plane on C. Substituting $x = x' + a$ and $y = y' + b$ in $y^2 - q(x)$ gives

$$(y' + b)^2 - q(x' + a). \tag{10}$$

when this quantity is multiplied out, the constant term is zero (since (a, b) satisfies $y^2 - q(x) = 0$) and y' has coefficient $2b$. Let s be the coefficient of x'. The proof of Theorem 4.10 shows that C is nonsingular at (a, b) if and only if s and $2b$ are not both zero and that, in this case, the tangent at (a, b) is

$$s(x - a) + 2b(y - b) = 0.$$

If $b \neq 0$, this shows that C is nonsingular at (a, b) and that its tangent there is not vertical. This gives part (i).

Any point on the x-axis the Euclidean plane has the form $(r, 0)$ for a real number r. This point lies on C if and only if $q(r) = 0$, which happens if and only if $x - r$ is a factor of $q(x)$ (by Theorem 1.10(ii)). In this case, we can write

$$q(x) = (x - r)h(x) \tag{11}$$

for a polynomial $h(x)$. Setting $a = r$ and $b = 0$ in (10) gives

$$y'^2 - q(x' + r). \tag{12}$$

Taking the expression for $q(x)$ from (11) and substituting it in (12) gives

$$y'^2 - x'h(x' + r).$$

When this quantity is multiplied out, x' has coefficient $-h(r)$ and y' has coefficient zero. Thus, C is nonsingular at $(r, 0)$ and has a vertical tangent there if $h(r) \neq 0$, and C is singular at $(r, 0)$ if $h(r) = 0$ (by the previous paragraph). By Theorem 1.10(ii), $h(r) = 0$ if and only if $x - r$ is a factor of $h(x)$, which happens if and only if $x - r$ is a repeated factor of $q(x)$ (by (11)). This gives (ii).

The equation of C in homogeneous coordinates is

$$y^2 z = x^3 + fx^2 z + gxz^2 + hz^3.$$

Setting $z = 0$ in this equation gives $x = 0$. Thus, $(0, 1, 0)$ is the only point at infinity on C, and (iii) holds (by Theorem 8.1(i)). □

We recall from single variable calculus that every polynomial $q(x)$ of degree 3 in one variable has a root. In fact, we can write

$$q(x) = ex^3 + fx^2 + gx + h$$

for $e \neq 0$. Factoring out ex^3 shows that

$$q(x) = ex^3\left(1 + \frac{f}{ex} + \frac{g}{ex^2} + \frac{h}{ex^3}\right). \tag{13}$$

As x goes to $+\infty$ or $-\infty$, so does x^3 (since the exponent 3 is odd), and the quantity in parentheses in (13) approaches 1 (since the last three terms inside the parentheses approach 0). Thus, $q(x)$ takes both positive and negative values. It follows that the graph of $y = q(x)$ crosses the x-axis at some point, and so $q(x)$ has a root. For the same reason, every polynomial of odd degree in one variable has a root in the real numbers.

We call a curve *singular* if it has a singular point in the sense of Definition 4.9. Other curves are *nonsingular*; these are the curves that have tangents at all of their points.

Combining Theorems 8.1(ii) and 8.2 gives one of the two main results of this section. It determines all nonsingular, irreducible cubics that have a flex.

Theorem 8.3
A cubic is nonsingular and irreducible and has a flex if and only if it can be transformed into

$$y^2 = x(x-1)(x-w) \tag{14}$$

or

$$y^2 = x(x^2 + kx + 1) \tag{15}$$

for $w > 1$ *and* $-2 < k < 2$.

Figures 8.3 and 8.4 show cubics given by (14) and (15), respectively. These figures illustrate several properties of the cubics. The cubics have points (x, y) for the values of x that make the right-hand sides of the equations nonnegative. For (14), this occurs for $0 \leq x \leq 1$ or $x \geq w$. For (15), this occurs for $x \geq 0$: since $-2 < k < 2$, the quadratic formula shows that $x^2 + kx + 1$ has no real roots and therefore takes positive values for all real numbers x. The x intercepts of the cubics are the roots of the right-hand sides of the equations: 0, 1, w for (14) and 0 for (15). These x intercepts are the points where the cubics have vertical tangents, as Theorem 8.2 states. The cubics are symmetric across the x-axis because

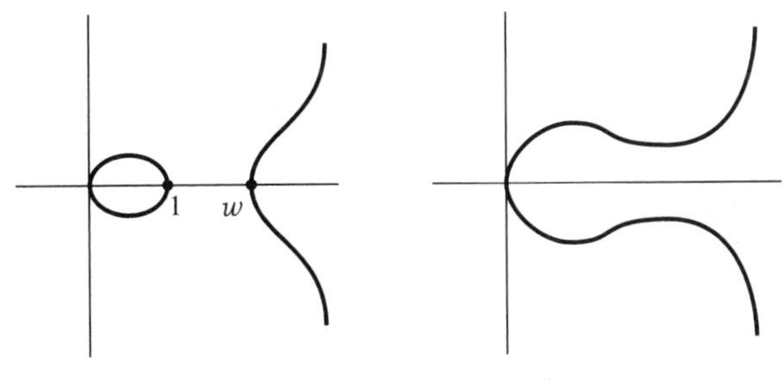

Figure 8.3 **Figure 8.4**

the equations are unchanged by replacing y with $-y$. The y coordinates of points on the cubics go to $\pm\infty$ as x goes to $+\infty$.

Proof

Assume first that C is a nonsingular, irreducible cubic that has a flex. C can be transformed into

$$y^2 = x^3 + fx^2 + gx + h, \tag{16}$$

(by Theorem 8.1(ii)). The right-hand side of this equation is a polynomial of degree 3 in x, and so it has a root r (by the discussion accompanying (13)). By Theorem 1.10(ii), we can rewrite (16) as

$$y^2 = (x - r)(x^2 + bx + c) \tag{17}$$

for real numbers b and c. Either the quadratic $x^2 + bx + c$ in (17) is irreducible or else we can write (17) as

$$y^2 = (x - r)(x - s)(x - t) \tag{18}$$

for real numbers s and t.

Because C is nonsingular, no two of the numbers r, s, t in (18) are equal (by Theorem 8.2). We label these numbers so that $r < s < t$. The transformation

$$x' = x - rz, \qquad y' = y, \qquad z' = z, \tag{19}$$

takes (18) to

$$y^2 = x(x - u)(x - v), \tag{20}$$

for $u = s - r$ and $v = t - r$. The inequalities $r < s < t$ imply that $0 < u < v$. The substitution

$$x = ux', \qquad y = u^{3/2}y', \qquad z = z',$$

arises from a transformation and takes (20) to

$$u^3y^2 = ux(ux - u)(ux - v).$$

Dividing both sides of this equation by u^3 gives (14) for $w = v/u > 1$.

If the quadratic $x^2 + bx + c$ in (17) is irreducible, the transformation in (19) takes (17) to

$$y^2 = x(x^2 + dx + e), \tag{21}$$

where the quadratic $x^2 + dx + e$ is irreducible. The quadratic formula shows that $d^2 - 4e$ is negative, and so e is positive. The substitutions

$$x = e^{1/2}x', \qquad y = e^{3/4}y', \qquad z = z',$$

arise from a transformation and take (21) to

$$e^{3/2}y^2 = e^{1/2}x(ex^2 + de^{1/2}x + e).$$

Dividing this equation by $e^{3/2}$ gives (15) for $k = de^{-1/2}$. Because the quadratic $x^2 + kx + 1$ is irreducible, the quadratic formula shows that $k^2 < 4$, and so $-2 < k < 2$.

Conversely, since the right-hand sides of (14) and (15) have no repeated factors, any cubic C that can be transformed into one of these equations is nonsingular (by Theorem 8.2 and the fact that transformations preserve singular points). Any such cubic C is irreducible and has a flex (by Theorem 8.1(ii)). □

We prove in Section 12 that every nonsingular, irreducible cubic has a flex. Thus, Theorem 8.3 actually determines all nonsingular, irreducible cubics. Theorem 8.4, the second main result of this section, determines all singular, irreducible cubics. The remaining sections (9 to 12) of this chapter focus on nonsingular, irreducible cubics and, except for some of their exercises, are independent of the rest of this section.

We could easily adapt the proof of Theorem 8.3 to show that a cubic C is singular, irreducible, and has a flex if and only if it can be transformed into one of the curves

$$y^2 = x^3, \qquad y^2 = x^2(x+1), \qquad y^2 = x^2(x-1).$$

Instead, by doing somewhat more work, we avoid assuming that C has a flex. That is, we prove that every singular, irreducible cubic C can be transformed into one of the three curves above. This is the second main result of this section. The fact that C has a singular point provides enough information to simplify its equation, just as, in Theorem 8.3, the fact that a cubic has a flex lets us simplify its equation.

Theorem 8.4

A cubic is singular and irreducible if and only if it can be transformed into one of the forms

$$y^2 = x^3, \tag{22}$$

$$y^2 = x^2(x+1), \tag{23}$$

$$y^2 = x^2(x-1). \tag{24}$$

Figures 8.5 to 8.7 show that cubics in (22)–(24), respectively, and illustrate several of their properties. The graphs have points for all values of x that make the right-hand sides of the equations nonnegative: $x \geq 0$ for (22), $x \geq -1$ for (23), and $x = 0$ or $x \geq 1$ for (24). In particular, the origin is an "isolated point" of the graph in Figure 8.7. Because x is a repeated factor of the right-hand sides of (22)–(24), the cubics are singular at the origin (by Theorem 8.2). The x intercepts other than the origin are the points where the graph has a vertical tangent (as in Theorem 8.2). The cubics are symmetric across the x-axis; and the y-coordinates of points on the graph go to $\pm\infty$ as x goes to $+\infty$.

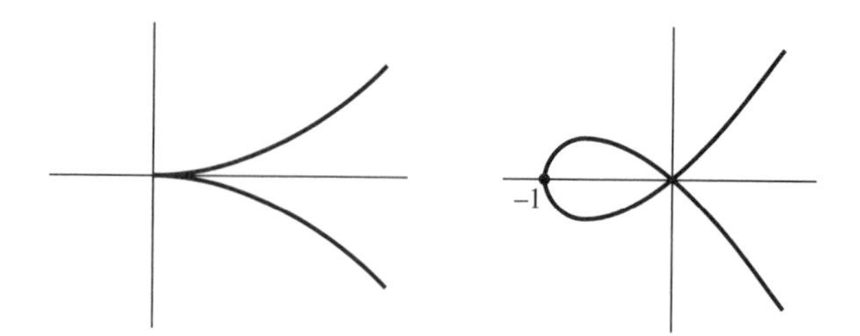

Figure 8.5 Figure 8.6

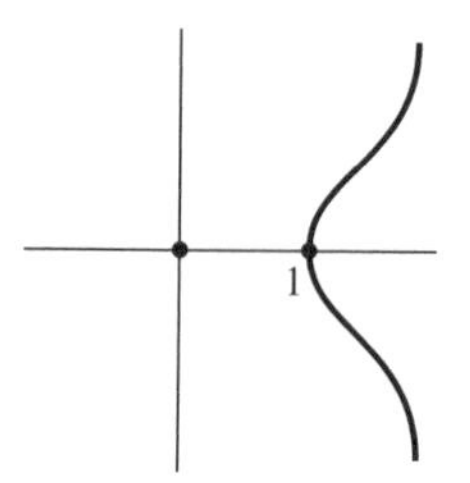

Figure 8.7

Proof

The cubics in (22)–(24) are singular at the origin (by Theorem 8.2), since x is a repeated factor of their right-hand sides. These cubics are irreducible (by Theorem 8.1(i)). Thus, any cubic that can be transformed into one of (22)–(24) is singular and irreducible.

Conversely, let C be a singular, irreducible cubic. We can use Theorem 3.4 to transform the singular point to the origin. Then C has equation

$$ax^2 + bxy + cy^2 = dx^3 + ex^2y + fxy^2 + gy^3 \tag{25}$$

in the Euclidean plane (by Theorem 4.7 and Definition 4.9), where we have collected terms of degree 2 on the left and degree 3 on the right. First, we prove that we can transform (25) so that $c = 1, b = 0$, and a is either 0, -1, or 1. Then we show that we can transform these curves into (22)–(24), where the three values of a correspond to the three equations (22)–(24).

If a, b, c were all zero, then d would be nonzero (since the irreducible cubic C does not have y as a factor). Then the polynomial

$$dx^3 + ex^2 + fx + g$$

would have a root r (as discussed before Theorem 8.3) and a factor $x - r$ (by Theorem 1.10(ii)). Then $x - ry$ would be a factor of the right-hand side of (25), and so it would be a factor of C (since we are assuming that a, b, c are all zero). This would contradict the fact that C is irreducible.

Thus, the coefficients a, b, c on the left-hand side of (25) are not all zero. If a and c are both zero, then b is nonzero, and the substitution

$$x = x' + y', \qquad y = y', \qquad z = z,$$

arises from a transformation and takes (25) to an equation of the same form with a nonzero y^2 term. Thus, we can assume that a or c is nonzero in (25). By interchanging x and y with a transformation, if necessary, we can assume that c is nonzero.

As discussed after Theorem 3.6, we can divide (25) by c and obtain

$$y^2 + bxy + ax^2 = dx^3 + ex^2y + fxy^2 + gy^3$$

for revised values of $a, b, d - g$. Completing the square in y on the left-hand side of this equation gives

$$\left(y + \tfrac{1}{2}bx\right)^2 + (a - \tfrac{1}{4}b^2)x^2 = dx^3 + ex^2y + fxy^2 + gy^3. \tag{26}$$

The transformation

$$x' = x, \qquad y' = y + \tfrac{1}{2}bx, \qquad z' = z,$$

takes (26) to

$$y^2 + ax^2 = dx^3 + ex^2y + fxy^2 + gy^3 \tag{27}$$

for revised values of $a, d - g$. In short, we have arranged to have $b = 0$ and $c = 1$ in (25).

Suppose first that $a = 0$. Because the cubic in (27) is irreducible, it does not have y as a factor, and so d is nonzero. Then

$$x' = d^{1/3}x, \qquad y' = y, \qquad z' = z,$$

is a transformation that takes (27) to

$$y^2 = x^3 + ex^2y + fxy^2 + gy^3 \tag{28}$$

for revised values of $e - g$. The substitutions

$$x = x' - \frac{e}{3}y', \qquad y = y', \qquad z = z',$$

eliminate the ex^2y term from (28), since the coefficient of x^2y in $(x - (e/3)y)^3$ is $-e$. Thus, the transformation

$$x' = x + \frac{e}{3}y, \qquad y' = y, \qquad z' = z,$$

takes (28) to

$$y^2 = x^3 + fxy^2 + gy^3$$

for revised values of f and g. The homogenization of this equation is

$$y^2z = x^3 + fxy^2 + gy^3,$$

which we can rewrite as

$$y^2(z - fx - gy) = x^3. \tag{29}$$

The transformation

$$x' = x, \qquad y' = y, \qquad z' = z - fx - gy,$$

takes (29) to $y^2z = x^3$, which is the homogenization of (22).

Henceforth we can assume that a is nonzero in (27). The transformation

$$x' = |a|^{1/2}x, \qquad y' = y, \qquad z' = z,$$

leaves x^2 with coefficient ± 1. Thus, we can assume that $a = \pm 1$ in (27).

Suppose first that $a = -1$. Then (27) becomes

$$(y^2 - x^2)z = dx^3 + ex^2y + fxy^2 + gy^3 \tag{30}$$

in homogeneous coordinates. The equations

$$x = \tfrac{1}{2}x' - \tfrac{1}{2}y', \qquad y = \tfrac{1}{2}x' + \tfrac{1}{2}y', \qquad z = z',$$

arise from a transformation because they can be solved for x', y', z' as follows:

$$x' = x + y, \qquad y' = -x + y, \qquad z' = z.$$

This transformation takes (30) to

$$xyz = dx^3 + ex^2y + fxy^2 + gy^3$$

for revised values of $d - g$. We can rewrite this equation as

$$xy(z - ex - fy) = dx^3 + gy^3,$$

and so the transformation

$$x' = x, \qquad y' = y, \qquad z' = z - ex - fy,$$

gives

$$xyz = dx^3 + gy^3.$$

Because C is irreducible, d and g are both nonzero. Then

$$x' = d^{1/3}x, \qquad y' = g^{1/3}y, \qquad z' = d^{-1/3}g^{-1/3}z,$$

is a transformation that gives the equation

$$xy = x^3 + y^3. \tag{31}$$

In particular, the cubic

$$y^2 - x^2 = x^3 \tag{32}$$

is irreducible (by Theorem 8.1(i)), and it has the form of (27) with $a = -1$. Accordingly, the previous paragraph shows that there is a transformation that takes (32) to (31). Reversing this gives a transformation that takes (31) to (32). Thus, by the previous paragraph, any irreducible cubic given by (27) with $a = -1$ can be transformed first into (31) and then into (32), which can be rewritten as (23).

Finally, suppose that $a = 1$ in (27). We can eliminate the ex^2y and the fxy^2 terms by rewriting (27) as

$$(y^2 + x^2)(z - fx - ey) = (d - f)x^3 + (g - e)y^3$$

in homogeneous coordinates. Then the transformation

$$x' = x, \qquad y' = y, \qquad z' = z - fx - ey,$$

gives an equation of the form

$$(y^2 + x^2)z = jx^3 + ky^3 \tag{33}$$

for real numbers j and k.

If $k = 0$, we must have $j \neq 0$, since C is irreducible. The transformation

$$x' = j^{1/3}x, \qquad y' = j^{1/3}y, \qquad z' = j^{-2/3}z,$$

takes (33) with $k = 0$ to

$$(y^2 + x^2)z = x^3,$$

which is the homogenization of (24). Thus, we can assume that $k \neq 0$ in (33).

Consider the substitutions

$$x = qx' + y', \qquad y = -x' + qy', \qquad z = z' + sx' + ty', \tag{34}$$

for real numbers q, s, t to be determined. If we multiply the second equation by q and add it to the first, we obtain

$$x + qy = (q^2 + 1)y'. \tag{35}$$

Since $q^2 + 1 \neq 0$ for any real number q, we can solve (35) for y' in terms of x and y. If we substitute the result into the second equation in (34), we can express x' in terms of x and y. We can then use the third equation in (34) to express z' in terms of x, y, z by substituting the expressions we have for x' and y'. Thus, the substitutions in (34) arise from a transformation.

The substitutions in (34) transform $y^2 + x^2$ into

$$(-x' + qy')^2 + (qx' + y')^2 = (q^2 + 1)(y'^2 + x'^2).$$

Accordingly, these substitutions transform (33) into

$$(q^2 + 1)(y^2 + x^2)(z + sx + ty) = j(qx + y)^3 + k(-x + qy)^3. \tag{36}$$

The x^2y terms on both sides of this equation will cancel if

$$(q^2+1)t = 3jq^2 + 3kq. \tag{37}$$

The xy^2 terms in (36) will cancel if

$$(q^2+1)s = 3jq - 3kq^2. \tag{38}$$

The y^3 terms in (36) will cancel if

$$(q^2+1)t = j + kq^3. \tag{39}$$

Combining this equation with (37) gives

$$3jq^2 + 3kq = j + kq^3. \tag{40}$$

Since $k \neq 0$ (by the paragraph after (33)), equation (40) is a polynomial of degree 3 in q. We can choose q to satisfy this equation (as discussed before Theorem 8.3). Since $q^2+1 \neq 0$, we can choose s and t so that (37) and (38) hold. Equation (39) follows from (37) and (40). In short, we have chosen q, s, and t so that the x^2y, xy^2, and y^3 terms in (36) all cancel. Thus, if we multiply out (36) and collect like terms, we obtain

$$(q^2+1)(y^2+x^2)z = ux^3 \tag{41}$$

for a real number u. Because $q^2+1 \neq 0$, equation (41) has the form of (33) with $k = 0$. Thus, we can transform (41) into (24), by the paragraph after (33). □

Theorems 8.4 and 8.1(ii) imply that *every singular, irreducible has a flex* because (22)–(24) are special cases of (6).

The cubics in (14) and (15) and (22)–(24) have flexes at the point $(0, 1, 0)$ at infinity on vertical lines (by Theorem 8.1(i)). To illustrate this, consider the homogenization $y^2z = x^3$ of (22). Interchanging y and z gives the curve $z^2y = x^3$ and takes the point $(0, 1, 0)$ to $(0, 0, 1)$. In Euclidean terms, we want to verify that $y = x^3$ has a flex at the origin $(0, 0)$, and we did so in the discussion accompanying Figure 8.2. In effect, the inflection point at the origin in Figure 8.2 shows how the two "ends" of the cubic in (22) and Figure 8.5 form a flex at infinity.

Theorems 8.3 and 8.4 characterize all irreducible cubics that have a flex or a singular point. We will prove in Section 12 that every nonsingular, irreducible cubic has a flex, and so Theorems 8.3 and 8.4 actually determine all irreducible cubics. Every nonsingular, irreducible cubic can be transformed into (14) for $w > 1$ or (15) for $-2 < k < 2$. Every singular, irreducible cubic can be transformed into one of the three equations (22)–(24).

Exercises

8.1. Let $q(x)$ be a polynomial in one variable, let $q'(x)$ be its derivative, and let r be a real number. Prove that $x - r$ is a repeated factor of $q(x)$ if and only if $q(r) = 0 = q'(r)$.

8.2. Find a polynomial $g(x)$ such that the curve $y = g(x)$ has a flex but not an inflection point at the origin.

8.3. Let $-2 < k < 2$, and consider $y = x^{1/2}(x^2 + kx + 1)^{1/2}$. Use single-variable calculus to prove that there are either two, one, or or zero values of x such that $dy/dx = 0$, and determine what values k in $(-2, 2)$ give each number. (The given function is the top half of the curve in (15). Figures 8.4, 8.8, and 8.9 sketch (15) in the three cases of this exercise.)

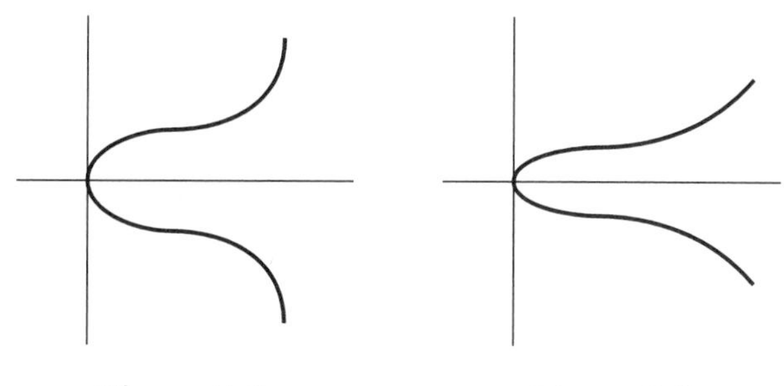

Figure 8.8 **Figure 8.9**

8.4. Sketch the graph of $y = x(x+1)^{1/2}$, analyzing it in terms of dy/dx and d^2y/dx^2 as in single-variable calculus. Deduce that Figure 8.6 is the graph of (23).

8.5. Let C be a nonsingular, irreducible cubic. Prove that any flex of C lies on the tangents of either one or three other points of C. Moreover, if a flex lies on the tangents of three other points of C, prove that the three points are collinear. (See Theorems 8.1–8.3.)

8.6. Let P and Q be two flexes of an irreducible cubic C. Prove that line PQ intersects C at a third point R, which is a flex of C other than P and Q.

(*Hint*: See Theorems 8.1 and 8.2 and take advantage of the symmetry of (6) across the x-axis. Why does no flex of (6) lie on the x-axis?)

8.7. Let P and Q be two flexes of an irreducible cubic. Prove that there is a transformation that interchanges P and Q and that does not change the equation of the cubic. (See Exercise 8.6 and its Hint.)

8.8. Conclude from Exercise 8.7 and Theorems 8.1–8.3 that no cubic can be transformed into both (14) for $w > 1$ and (15) for $-2 < k < 2$.

8.9. Consider a transformation of the projective plane. Let f, g, h be real numbers. Prove that the following conditions are equivalent, i.e., that one holds if and only if the other does.

(i) The transformation fixes $(0, 1, 0)$ and maps $y^2 = x^3 + fx^2 + gx + h$ to $y^2 = x^3 + qx^2 + rx + s$ for real numbers q, r, s.

(ii) There are real numbers c and $e \neq 0$ such that the transformation maps any point (x, y, z) to the point (x', y', z') given by the equations

$$x' = x + cz, \qquad y' = ey, \qquad z' = z/e^2. \tag{42}$$

8.10. (a) Let k be a number such that $-2 < k < 2$. Consider a transformation that fixes $(0,1,0)$ and maps (15) to

$$y^2 = x(x^2 + jx + 1)$$

for a number j such that $-2 < j < 2$. Conclude from Exercise 8.9 that, for $e = 1$ or $e = -1$, the transformation maps any point (x, y, z) to the point (x', y', z') given by the equations $x' = x$, $y' = ey$, and $z' = z$. Deduce that $j = k$.
(b) If a cubic can be transformed into (15) for $-2 < k < 2$, conclude from (a), Exercise 8.7, and Theorem 8.2 that the value of k is unique.

8.11. (a) Let $x_1 < x_2$ be real numbers. Prove that the transformation in (42) maps $(x_1, 0, 1)$ and $(x_2, 0, 1)$ to points $(x_3, 0, 1)$ and $(x_4, 0, 1)$ for real numbers x_3 and x_4 such that $x_3 < x_4$.
(b) Let w be a real number greater than 1. Consider a transformation that fixes $(0,1,0)$ and maps (14) to

$$y^2 = x(x-1)(x-v)$$

for a real number $v > 1$. Conclude from (a), Theorem 8.2, and Exercise 8.9 that, for $e = 1$ or $e = -1$, the transformation maps any point (x, y, z) to the point (x', y', z') given by the equations $x' = x$, $y' = ey$, and $z' = z$. Deduce that $v = w$.
(c) If a cubic can be transformed into (14) for $w > 1$, conclude from (b) and Exercise 8.7 that the value of w is uniquely determined.

(Together with Theorem 8.3 and Exercises 8.8 and 8.10, this exercise proves that *every nonsingular, irreducible cubic that has a flex can be transformed into exactly one of the cubics in* (14) *and* (15) *for* $w > 1$ *and* $-2 < k < 2$. In fact, this holds for all nonsingular, irreducible cubics because all such cubics have flexes, as we prove in Section 12.)

8.12. A cubic C and a point P are given in each part of this exercise. First prove that P is a flex of C by finding the tangent at P and proving that it intersects C three times at P. Then transform C into a cubic C' that has the form of (5) by taking the image of C under a transformation that maps P to $(0,1,0)$ and maps the tangent at P to the z-axis. Finally, transform C' into (14) for $w > 1$ or (15) for $-2 < k < 2$. (It follows from Theorem 8.3 that C is nonsingular and irreducible.)
(a) $y^3 = x^3 + 3x$, the origin.
(b) $y^3 = 3x^3 + 4x^2 + x$, the origin.
(c) $y^2 = x^2y + 4$, the point at infinity on horizontal lines.
(d) $x^3 = xy^2 + 2y$, the origin.
(e) $y^2x + y^2 = x^2 - x$, the point at infinity on vertical lines.
(f) $x^2y + xy^2 = 1$, the point at infinity on vertical lines.

8.13. Graph the cubics in Exercise 8.12.

8.14. (a) For each of the cubics in (22)–(24), determine how many lines through the origin intersect the cubic three times there. (Note that these are the lines that best approximate the cubic near the origin in Figures 8.5–8.7.)

(b) Conclude from part (a) and Theorems 8.2 and 8.4 that every singular, irreducible cubic can be transformed into exactly one of the equations (22)–(24).

8.15. A cubic C is given in each part of this exercise. Graph C, and prove that it is irreducible. Prove that C is singular at the point Q at infinity on vertical lines. Determine how many lines through Q intersect C three times there. Use this information, Exercise 8.14, and Theorem 8.4 to determine which of the equations (22)–(24) C can be transformed into:
(a) $x^2y = x^3 + 1$.
(b) $x^2y + y = x$.
(c) $x^2y = x + y$.
(d) $y = x - x^3$.
(e) $xy = x^3 + 1$.
(f) $x^2y + 4y = x^2 - 1$.
(g) $x^2y = x^2 - 1$.

8.16. In each part of Exercise 8.15, find a sequence of transformations that maps the given cubic to one of the equations (22)–(24). (See the proof of Theorem 8.4.)

8.17. Let $L = 0$ be the equation in homogeneous coordinates of a line other than the line at infinity, and let P be the point at infinity on L. Prove that a cubic has a flex at P and is tangent to L at P if and only if the cubic has equation $LG = uz^3$ where u is a real number and G is a homogeneous polynomial of degree 2 such that the curve $G = 0$ does not contain P.

(*Hint*: One possible approach is to show that there is a transformation that maps P to the origin and maps the lines $L = 0$ and $z = 0$ to the lines $y = 0$ and $x = 0$, respectively.

8.18. (a) Prove that the lines $y = -1$, $y = 3^{1/2}x + 2$, and $y = -3^{1/2}x + 2$ are the sides of an equilateral triangle centered at the origin.
(b) Use Exercise 8.17 and Theorem 1.9 to deduce that a cubic C is tangent to the lines in part (a) at their points at infinity and has these points as flexes if and only if C has equation

$$(y + 1)(y - 3^{1/2}x - 2)(y + 3^{1/2}x - 2) = u \tag{43}$$

for a real number u. (The cubic in (43) has the lines in part (a) as asymptotes. Figures 8.10 and 8.11 show (43) for $u = -10$ and $u = 2$.)
(c) Prove that the cubic in (46) maps to itself when the Euclidean plane is rotated 120° about the origin. (Thus, the graph has three-fold rotational symmetry about the origin.)
(d) Prove that a cubic has three collinear flexes at which the tangents are not concurrent if and only if it can be transformed into (43) for some value of u.

8.19. (a) Consider the lines through the origin parallel to the lines in Exercise 8.18(a). Prove that a cubic C is tangent to these lines at the points at infinity they contain and has these points as flexes if and only if C has the equation

$$y(y - 3^{1/2}x)(y + 3^{1/2}x) = v$$

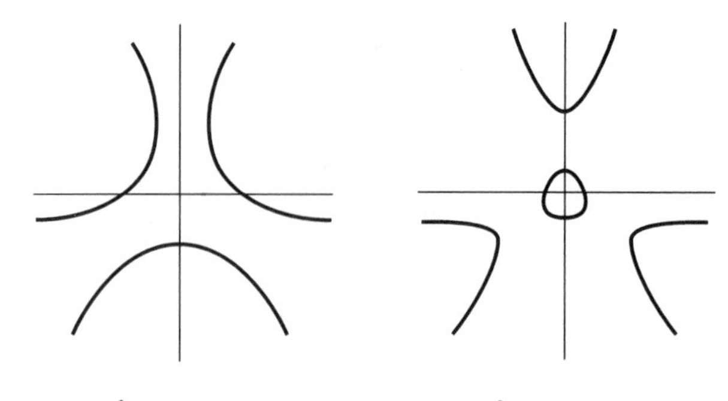

Figure 8.10 **Figure 8.11**

for a real number v. (See Exercise 8.17 and Theorem 1.9. Setting $v = 1$ gives

$$y(y - 3^{1/2}x)(y + 3^{1/2}x) = 1, \tag{44}$$

which is shown in Figure 8.12. The tangents at points at infinity are asymptotes.)

(b) If a cubic C is irreducible and has three collinear flexes at which the tangents are concurrent, prove that C can be transformed into the cubic in (44).

(c) Prove that (44) maps to itself when the Euclidean plane is rotated 120° about the origin.

(Exercise 12.8 shows that every nonsingular, irreducible cubic has three collinear flexes. Thus, Exercises 8.18 and 8.19 imply that *every nonsingular, irreducible cubic can be transformed so that it has three-fold symmetry*. Exercises 8.31 and 12.10 provide additional information about these cubics.)

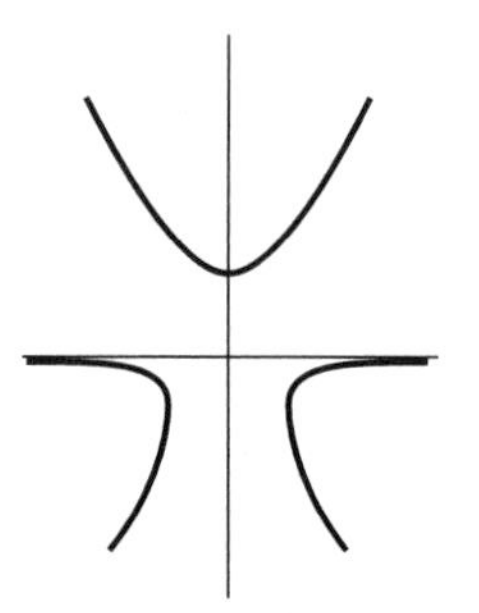

Figure 8.12

8.20. (a) For any real number u, prove that there is a transformation that takes (43) to

$$xyz = p(x + y + z)^3 \tag{45}$$

for $p = u/108$. (Setting $z = 1$ in (45) gives

$$(x + y + 1)^3 = wxy \tag{46}$$

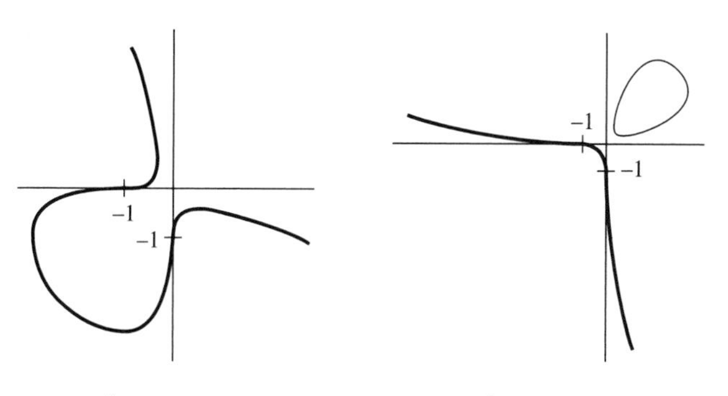

Figure 8.13 **Figure 8.14**

for $w = 1/p$ when $p \neq 0$. Figures 8.13 and 8.14 show (46) for $w = -8$ and $w = 40$.)

(b) Prove that a cubic C has flexes at $(0, -1, 1)$, $(-1, 0, 1)$, $(-1, 1, 0)$ and has tangents $x = 0$, $y = 0$, and $z = 0$ there if and only if C is given by (45) for a real number p. (See part (a) and Exercise 8.18(b).)

(c) If a transformation maps (45) to

$$xyz = q(x + y + z)^3$$

for real numbers p and q and fixes $(0, -1, 1)$, $(-1, 0, 1)$, and $(-1, 1, 0)$, prove that $p = q$ and the transformation fixes every point.

8.21. For any real number t, prove that

$$x^3 + y^3 + z^3 = txyz \tag{47}$$

has flexes at $(0, -1, 1)$, $(-1, 0, 1)$, $(-1, 1, 0)$ at which the tangents are $tx + 3y + 3z = 0$, $3x + ty + 3z = 0$, and $3x + 3y + tz = 0$. (Setting $z = 1$ in (47) gives

$$x^3 + y^3 + 1 = txy. \tag{48}$$

Figures 8.15 and 8.16 show (48) for $t = -1$ and $t = 8$.)

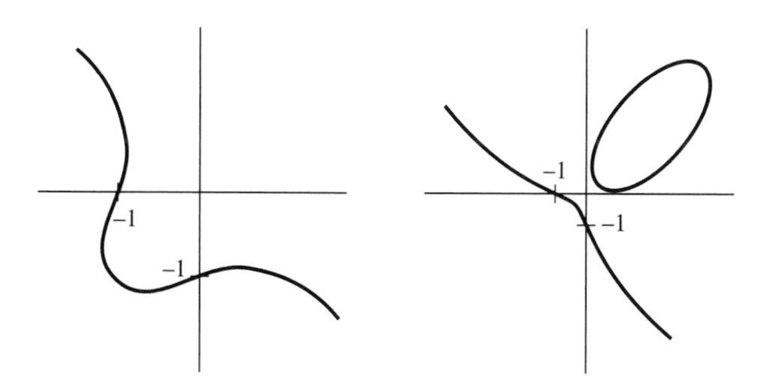

Figure 8.15 **Figure 8.16**

8.22. Comparing Exercises 8.20 and 8.21 suggests considering the equations

$$x' = tx + 3y + 3z,$$
$$y' = 3x + ty + 3z,$$
$$z' = 3x + 3y + tz.$$

Prove that these equations give a transformation if and only if t is not 3 or -6.

8.23. If t is not 3 or -6, prove that the transformation in Exercise 8.22 maps (47) to (45) for

$$p = \frac{t^2 + 3t + 9}{(t+6)^3}. \tag{49}$$

8.24. Taking the reciprocal of the right-hand side of (49) gives the expression

$$w = \frac{(t+6)^3}{t^2 + 3t + 9}. \tag{50}$$

This expression defines w as a function of t. Prove that this function is continuous and increasing for all real numbers t. Prove that the function takes arbitrarily large positive and arbitrarily negative values. Why does it follow that every real number w arises from exactly one real number t via equation (50)?

8.25. If $t = -6$, prove that the cubic in Exercise 8.21 has concurrent tangents at its flexes $(0, -1, 1)$, $(-1, 0, 1)$, and $(-1, 1, 0)$. (Figure 8.17 shows the cubic in (48) for $t = -6$.)

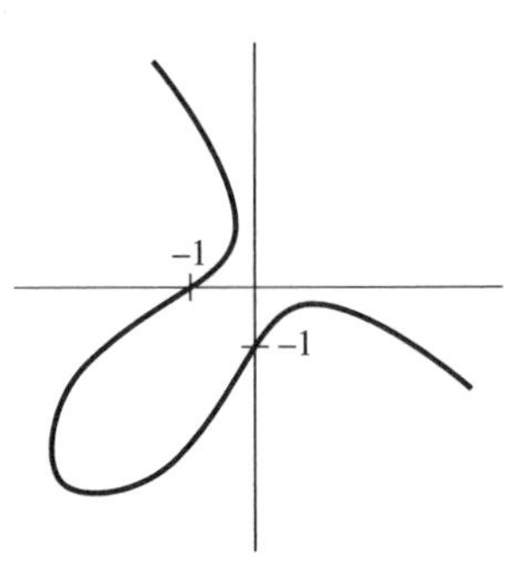

Figure 8.17

8.26. (a) Prove that a cubic C has flexes at $(1, 0, 0)$ and $(0, 1, 0)$ with tangent lines intersecting at $(0, 0, 1)$ if and only if C has the equation

$$xy(px + qy + mz) = nz^3 \tag{51}$$

for real numbers $p \neq 0$, $q \neq 0$, m, and n. (See Exercise 8.17.)
(b) Use part (a), Exercise 8.17, and Theorem 3.4 to do Exercise 8.6.

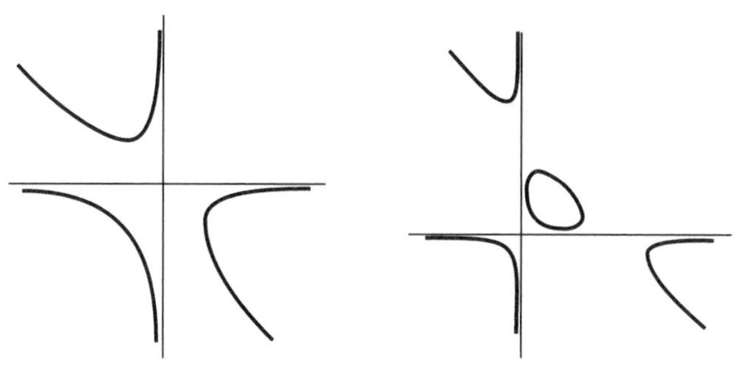

Figure 8.18 **Figure 8.19**

8.27. If the cubic in (51) is irreducible, prove that it can be transformed into

$$x^2y + xy^2 + z^3 = mxyz \tag{52}$$

for a real number m. (Setting $z = 1$ in (52) gives

$$x^2y + xy^2 + 1 = mxy.$$

Figures 8.18 and 8.19 show this cubic for $m = -1$ and $m = 4$.)
(b) Use part (a) and Exercise 8.26(a) to do Exercise 8.7.

8.28. If a transformation maps (52) to

$$x^2y + xy^2 + z^3 = sxyz$$

for a real number s and fixes $(1, 0, 0)$ and $(0, 1, 0)$, prove that $s = m$. (See Exercise 8.26(a).)

8.29. Prove that no irreducible curve of degree 2 has a flex.

8.30. Let $L = 0$ be the tangent line to a cubic $C = 0$ at a flex P. If C is reducible, prove that L is a factor of C.

(*Hint*: One possible approach is to combine Theorem 8.1(i) with the discussion after (3). Another approach is to deduce from Theorem 3.6 that P is a flex of an irreducible factor of C and to use Exercise 8.29.)

8.31. Use Exercise 8.30 and the exercises discussing the following cubics to prove that these cubics are irreducible.
(a) Equation (43) for $u \neq 0$.
(b) Equation (44).
(c) Equation (45) for $p \neq 0$.
(d) Equation (47) for $t \neq 3$.
(e) Equation (52) for all real numbers m.

§9. Addition on Cubics

We interrupt our work classifying cubics to devote this section to one of their most important properties. We give a geometric construction for adding the points of a nonsingular, irreducible cubic C with respect to a flex O. It is easy to see that addition is commutative, O is an identity element for addition, and every point of C has an additive inverse. The key property to be proved is the associative law of addition, which follows from Theorem 6.4 on "peeling off a line."

An elliptic curve is a nonsingular cubic of the form

$$y^2 = x^3 + ax^2 + bx + c \tag{1}$$

for rational numbers a, b, c. We describe addition algebraically for elliptic curves. A major open question in number theory is to determine all pairs of rational numbers x and y that satisfy (1). Work on this question is based on addition of points of C.

Elliptic curves are also important in number theory because of their role in the 1995 proof of *Fermat's Last Theorem*. This theorem, originally conjectured by Pierre de Fermat in 1665, states that the equation

$$x^n + y^n = z^n$$

has no solution in nonzero integers x, y, z when n is an integer greater than or equal to 3. In fact, any such solution would imply that there are nonzero integers a, b, c and a prime $p \geq 5$ such that

$$a^p + b^p = c^p,$$

where a is even and b is an integer 3 more than a multiple of 4. G. Frey observed in 1985 that the corresponding elliptic curve

$$y^2 = x(x + a^p)(x - b^p)$$

would have very unusual properties. Andrew Wiles proved in 1995 that no elliptic curve can have these properties, and therefore Fermat's Last Theorem holds.

The definition of addition of points on a nonsingular, irreducible cubic C is based on the intersections of lines with C. We start by analyzing these intersections.

Theorem 9.1

Let l be a line that intersects an irreducible cubic C at least twice, counting multiplicities. Then l intersects C exactly three times, counting multiplicities.

Proof

There is a transformation that maps two points of l to two points on the x-axis $y = 0$ (by Theorem 3.4). This transformation maps l to $y = 0$, maps C to another irreducible cubic, and preserves intersection multiplicities (by the remarks after the proofs of Theorems 3.4 and 4.4 and by Property 3.5). By replacing l and C with their images under this transformation, we can assume that l is the line $y = 0$.

C does not have y as a factor because it is irreducible. By Theorem 4.4, the number of times, counting multiplicities, that $y = 0$ intersects C in the projective plane is the degree 3 of C minus the degree of a polynomial $r(x)$ that has no real roots. Because $y = 0$ intersects C at least twice, $r(x)$ has degree at most 1. On the other hand, every polynomial in one variable of degree 1 has a root: $sx + t$ has root $-t/s$ for any real numbers $s \neq 0$ and t. Thus, $r(x)$ has degree 0; that is, it is a constant. Then $y = 0$ intersects C three times. □

As noted before Theorem 3.4, we call points distinct when no two of them are equal. We say that the intersections of curves F and G are *listed by multiplicity* if each point appears in the list as many times as F and G intersect there. For example, if the list is P, P, P, Q, R, R, S for distinct points P–S, then F and G intersect three times at P, twice at R, and once at each of the points Q and S.

Let C be a nonsingular, irreducible cubic. We define *line PQ* for any points P and Q on C, as follows. If $P \neq Q$, line PQ is the unique line through P and Q (by Theorem 2.2), as always. If $P = Q$, line PP is the tangent at P (as discussed before (9) of Section 6).

If $P \neq Q$, then line PQ intersects C at least once at P and at least once at Q. If $P = Q$, then line $PP = \tan P$ intersects C at least twice at P (by Definition 4.9). Thus, for any points P and Q of C, the intersections of line PQ and C, listed by multiplicity, include P and Q. Then line PQ intersects C exactly three times, counting multiplicities (by Theorem 9.1). The *third intersection* of PQ and C is the point R such that line PQ intersects C at P, Q, R, listed by multiplicity.

Figure 9.1–9.5 illustrate cases where R is the third intersection of line PQ. We have three distinct points P, Q, R in Figure 9.1, $P = Q \neq R$ in Figure 9.2, $P = R \neq Q$ in Figure 9.3, $Q = R \neq P$ in Figure 9.4, and $P = Q = R$ in Figure 9.5. We use intersection multiplicities to handle these cases simultaneously.

Figures 9.1–9.5 suggest that the condition that R is the third intersection of line PQ is symmetric in the points P, Q, R. The next theorem shows that this is so.

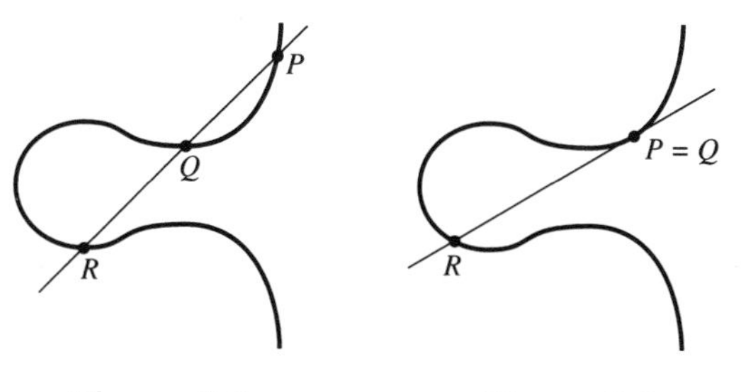

Figure 9.1 **Figure 9.2**

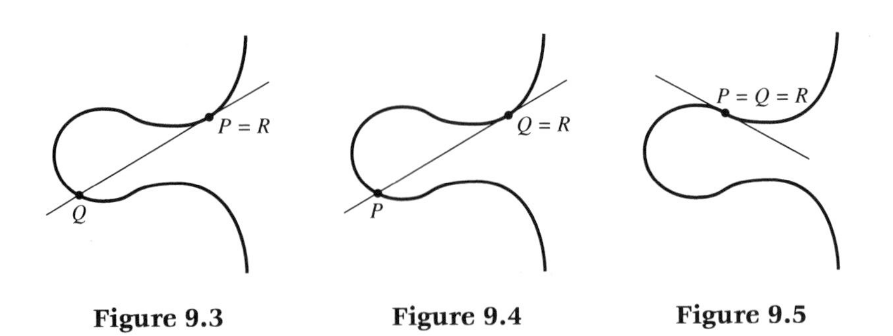

Figure 9.3 Figure 9.4 Figure 9.5

Theorem 9.2
Let C be a nonsingular, irreducible cubic. Let P, Q, R be points on C that are not necessarily distinct (*Figures* 9.1–9.5).

(i) *R is the third intersection of line PQ if and only if there is a line l that intersects C at P, Q, R, listed by multiplicity.*
(ii) *If R is the third intersection of line PQ, then Q is the third intersection of line PR, and R is the third intersection of line PQ.*

Proof
(i) If R is the third intersection of line PQ, then PQ is a line that intersects C at P, Q, R, listed by multiplicity. Conversely, let l be a line that intersects C at P, Q, R, listed by multiplicity. If $P \neq Q$, then l is the unique line PQ through P and Q (by Theorem 2.2). If $P = Q$, then l intersects C at least twice at P, and l is $\tan P = PP$ (by Definition 4.9). In short, l is line PQ whether or not P and Q are distinct. Since l intersects C at P, Q, R, counting multiplicities, R is the third intersection of line $PQ = l$.

(ii) The condition that a line l intersects C at P, Q, R, counting multiplicities, is symmetric in the points P, Q, R. Thus, part (ii) follows from part (i). □

We can now define addition on cubics.

Definition 9.3
Let C be a nonsingular, irreducible cubic with a flex O. Let P and Q be points of C that are not necessarily distinct. Then $P + Q$ is the point of C determined as follows: if S is the third intersection of line PQ, then $P + Q$ is the third intersection of line OS (Figure 9.6). □

Definition 9.3 actually applies to all nonsingular, irreducible cubics because all such cubics have flexes, as we show in Section 12. *We assume in the rest of this section that C is a nonsingular, irreducible cubic with flex O, and that addition on C is given by Definition* 9.3. Figure 9.6 depicts O as an inflection point because of the requirement that O be a flex.

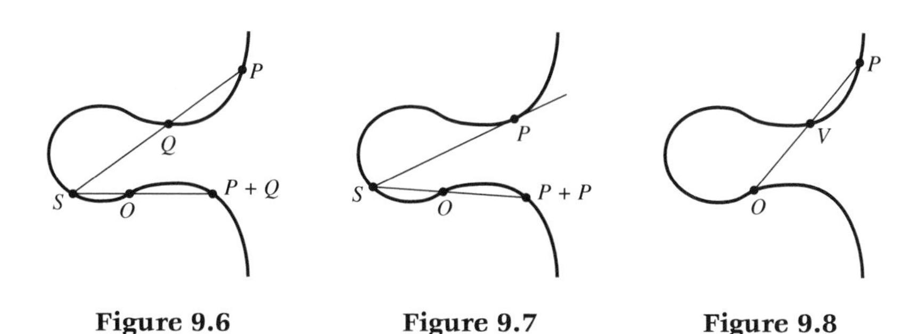

Figure 9.6 Figure 9.7 Figure 9.8

Definition 9.3 includes a number of special cases where the points P, Q, S, O, $P+Q$ are not all distinct. In particular, taking $P=Q$ in Definition 9.3 shows that $P+P$ is the third intersection of line OS, where S is the third intersection of line $PP = \tan P$ (Figure 9.7).

Because line PQ is unchanged if we switch P and Q, Definition 9.3 is symmetric in P and Q. Thus, we have the *commutative law*

$$P+Q=Q+P \tag{2}$$

for any points P and Q on C. In other words, the order in which we add points does not matter.

For any point P on C, let V be the third intersection of line PO (Figure 9.8). Taking $Q=O$ in Definition 9.3 gives $S=V$. Then $P+O$ is the third intersection of line $OS=OV$ (by Definition 9.3), and that point is P (by Theorem 9.2(ii) and the choice of V as the third intersection of line PO). In short, we have

$$P+O=P \tag{3}$$

for any point P of C. We call O the *identity element* for addition as a shorthand way to say that (3) holds for every point P of C. Equation (3) shows that adding the identity element to any point of C gives the same point back. Definition 9.3 assigns a special role to the point O precisely to make O the identity element.

In the notation of the previous paragraph, O is the third point of intersection of line PV (by Theorem 9.2(ii), since V is the third intersection of line PO). Thus, taking $Q=V$ in Definition 9.3 gives $S=O$ (Figure 9.9). Then $P+V$ is the third intersection of line $OS=OO=\tan O$ (by Definition 9.3), and that point is O (since $\tan O$ intersects C three times at O because O is a flex). If we write V as $-P$, we have shown the following: for any point P of C, there is a point $-P$ such that

$$P+(-P)=O. \tag{4}$$

We call $-P$ the *additive inverse* of P; it is a point that gives the identity element when it is added to P. We have shown that the inverse of any point P of C is the third intersection of line PO. Definition 9.3 specifies

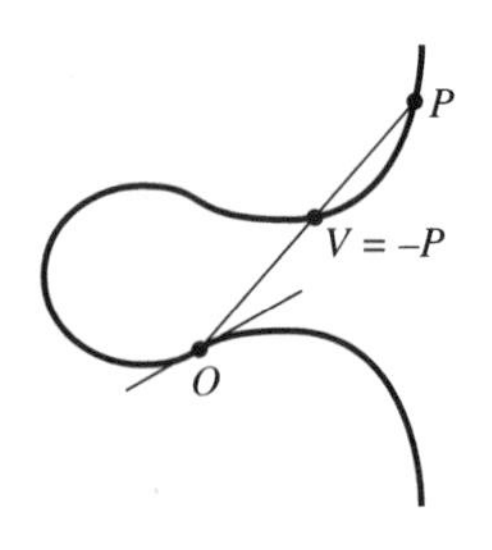

Figure 9.9

that O must be a flex precisely to permit this simple construction of inverses.

The key property of addition is the *associative law*

$$(P+Q)+R = P+(Q+R)$$

for any points P, Q, R of C. The proof depends on the following result, which relates the number of times that two curves intersect at a point to the number of times they each intersect a line there:

Theorem 9.4

Let $G=0$ and $H=0$ be curves, and let $L=0$ be a line such that L is not a factor of G. Let A be a point at which G is nonsingular. If

$$I_A(L,G) > I_A(L,H), \tag{5}$$

then we have

$$I_A(G,H) = I_A(L,H). \tag{6}$$

We can paraphrase as follows the fact that (5) implies (6): if L approaches G more closely than H at A, then L and G approach H with the same degree of closeness. Figure 9.10 illustrates this when $I_A(L,H)=1$.

Proof

Because $I_A(L,H) \geq 0$, inequality (5) implies that $I_A(L,G) \geq 1$. Assume first that $I_A(L,G)=1$. Then $I_A(L,H)=0$ (by inequality (5)), and it follows

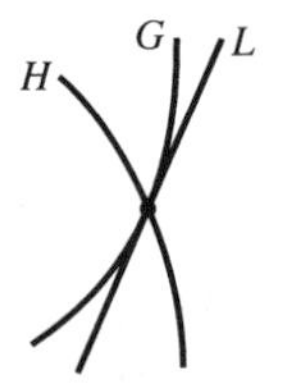

Figure 9.10

that A lies on L but not H (by Theorem 3.6(iii)). This implies that $I_A(G,H) = 0$ (by Theorem 3.6(iii)), and so (6) holds in this case.

Thus, we can assume that $I_A(L,G) \geq 2$. Then G is tangent to L at A (by Definition 4.9, since G is nonsingular at A).

There is a transformation that maps A to the origin O and maps a second point of L to a second point on the x-axis (by Theorem 3.4). Because transformations preserve intersection multiplicities and factorizations (by Property 3.5 and the discussion before Theorem 4.5), we can assume that A is the origin O and that L is the x-axis $y = 0$. By Property 3.1, we can replace G and H with their restrictions $g(x,y) = G(x,y,1)$ and $h(x,y) = H(x,y,1)$ to the Euclidean plane in computing intersection multiplicities at the origin.

The assumption that L is not a factor of G means that y is not a factor of g. If x^r is the smallest power of x appearing in the terms of g that do not have y as a factor, we can write

$$g(x,y) = yu(x,y) + x^r p(x) \tag{7}$$

for polynomials u and p such that $p(0) \neq 0$: $yu(x,y)$ is the sum of the terms of $g(x,y)$ in which y appears, and $x^r p(x)$ is the sum of the other terms. Setting $y = 0$ gives

$$g(x,0) = x^r p(x),$$

and so we have

$$I_O(y,g) = r, \tag{8}$$

(by Theorem 4.2). Moreover, because g is tangent to $y = 0$ at the origin, the coefficient of y in g is nonzero (by Theorem 4.7 and Definition 4.9). Thus, $u(x,y)$ has nonzero constant term, and so

$$u(0,0) \neq 0. \tag{9}$$

Inequality (5) shows that $I_O(y,h)$ is finite, and so y is not a factor of h (by Theorem 1.7). It follows, as in the previous paragraph, that we can write

$$h(x,y) = yv(x,y) + x^S q(x) \tag{10}$$

for polynomials v and q, where

$$q(0) \neq 0 \tag{11}$$

and

$$I_O(y,h) = s. \tag{12}$$

Equations (7) and (10) show that

$$I_O(g,h) = I_O(yu + x^r p, yv + x^s q).$$

We can multiply the last polynomial by u (by inequality (9) and Theorem

1.8). This gives

$$I_O(yu + x^r p, yuv + x^s qu).$$

We can cancel yuv by subtracting v times the first polynomial from the second (by Property 1.5). This leaves

$$I_O(yu + x^r p, -x^r pv + x^s qu). \tag{13}$$

We have

$$r > s \tag{14}$$

(by inequality (5) and equations (8) and (12)), and so we can factor x^s out of the second polynomial in (13). This gives

$$I_O(yu + x^r p, x^s w) \tag{15}$$

for

$$w(x,y) = -x^{r-s} p(x) v(x,y) + q(x) u(x,y).$$

Setting $x = 0$ and $y = 0$ in $w(x,y)$ makes x^{r-s} zero (by inequality (14)) and $q(x)u(x,y)$ nonzero (by inequalities (9) and (11)), and so $w(0,0)$ is nonzero. Thus, we can drop $w(x,y)$ from (15) (by Theorem 1.8) and leave

$$I_O(yu + x^r p, x^s).$$

This quantity equals

$$sI_O(yu + x^r p, x) \tag{16}$$

(by Property 1.6 if $s > 0$ and by Properties 1.1 and 1.3 if $s = 0$). Since $r > 0$ (by inequality (14)), $x^r p$ is a multiple of x, and so it can be omitted from (16) (by Properties 1.2 and 1.5). Thus, the quantity in (16) equals

$$\begin{aligned} sI_O(yu, x) &= sI_O(x, yu) && \text{(by Property 1.2)} \\ &= sI_O(x, y) && \text{(by inequality (9) and Theorem 1.8)} \\ &= s && \text{(by Property 1.4).} \end{aligned}$$

Together with (12), this establishes (6). □

In Theorem 9.4, the condition that

$$I_A(L, G) > I_A(L, H) \tag{17}$$

implies that $I_A(G, H) = I_A(L, H)$, and so we have

$$I_A(L, G) > I_A(G, H). \tag{18}$$

Thus, if inequality (18) does not hold, neither does inequality (17). This gives the following result:

Theorem 9.5
Let $G = 0$ and $H = 0$ be curves, and let $L = 0$ be a line such that L is not a factor of G. Let A be a point at which G is nonsingular. If

$$I_A(G, H) \geq I_A(L, G),$$

then we have

$$I_A(L, H) \geq I_A(L, G). \qquad \square$$

The next result is the geometric form of the associative law of addition on cubics, as the proof of Theorem 9.7 will show. The key to proving the next result is Theorem 6.4 on "peeling off a line." Theorem 9.5 equips us to handle multiple intersections smoothly.

Theorem 9.6
Let C be nonsingular, irreducible cubic. Let E, F, G, H be points of C that are not necessarily distinct. Let W and X be the third intersections of lines EF and GH, and let Y and Z be the third intersections of lines EG and FH. Then the third intersections of the lines WX and YZ are the same point (*Figure* 9.11).

Proof
Let T be the third intersection of line YZ (Figure 9.12). It suffices to prove that there is a line that intersects C at the points W, X, T, listed by multiplicity; if so, T is the third intersection of line WX, as desired (by Theorem 9.2(i)).

The lines

$$EF, \quad GH, \quad YZ \tag{19}$$

are given by homogeneous polynomials of degree 1. The product of these polynomials is a cubic D, which consists of the three heavy lines in Figure 9.12. Because C intersects line EF at E, F, W, line GH at G, H, X, and line YZ at Y, Z, T, listing points by multiplicity, C and D intersect at

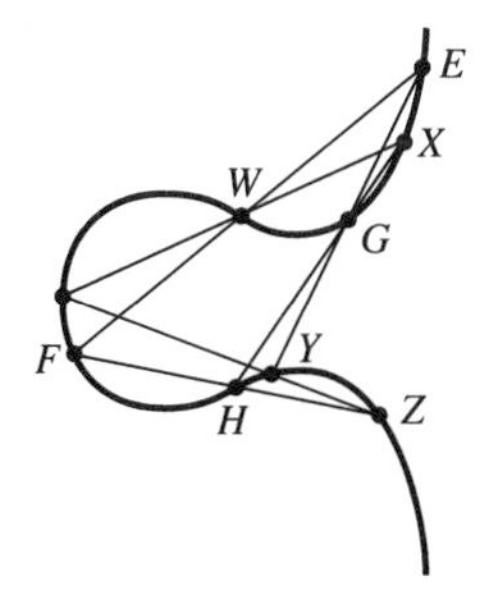

Figure 9.11

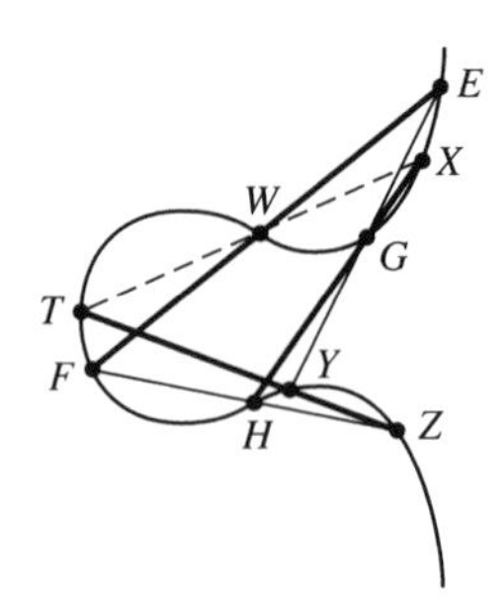

Figure 9.12

the nine points

$$E, F, G, H, W, X, Y, Z, T, \tag{20}$$

listed by multiplicity (by Theorem 3.6(v)).

Assume first that line EG does not equal any of the lines in (19). Then EG intersects each of these lines exactly once (by Theorem 4.1). It is clear that EG intersects line EF at E, line GH at G, and line YZ at Y. Thus, line EG intersects the cubic D at E, G, Y, listed by multiplicity (by Theorem 3.6(v)). EG also intersects the cubic C at E, G, Y, listed by multiplicity. Thus, by Theorem 6.4 on "peeling off a line," there is a curve $K = 0$ of degree 2 that intersects C in the six points

$$F, H, W, X, Z, T, \tag{21}$$

listed by multiplicity, that are left after removing E, G, Y from the list of points in (20).

On the other hand, suppose that EG is one of the lines in (19). The other two lines are given by homogeneous polynomials of degree 1, and we let K be the product of these two polynomials. Then $K = 0$ is again a curve of degree 2 that intersects C in the six points in (21), listed by multiplicity (by Theorem 3.6(v)).

The last two paragraphs show that there is always a curve $K = 0$ of degree 2 that intersects C at the six points in (21), listed by multiplicity. Let $L = 0$ be line FH. L intersects C at the points F, H, Z, listed by multiplicity, and these are among the points in (21) where K intersects C, listed by multiplicity. Thus, the relation

$$I_A(C, K) \geq I_A(L, C) \tag{22}$$

holds for every point A of C. By assumption, C is nonsingular, and, because it is irreducible, it does not have L as a factor. Then Theorem 9.5 and inequality (22) imply that

$$I_A(L, K) \geq I_A(L, C) \tag{23}$$

for every point A of C. Because L intersects C three times, counting multiplicities, inequality (25) shows that L intersects K at least three times, counting multiplicities. Since K has degree 2, it follows that L is a factor of K, by Theorem 4.5.

We write $K = LM$ for a homogeneous polynomial M. Since K has degree 2, M has degree 1 and so $M = 0$ is a line. By Theorem 3.6(v), the list of points in (21) where K intersects C consists of the points F, H, Z where L intersects C together with the points where M intersects C. Thus, M intersects C at the points W, X, T, listed by multiplicity. We are done by the first paragraph of the proof. □

If we take the point H in the previous result to be the flex O used to define addition on the cubic C, we obtain the associative law.

Theorem 9.7

Let C be a nonsingular, irreducible cubic that has a flex O. If P, Q, R are points of C that are not necessarily distinct, then we have

$$(P + Q) + R = P + (Q + R). \tag{24}$$

Proof

Definition 9.3 determines each side of (24) geometrically. If S is the third intersection of line PQ, then $P + Q$ is the third intersection of line OS (Figure 9.13). If T is the third intersection of line $(P + Q)R$, then $(P + Q) + R$ is the third intersection of line OT.

Likewise, if U is the third intersection of line QR, then $Q + R$ is the third intersection of line OU (Figure 9.14). If V is the third intersection of line $P(Q + R)$, then $P + (Q + R)$ is the third intersection of line OV.

Because the quantities $(P + Q) + R$ and $P + (Q + R)$ in (24) are the third intersections of the lines OT and OV, respectively, (24) is equivalent to the equation $T = V$. Accordingly, it suffices to prove that the third intersections of the lines $(P + Q)R$ and $P(Q + R)$ are the same point.

We apply Theorem 9.6, taking E, F, G, H to be the points Q, S, U, O, respectively (Figure 9.15). By Theorem 9.2(ii), the third intersections of

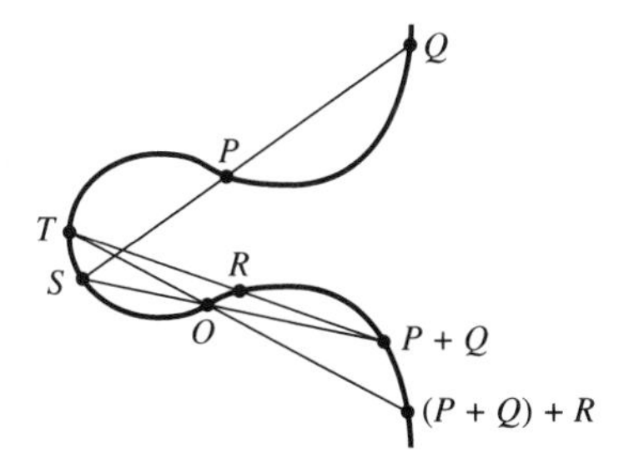

Figure 9.13

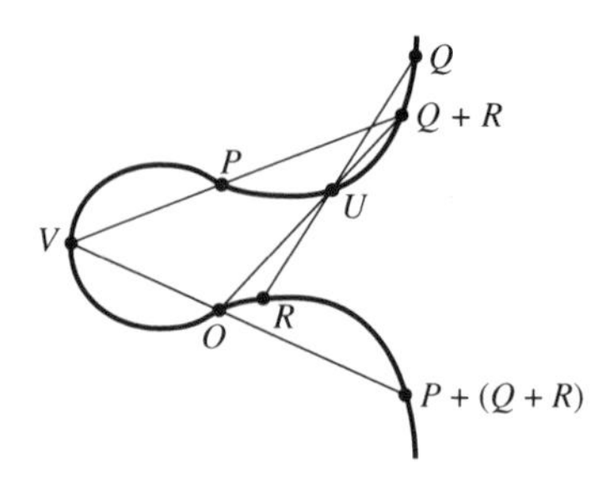

Figure 9.14

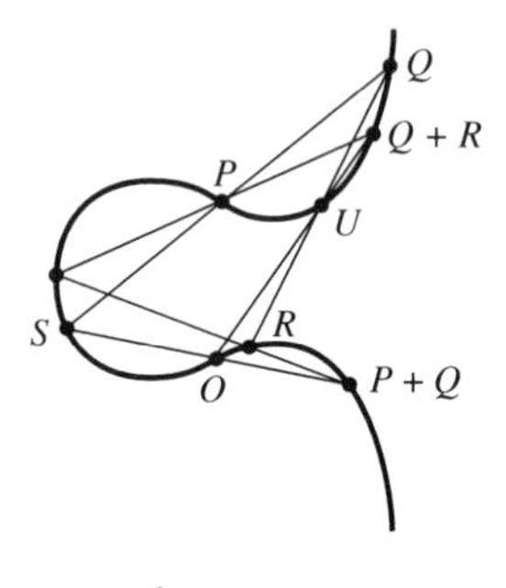

Figure 9.15

lines QS and UO are P and $Q+R$, and so the points W and X in Theorem 9.6 are now P and $Q+R$. The third intersections of lines QU and SO are R and $P+Q$ (by Theorem 9.2(ii)), and so the points Y and Z in Theorem 9.6 are now R and $P+Q$. Then Theorem 9.6 shows that the third intersections of the lines $P(Q+R)$ and $(P+Q)R$ are the same point, as desired. □

Let C be a nonsingular, irreducible cubic with a flex O. Readers familiar with abstract algebra will recognize that we have made C an *abelian group*. This means that C is a set with an operation on its elements called addition such that the sum of two elements of C is an element of C, the commutative and associative laws hold, there is an identity element for addition, and every element of C has an additive inverse. We have established these properties in Definition 9.3, equations (2), (3), and (4), and Theorem 9.7.

Let P be a point of C, and let k be a positive integer. We define kP to be the sum $P+\cdots+P$ of k terms equal to P. We do not need to use parentheses in the sum because the associative law implies that we can group the terms in any way. We say that P has *finite order* n if n is the least positive integer such that $nP = O$. If P does not have finite order, we say that it has *infinite order*, which means that $kP \neq O$ for every positive integer k.

We assume throughout the rest of this section that the cubic C has the form

$$y^2 = x^3 + ax^2 + bx + c, \tag{25}$$

where a, b, c are rational numbers and the right-hand side of this equation has no repeated factors of the form $x - r$ for a real number r. A cubic of this form is called an *elliptic curve*. C is nonsingular (by Theorem 8.2), and it is irreducible and has a flex at $(0, 1, 0)$ (by Theorem 8.1(i)). *We take* $(0, 1, 0)$ *to be the identity element O for addition on C.* O is the point at infinity on vertical lines, and it is the only point at infinity on C (by Theorem 8.2(iii)).

Let $P = (t, u)$ be any point of C in the Euclidean plane, and let

$$V = (t, -u) \tag{26}$$

be the reflection of P across the x-axis. Since OP is the vertical line through P, it contains V. V lines on C, since (25) is unchanged by replacing y with $-y$. If $u \neq 0$, then $P \neq V$, and V is the third intersection of OP (Figure 9.16). If $u = 0$, then P equals V (Figure 9.17). In this case, the

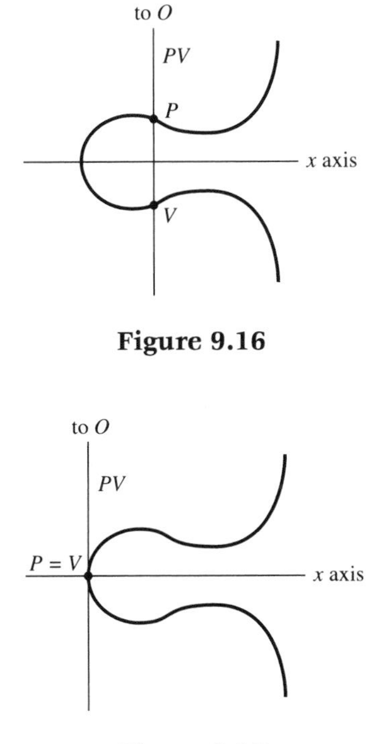

Figure 9.16

Figure 9.17

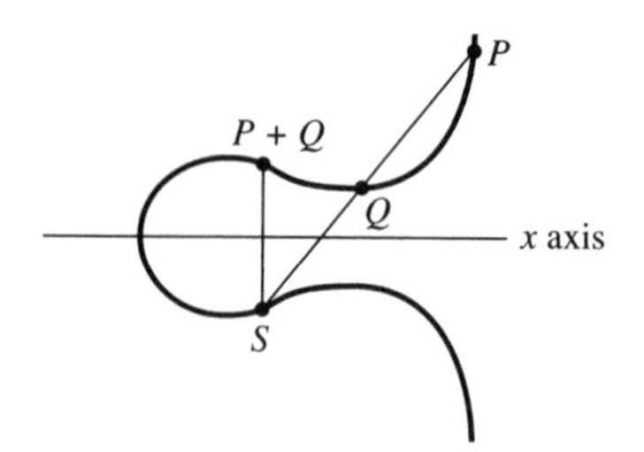

Figure 9.18

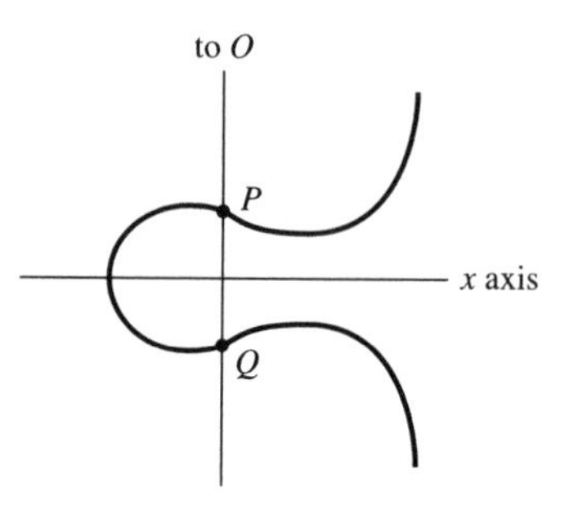

Figure 9.19

tangent to C at P is vertical (by Theorem 8.2(ii)), OP intersects C twice at P (by Definition 4.9), and so P is the third intersection of OP. In short, *for every point P of C in the Euclidean plane, the third intersection of OP is the reflection V of P across the x-axis*. V equals P exactly when P lies on the x-axis (Figures 9.16 and 9.17).

This simplifies the geometric construction of addition on C. For any points P and Q in the Euclidean plane, not necessarily distinct, let S be the third intersection of line PQ. If S lies in the Euclidean plane, then $P+Q$ is the reflection of S across the x-axis (Figure 9.18) (by the previous paragraph and Definition 9.3). If $S = O$, then $P+Q$ equals O (Figure 9.19), since $P+Q$ is the third intersection of $OS = OO = \tan O$, and that point is O (because O is a flex).

We can simplify the geometric construction of additive inverses in a similar way. For any point P of C in the Euclidean plane, $-P$ is the third intersection of line OP (by the discussion accompanying (4)). Together with the discussion after (26), this shows that *$-P$ is the reflection of P across the x-axis*. We can rewrite this algebraically as

$$-(t, u) = (t, -u). \tag{27}$$

We also have

$$-O = O \tag{28}$$

because $-O$ is the third intersection of line $OO = \tan O$ (by the discussion accompanying (4)), and that point is O (since O is a flex).

We use the results of the last two paragraphs to add points on elliptic curves algebraically. Since

$$O + X = X \tag{29}$$

for any point X on C, we need only evaluate $P + Q$ when

$$P = (t, u) \qquad \text{and} \qquad Q = (v, w) \tag{30}$$

are points of C in the Euclidean plane. We can assume that line PQ is not vertical, since we have

$$P + Q = O \tag{31}$$

when line PQ is vertical (by the discussion accompanying Figure 9.19).

Suppose first that $P \neq Q$. Since we are assuming that the line PQ through the points in (30) is not vertical, it has slope

$$m = \frac{u - w}{t - v}. \tag{32}$$

and equation

$$y = m(x - t) + u. \tag{33}$$

Substituting this expression for y in the difference between the two sides of (25) gives

$$x^3 + ax^2 + bx + c - (mx - mt + u)^2. \tag{34}$$

The factors of this quantity that have degree 1 correspond to the intersections of line PQ and C in the Euclidean plane, counted with multiplicity (by Theorem 4.3). Since PQ intersects C at the points P and Q with x-coordinates t and v, the quantity in (34) factors as

$$(x - t)(x - v)(x - g) \tag{35}$$

for a real number g, and g is the x-coordinate of the third intersection of line PQ (by Theorem 4.3). Comparing the coefficients of x^2 when we multiply out (34) and (35) gives

$$a - m^2 = -t - v - g.$$

Solving this equation for g gives

$$g = m^2 - a - t - v. \tag{36}$$

This is the x-coordinate of the third intersection of line PQ. We find the y-coordinate by setting $x = g$ in (33). Multiplying the result by -1 gives the y-coordinate of $P + Q$ (by the discussion accompanying Figure 9.18). In short, $P + Q$ is the point (g, h) for g given by (36) and for

$$h = m(t - g) - u. \tag{37}$$

On the other hand, suppose that $P = Q$. We must determine the point $P + P = 2P$, which is the reflection across the x-axis of the third intersection of line $PP = \tan P$. Differentiating (25) implicitly with respect to x gives

$$2y\,\frac{dy}{dx} = 3x^2 + 2ax + b. \tag{38}$$

Because the tangent at $P = (t, u)$ is not vertical (as we assumed after (30)), u is nonzero (by Theorem 8.2(ii)). Thus, we can substitute (t, u) for (x, y) in (38) and solve the result for dy/dx. This gives the slope $m = dy/dx$ of the tangent at P (as discussed between Theorems 4.10 and 4.11), and so we have

$$m = \frac{3t^2 + 2at + b}{2u}. \tag{39}$$

As in the previous paragraph, the intersections of line $PP = \tan P$ and C, counted with multiplicity, correspond to the factors of the quantity in (34) that have degree 1. The third intersection of line PP has x-coordinate g, where (34) factors as

$$(x - t)^2(x - g).$$

This quantity arises from (35) by setting $v = t$, which corresponds to taking $Q = P$. It follows, as in the previous paragraph, that $P + P = 2P$ is the point (g, h) given by (36) and (37) when m is given by (39) and we replace v by t in (36).

The fundamental question about an elliptic curve C is to determine all points (x, y) on C that have rational coordinates. In other words, we want to find all solutions of (25) in rational numbers. Addition of points is the key to attacking this problem.

Let C^* be the subset of C composed of the point O at infinity on C and all points (x, y) of C that have rational coordinates x and y. If the points P and Q in (30) have rational coordinates t, u, v, w, then the values of m, g, h given by (32), (36), (37), and (39) are also rational (because the coefficients a–c of C are rational, by assumption). Together with (27)–(29) and (31), this shows that *$P + Q$ and $-P$ belong to C^* for any points P and Q in C^**. In other words, sums and additive inverses of points of C with rational coordinates are again points of C with rational coordinates. Readers familiar with abstract algebra will note that we have shown that C^* is a *subgroup* of C; this means that C^* is a subset of C that contains the identity element and is closed under addition and taking inverses.

The most basic way to use addition of points to produce elements of C^* is this: given a point P of C^*, the points kP also belong to C^* for all positive integers k, by the previous paragraph. For example, consider the cubic

$$y^2 = x^3 + 3x. \tag{40}$$

This is an elliptic curve because the coefficients are rational and the factorization

$$x(x^2 + 3)$$

of the right-hand side over the real numbers has no repeated factors of degree 1. $P = (1, 2)$ is an obvious point on (40) with rational coordinates. We find another by computing $2P$. We take $P = Q = (1, 2) = (t, u) = (v, w)$ in (30) and $a = 0$, $b = 3$, and $c = 0$ in (25). Equation (39) becomes

$$m = \frac{3 \cdot 1^2 + 2 \cdot 0 \cdot 1 + 3}{2 \cdot 2} = \frac{6}{4} = \frac{3}{2}.$$

Substituting $m = \frac{3}{2}$, $a = 0$, and $t = v = 1$ in (36) gives

$$g = (\tfrac{3}{2})^2 - 0 - 2 \cdot 1 = \tfrac{1}{4},$$

and (37) gives

$$h = \tfrac{3}{2}(1 - \tfrac{1}{4}) - 2 = \tfrac{9}{8} - 2 = -\tfrac{7}{8}.$$

Thus, $2P = (\frac{1}{4}, -\frac{7}{8})$ is also a point of C with rational coordinates. We can check that this point satisfies (40).

Similarly, we can find another point of C with rational coordinates by adding P and $2P$. We take

$$P = (t, u) = (1, 2) \qquad \text{and} \qquad Q = 2P = (v, w) = (\tfrac{1}{4}, -\tfrac{7}{8})$$

in (30). Equation (32) gives

$$m = \frac{2 + \frac{7}{8}}{1 - \frac{1}{4}} = \frac{\frac{23}{8}}{\frac{3}{4}} = \tfrac{23}{6},$$

equation (36) gives

$$g = \left(\frac{23}{6}\right)^2 - 0 - 1 - \tfrac{1}{4} = \frac{484}{36} = \frac{121}{9},$$

and (37) gives

$$h = \frac{23}{6}\left(1 - \frac{121}{9}\right) - 2 = -\frac{1342}{27}.$$

Thus, $3P = (121/9, -1342/27)$ is another point of C with rational coordinates. We can check that this point satisfies (40).

There are also much deeper connections between C^* and addition of points. Barry Mazur proved in 1976 that every element of C^* of finite order has order $1, 2, \ldots, 10$, or 12, and he determined all possibilities for the subgroup formed by the elements of C^* of finite order. Much less is known about the elements of C^* of infinite order, and important conjectures remain open. We say that points $P_1, \ldots, P_k$ of C^* *generate* C^* if we can obtain every element of C^* by adding these points and their inverses any number of times. L.J. Mordell proved in 1922 that C^* is generated by a finite number of points. It remains a major unsolved problem in number theory to determine the least number of points needed to generate C^*.

Exercises

9.1. A cubic C and a point P are given in each part of this exercise. Check that C is an elliptic curve and that P is a point of C^*. Find the coordinates of the points $2P$ and $3P$ of C^*. Check your work by verifying that these coordinates satisfy the given equation.
(a) $y^2 = x^3 + 8; (2, 4)$. (b) $y^2 = x^3 - 2x; (2, 2)$.
(c) $y^2 = x^3 + x + 1; (0, 1)$. (d) $y^2 = x^3 - x^2 - 3x; (-1, 1)$.
(e) $y^2 = x^3 + x - 1; (1, 1)$. (f) $y^2 = x^3 - 4; (2, 2)$.
(g) $y^2 = x^3 - 2x; (-1, 1)$. (h) $y^2 = x^3 + 5x^2 + 3x; (1, 3)$.

9.2. Let C be a nonsingular, irreducible cubic with a flex O. Add points of C with respect to O as in Definition 9.3.
(a) Let P, Q, R be points of C that are not necessarily distinct. Prove that there is a line that intersects C at P, Q, R, counting multiplicities if and only if

$$P + Q + R = O.$$

(b) Let P be a point of C. Prove that P is a flex of C if and only if $3P = O$.
(c) Let P and Q be distinct flexes of C. Use parts (a) and (b) to prove that the third intersection of line PQ and C is a flex of C that is distinct from P and Q and collinear with them. (This gives another proof of Exercise 8.6.)

9.3. Let C be an elliptic curve.
(a) Prove that C^* has either zero, one, or three points of order 2. For each of the numbers 0, 1, 3, give the equation of an elliptic curve such that C^* has the specified number of points of order 2. (See Theorem 8.2 and the discussion accompanying Figure 9.19.)
(b) If P and Q are distinct points of order 2 on C^*, prove that $P + Q$ is a point of order 2 on C^* that is distinct from both P and Q.

9.4. Let P be a point of an elliptic curve C.
(a) Prove that P has order 3 if and only if P is a flex of C that lies in the Euclidean plane.
(b) Prove that P has order 3 if and only if P and $2P$ are points of the Euclidean plane that have the same x-coordinate.

9.5. Let P be a point of an elliptic curve C.
(a) Prove that P has order 4 if and only if the tangent at P intersects C at a point on the x-axis not equal to P.

(b) If P has order 4, describe the relative positions of the points P, $2P$, and $3P$, and use the discussion accompanying Figures 9.18 and 9.19 to justify your answer. Illustrate your answer with a figure that shows an elliptic curve C, a point P of C of order 4, and the points $2P$ and $3P$.

9.6. Let C be a nonsingular, irreducible cubic with a flex O. Add points of C with respect to O as in Definition 9.3. Let P be a point of C of order 6.

(a) Prove that $2P$ is a flex of C collinear with P and $3P$. Prove that $4P$ is a flex of C collinear with $5P$ and $3P$. (See Exercise 9.2.)

(b) Prove that the tangent at P intersects C twice at P and once at $4P$. Prove that the tangent at $5P$ intersects C twice at $5P$ and once at $2P$. (See Exercise 9.2(a).)

(c) Illustrate parts (a) and (b) with a figure that shows an elliptic curve C, the x-axis, a point P of order 6, the points $2P$, $3P$, $4P$, $5P$, the lines through the two triples of collinear points in (a), and the two tangent lines in (b).

9.7. Let P be a point on an elliptic curve C.

(a) Prove that P has order 6 if and only if it is the third intersection of the line determined by a flex of C in the Euclidean plane and a point of C on the x-axis. (See Exercises 9.6(a) and 9.2.)

(b) Prove that P has order 6 if and only if the tangent at P contains a flex of C that lies in the Euclidean plane and is not equal to P. (See Exercises 9.6(b) and 9.2.)

9.8. Illustrate Theorem 9.6 with a figure in each of the following cases. Restate the theorem in terms of tangents and flexes as appropriate in each case.

(a) $E = G$ and $F = H$.

(b) $E = G$ and $H = Z$.

(c) $E = G$ and $Y = Z$.

(d) $E = G = Y$ and $F = H$.

(e) $E = G = Y$.

(f) $E = G$.

(g) $E = G = Y$ and $F = H = Z$.

9.9. Let C be a nonsingular, irreducible cubic. Let P be a point of C that lies on the tangents at three collinear points R, S, T of C other than P. Prove that P is a flex of C and that P, R, S, T are the only points of C whose tangents contain P. (This is the converse of the third sentence of Exercise 8.5. Exercise 9.8(a) and Theorems 8.1 and 8.2 may be helpful.)

9.10. (a) Prove that C is an elliptic curve and P is a point of C^* such that $2P = (0, 0)$ if and only if there are nonzero rational numbers t and m such that

$$P = (t, mt), \tag{41}$$

C has equation

$$y^2 = x^3 + (m^2 - 2t)x^2 + t^2x, \tag{42}$$

and the inequality

$$m^2 \neq 4t \tag{43}$$

holds.

(*Hint*: Let C be given by (25). One possible approach is to observe that the origin is the third intersection of the tangent to C at the point P given by (41) if and only if t and m are both nonzero and the polynomial

$$x^3 + ax^2 + bx + c - (mx)^2$$

factors as $x(x-t)^2$. Show in this case that inequality (43) is equivalent to the condition that the right-hand side of (42) has no repeated factors of degree 1.)

(b) If P and C are as in part (a), prove that the point P of C^* has order 4.

9.11. Let C, P, t, m be as in Exercise 9.10(a).

(a) Prove that there is a point Q of C^* such that $\tan Q$ has slope 1 and has P as its third point of intersection if and only if the equation

$$(m^2 - t - 1)^2 = 4t(m-1)^2 \tag{44}$$

holds.

(*Hint:* One possible approach is to show that a point Q of C^* with x-coordinate v has the required properties if and only if the polynomial

$$x^3 + (m^2 - 2t)x^2 + t^2x - (x - t + mt)^2 \tag{45}$$

factors as $(x-t)(x-v)^2$.)

(b) Prove that the point Q in part (a) has order 8.

(c) Prove that rational numbers t and m satisfy (44) if and only if

$$(2m^2 - 2m)^{1/2} \tag{46}$$

is a rational number and if $t = n^2$ for

$$n = -m + 1 \pm (2m^2 - 2m)^{1/2}.$$

(*Hint*: One possible approach is to prove that (44) holds if and only if there is a rational number n such that $t = n^2$ and

$$m^2 - n^2 - 1 = 2n(m-1).$$

Use the quadratic formula to solve this equation for n in terms of m.)

(d) Prove that the nonzero rational numbers m, such that the quantity in (46) is also rational, are exactly the numbers

$$m = \frac{2p^2}{2p^2 - q^2}$$

as p and q vary over all pairs of integers such that $p \neq 0$.

(*Hint*: One possible approach is to set

$$2m^2 - 2m = k^2$$

for a rational number k, substitute $m = p/r$ and $k = q/r$ for integers q and r, solve for r in terms of p and q, and then express m in terms of p and q.)

(e) Use parts (a)–(d) and Exercise 9.10 to find two elliptic curves C such that C^* contains an element of order 8. Be sure to check that the conditions $t \neq 0$ and $t \neq m^2/4$ in Exercise 9.10 hold.

9.12. (a) Prove that the elliptic curves C that contain the origin and are such that C^* has a point of order 3 with x-coordinate 1 are exactly the curves

$$y^2 = x^3 + (r^2 - 3)x^2 + (2r + 3)x \tag{47}$$

for all rational numbers r except 3, -1, and $-\frac{3}{2}$.

(*Hint*: Let C be given by (25) with $c = 0$. One possible approach is to prove that C has a flex at a point with x-coordinate 1 and has tangent $y = rx + s$ at that point if and only if

$$x^3 + ax^2 + bx - (rx + s)^2$$

factors as $(x - 1)^3$. Ensure that the right-hand side of (47) has no repeated factors of degree 1, and then apply Exercise 9.2(b).)

(b) Prove that $(0, 0) + (1, r + 1)$ is an element of C^* of order 6. What are the (x, y) coordinates of this point?

9.13. (a) Find an elliptic curve C such that C^* has a point P of order 4 and a point Q of order 3.

(*Hint*: One possible approach is to set $r = m$ in (47) and find values of t and m so that (42) and (47) coincide. Apply Exercises 9.10 and 9.12 after checking that t, m, and r satisfy the conditions in these exercises.)

(b) Prove that $P + Q$ is an element of order 12 in C^*.

9.14. Let C be an elliptic curve, and let P be a point of the Euclidean plane on C^*.

(a) If P has odd order, prove that there is a point Q of C^* such that $2Q = P$.

(b) If C^* has no points Q such that $2Q = P$, and if C^* has no points of order 2, prove that the order of P is infinite.

9.15. (a) Why is $y^2 = x^3 + 3$ an elliptic curve C, and why is $P = (1, 2)$ a point of C^*?

(b) Why does C^* have no elements of order 2?

(c) If P were the third intersection of the tangent at a point of C^*, prove that the slope m of the tangent would be a rational number m such that

$$(m^2 - 1)^2 = 4(m^2 - 4m + 1). \tag{48}$$

(*Hint*: One possible approach is to prove that, if Q existed, the polynomial

$$x^3 + 3 - (m(x - 1) + 2)^2$$

would factor as

$$(x - 1)(x - v)^2$$

for some value of v.)

(d) Prove that there is no rational number m that satisfies (48).

(*Hint*: Recall the *Rational Root Theorem*: if a polynomial

$$x^n + a_{n-1}x^{n-1} + \cdots + a_1x + a_0$$

with integer coefficients a_i has a rational root r, then r is an integer and a positive or negative factor of a_0.)

(e) Conclude that P is a point of C^* that has infinite order. (See Exercise 9.14.)

9.16. Adapt the approach of Exercise 9.15 to show that $y^2 = x^3 + x - 1$ is an elliptic curve C and that $(1, 1)$ is an element of C^* that has infinite order.

9.17. Let $q(x) = x^3 + ax^2 + bx + c$ be a polynomial of degree 3 in one variable x with leading coefficient 1. The quantity

$$\Delta = 4b^3 + 27c^2 - a^2b^2 + 4a^3c - 18abc \tag{49}$$

is called the *discriminant* of q. This exercise shows that q has a repeated factor of degree 1 if and only if the discriminant is zero. By Theorem 8.2, this makes it easy to check whether the cubic $y^2 = q(x)$ is nonsingular: it is nonsingular if and only if $\Delta \neq 0$.

(a) Prove that there is a unique real number k such that

$$q(x + k) = x^3 + sx + t \tag{50}$$

for real numbers s and t. Express k, s, t in terms of a, b, c.

(b) Use Exercise 8.1 to prove that $x^3 + sx + t$ has a repeated factor if and only if

$$4s^3 + 27t^2 = 0.$$

(c) Use the expressions for s and t from part (a) to prove that $4s^3 + 27t^2$ equals the quantity on the right-hand side of (49).

(d) Conclude that $q(x)$ has a repeated factor of degree 1 if and only if $\Delta = 0$

9.18. An elliptic curve that contains the point $P = (0, 1)$ has the form $y^2 = q(x)$, where

$$q(x) = x^3 + ax^2 + bx + 1 \tag{51}$$

for rational numbers a and b. Set

$$e = \frac{b^2}{4} - a. \tag{52}$$

(a) Prove that P has order 5 if and only if $2P$ and $3P$ are points of the Euclidean plane that have the same x-coordinates.

(b) If $3P \neq O$, prove that $2P$ has x-coordinate $e \neq 0$ and that $3P$ has x-coordinate $(2be + 4)/e^2$.

(c) Conclude from parts (a) and (b) that P has order 5 if and only if e is nonzero and

$$b = \frac{e^3 - 4}{2e}. \tag{53}$$

9.19. (a) Find an elliptic curve C such that $(0, 1)$ is a point of C^* of order 5 by taking $e = 2$ in Exercise 9.18, using (53) to determine b, and using (52) to determine a. Check that C is nonsingular by using Exercises 9.17 or 8.1.

(b) Repeat part (a) with $e = 1$.

(c) Repeat part (a) with $e = -2$.

§10. Complex Numbers

The complex numbers are formed by adding a square root of -1 to the real numbers. The Fundamental Theorem of Algebra states that every polynomial in one variable factors over the complex numbers as a product of polynomials of degree 1. We introduce the complex numbers in this section, derive their basic properties, and prove the Fundamental Theorem.

Over the real numbers, some curves intersect fewer times than others that have the same degrees. For example, a line and a circle may intersect either twice or not at all. Theorems 4.4 and 5.8 suggest that this happens because some polynomials in one variable, such as $x^2 + 1$, do not factor over the real numbers into polynomials of degree 1. The Fundamental Theorem shows that this does not happen over the complex numbers. We use the Fundamental Theorem in Section 11 to prove Bezout's Theorem, which states that curves of degrees m and n without common factors intersect exactly mn times, counting multiplicities, over the complex numbers. We use Bezout's Theorem in Section 12 to complete the classification of cubics that we began in Section 8.

We construct the complex numbers from the real numbers by adding a quantity i whose square is -1. In other words, because $x^2 + 1$ has no roots in the real numbers, we add one. Formally, a *complex number* is a quantity of the form $a + bi$, where a and b are real numbers. We want the commutative and associative laws of addition and multiplication and the distributive law to generalize from the real to the complex numbers, and we want the relation $i^2 = -1$ to hold. This leads us to define addition and multiplication of complex numbers as follows:

$$(a + bi) + (c + di) = (a + c) + (b + d)i, \tag{1}$$

$$(a + bi)(c + di) = (ac - bd) + (ad + bc)i, \tag{2}$$

for all real numbers a–d. Equation (2) arises from the desire to have

$$(bi)(di) = bd(i^2) = bd(-1) = -bd.$$

Consider complex numbers

$$z = a + bi, \qquad w = c + di, \qquad v = e + fi, \tag{3}$$

for real numbers a–f. The commutative laws for adding and multiplying complex numbers

$$z + w = w + z \qquad \text{and} \qquad zw = wz \tag{4}$$

hold because the right-hand sides of (1) and (2) are unaffected by interchanging a with c and b with d. The associative law for adding complex numbers

$$(z + w) + v = z + (w + v) \tag{5}$$

holds because both sides of this equation equal

$$(a+c+e)+(b+d+f)i.$$

The associative law for multiplying complex numbers

$$(zw)v=z(wv) \tag{6}$$

holds because both sides of this equation equal

$$(ace-adf-bcf-bde)+(bce+ade+acf-bdf)i.$$

The distributive law for complex numbers

$$z(w+v)=zw+zv \tag{7}$$

holds because both sides of this equation equal

$$(ac-bd+ae-bf)+(ad+bc+af+be)i.$$

The associative law of addition (5) ensures that we can write sums of complex numbers without parentheses. The associative law of multiplication (6) lets us define z^n as the product of n factors of z for any complex number z and any positive integer n. Equations (4)–(7) imply that we can work with polynomials with complex coefficients and evaluate them by substituting complex numbers for the variables just as we do over the real numbers.

We identify each real number a with the complex number $a+0i$. The addition and multiplication of complex numbers in (1) and (2) give the usual addition and multiplication of real numbers. Thus, we can think of the complex numbers as containing the real numbers. Equations (1) and (2) imply that 0 and 1 satisfy their usual properties $0+w=w, 0w=0$, and $1w=w$ for all complex numbers w.

We define $-w$ to be $(-1)w$, and we define $z-w$ to be $z+(-w)$, for any complex numbers w and z. In the notation of (3), we have

$$-w=(-c)+(-d)i, \tag{8}$$

$$z-w=(a-c)+(b-d)i, \tag{9}$$

$$w-w=w+(-w)=0. \tag{10}$$

We match up the complex numbers with the points of the Euclidean plane by associating the complex number $a+bi$ with the point (a,b) in standard (x,y) coordinates for all real numbers a and b (Figure 10.1). We define the *modulus* $|a+bi|$ of $a+bi$ to be the distance from this point to the origin. By the Pythagorean Theorem, we have

$$|a+bi|=(a^2+b^2)^{1/2}. \tag{11}$$

Because the modulus $|z|$ of a complex number z is the distance from z to

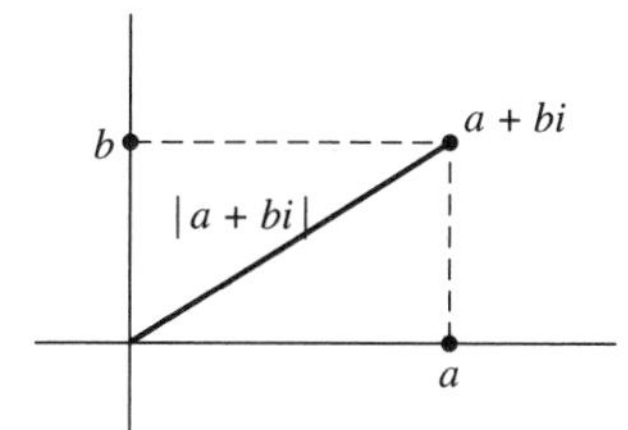

Figure 10.1

zero, it represents the "size" of z. It is clear that

$$|z| \geq 0 \tag{12}$$

for every complex number z and that

$$|z| = 0 \quad \text{if and only if} \quad z = 0. \tag{13}$$

Because the right-hand side of (11) is unchanged if we replace a and b with $-a$ and $-b$, equations (8) and (11) imply that

$$|-z| = |z|. \tag{14}$$

In the notation of (3), we have

$$|z - w| = [(a - c)^2 + (b - d)^2]^{1/2},$$

by (9) and (11). Thus, $|z - w|$ *is the distance between the points* z *and* w *in the plane*. In particular, it follows that

$$|z - w| = |w - z|. \tag{15}$$

Equation (1) shows that the x- and y-coordinates of $z + w$ are the sums of the x- and y-coordinates of z and w. It follows that $z + w$ is the fourth vertex of a parallelogram that has $0, z, w$ as the other three vertices (as in Figure 10.2). (The parallelogram collapses when $0, z, w$ lie on a line.) Because a straight line is the shortest distance between two points, it

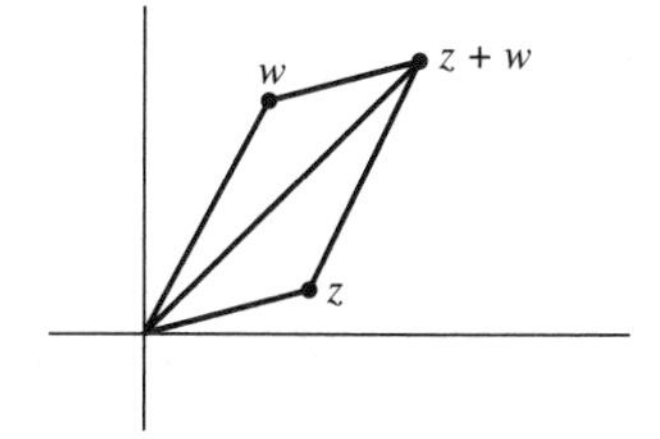

Figure 10.2

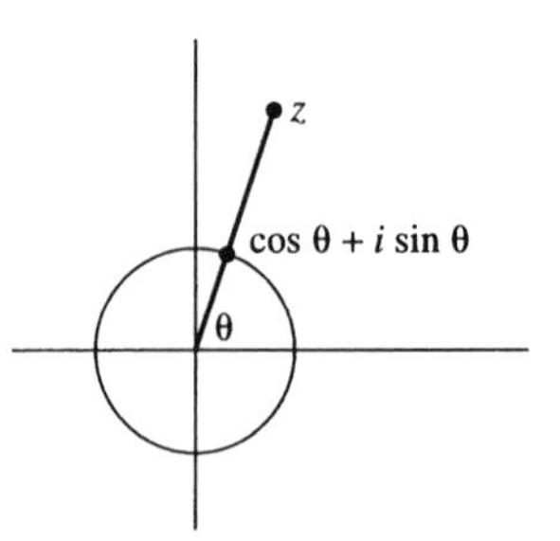

Figure 10.3

follows that

$$|z+w| \leq |z| + |w| : \tag{16}$$

the left-hand side is the distance from 0 to $z+w$, and the right-hand side is the sum of the distances from 0 to z and from z to $z+w$.

If we substitute $z+w$ for z and $-w$ for w in (16), we obtain

$$|z| \leq |z+w| + |-w|, \tag{17}$$

since (5) and (10) imply that $(z+w)+(-w) = z+0 = z$. By (14), we can rewrite inequality (17) as

$$|z+w| \geq |z| - |w|. \tag{18}$$

Let z be a nonzero complex number, and let θ be the angle that lies counterclockwise after the positive x-axis and before the ray from the origin O through z (Figure 10.3). This ray intersects the unit circle at the point that has (x, y)-coordinates $(\cos\theta, \sin\theta)$ and that corresponds to the complex number $\cos\theta + i\sin\theta$. If we multiply the x- and y-coordinates of this point by $|z|$, we obtain the x- and y-coordinates of z. Thus, we have

$$z = |z|(\cos\theta + i\sin\theta). \tag{19}$$

We call this the *polar form* of z because $|z|$ and θ are the polar coordinates of the point z. We can also write 0 in the form of (19) for any angle θ, since $|0| = 0$.

For any complex number z given by (19), $-z$ has the same modulus as z (by (14)), and it corresponds to the angle $\theta + \pi$ (since $-z$ and z lie in diametrically opposite directions from the origin, as in Figure 10.4). Thus, we have

$$-z = |z|(\cos(\theta+\pi) + i\sin(\theta+\pi)). \tag{20}$$

Let z and w be complex numbers given in polar form by (19) and the equation

$$w = |w|(\cos\psi + i\sin\psi). \tag{21}$$

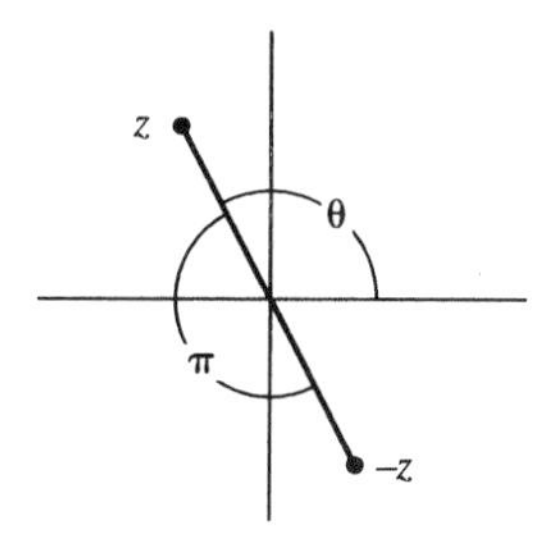

Figure 10.4

Equation (2) shows that

$$\begin{aligned}(\cos\theta + i\sin\theta)(\cos\psi + i\sin\psi) \\ &= (\cos\theta\cos\psi - \sin\theta\sin\psi) + i(\sin\theta\cos\psi + \cos\theta\sin\psi) \\ &= \cos(\theta+\psi) + i\sin(\theta+\psi)\end{aligned}$$

(by the angle-addition formulas of trigonometry). Multiplying these quantities by $|z|\,|w|$ shows that

$$zw = |z|\,|w|[\cos(\theta+\psi) + i\sin(\theta+\psi)] \tag{22}$$

(by (19), (21), (4), and (6)). Thus, *we multiply complex numbers in polar form by multiplying their moduli and adding their angles.*

Because (22) express zw in polar form, it follows that

$$|zw| = |z|\,|w|. \tag{23}$$

Equations (13) and (23) imply that

$$zw = 0 \quad \text{if and only if} \quad z = 0 \quad \text{or} \quad w = 0. \tag{24}$$

For any positive integer n, it follows from (22) that

$$z^n = |z|^n(\cos(n\theta) + i\sin(n\theta)). \tag{25}$$

Any complex number w has an nth root z for any positive integer n: if (21) gives w in polar form, then

$$z = |w|^{1/n}(\cos(\psi/n) + i\sin(\psi/n))$$

satisfies $z^n = w$ (by (25)).

Let z be a nonzero complex number given in polar form by (19). Since $|z| \neq 0$ (by (13)), we can set

$$z^{-1} = \frac{1}{|z|}(\cos(-\theta) + i\sin(-\theta)). \tag{26}$$

This equation shows that

$$|z^{-1}| = \frac{1}{|z|}. \tag{27}$$

If we take $w = z^{-1}$, (22) becomes

$$zz^{-1} = 1, \tag{28}$$

because $|z|\,|w| = |z|\,|z^{-1}| = 1$ (by (27)), $\theta + \psi = \theta - \theta = 0$, and $\cos 0 + i \sin 0 = 1 + i0 = 1$. Equation (28) shows that we can think of z^{-1} as the reciprocal of z and think of multiplication by z^{-1} as "division by z." Of course, we cannot divide by zero.

For any real numbers a and b, we define the *conjugate* $\bar{z}$ of the complex number $z = a + bi$ by setting

$$\bar{z} = a - bi.$$

It is clear that

$$\bar{z} = z \quad \text{if and only if } z \text{ is real.} \tag{29}$$

The conjugate of $a - bi$ is $a - (-b)i = a + bi$, and so

$$w = \bar{z} \quad \text{if and only if} \quad z = \bar{w}. \tag{30}$$

In other words, conjugation interchanges complex numbers in pairs, pairing each real number with itself.

For any complex numbers $z = a + bi$ and $w = c + bi$, we have

$$\overline{z + w} = \bar{z} + \bar{w}, \tag{31}$$

since both sides of this equation equal

$$(a + c) + (-b - d)i.$$

We also have

$$\overline{zw} = \bar{z} \cdot \bar{w} : \tag{32}$$

conjugating both sides of (2) shows that $\overline{zw}$ equals

$$(ac - bd) - (ad + bc)i,$$

and replacing b and d with $-b$ and $-d$ in (2) shows that $\bar{z} \cdot \bar{w}$ has the same value. Equations (31) and (32) show that conjugation preserves addition and multiplication of complex numbers.

We have formed the complex numbers from the real numbers by adding a root i of the polynomial $x^2 + 1$. Amazingly, this is enough to ensure that every nonconstant polynomial in one variable has a root in the complex numbers. Equivalently, every nonconstant polynomial in one variable factors over the complex numbers as a product of polynomials of degree 1. This is the Fundamental Theorem of Algebra, and we devote the rest of the section to its proof. The self-contained proof we present is adapted from the article by Charles Fefferman listed in the References at the end of the book.

For the rest of this section, we let $f(x)$ be a nonconstant polynomial in one variable with complex coefficients. We write

$$f(x) = a_n x^n + \cdots + a_1 x + a_0 \tag{33}$$

for complex numbers $a_n, \ldots, a_0$, where $a_n \neq 0$ and $n \geq 1$.

To prove the Fundamental Theorem of Algebra, we must prove that $f(w) = 0$ for some complex number w. We do so by proving that $|f(x)|$ takes a minimum value at a complex number w and deducing that $|f(w)|$ must be zero. We divide the argument into a sequence of six claims.

Claim 1

Let w be a complex number, and let ε be a positive real number. Then there is a real number $\delta > 0$ such that $|f(z) - f(w)| < \varepsilon$ for all complex numbers z such that $|z - w| < \delta$.

This claim means that $f(z)$ approaches $f(w)$ as z approaches w. More precisely, it shows that we can make $f(z)$ as close as we wish to $f(w)$ by choosing z close enough to w. It is the analogue over the complex numbers of the result in single-variable calculus that polynomials are continuous.

To prove the claim, we assume first that $w = 0$. Since $f(0) = a_0$ (by (33)), we must find a real number $\delta > 0$ such that $|f(z) - a_0| < \varepsilon$ for all complex numbers z such that $|z| < \delta$. Equation (33) shows that

$$\begin{aligned} |f(z) - a_0| &= |a_n z^n + \cdots + a_1 z| \\ &\leq |a_n z^n| + \cdots + |a_1 z| \quad \text{(by (16))} \\ &= |a_n|\,|z|^n + \cdots + |a_1|\,|z| \end{aligned} \tag{34}$$

(by (23)). If $|z| \leq 1$, we have $|z|^j \leq |z|$ for each integer $j \geq 1$, and the quantity in (34) is less than or equal to

$$(|a_n| + \cdots + |a_1|)|z|.$$

This quantity is less than ε if δ is no larger than

$$\frac{\varepsilon}{|a_n| + \cdots + |a_1|}. \tag{35}$$

In short, if we take δ to be the smaller of 1 and the quantity in (35), then δ is a positive real number such that

$$|f(z) - f(0)| < \varepsilon \qquad \text{if} \quad |z| < \delta.$$

This establishes Claim 1 when $w = 0$.

To prove the claim when w is any complex number, we define a new polynomial $g(t)$ with complex coefficients by setting

$$g(t) = f(t + w). \tag{36}$$

The previous paragraph shows that there is a real number $\delta > 0$ such that

$$|g(t) - g(0)| < \varepsilon \qquad \text{if} \quad |t| < \delta.$$

Substituting from (36) shows that

$$|f(t+w) - f(w)| < \varepsilon \qquad \text{if} \quad |t| < \delta.$$

Finally, setting $t = z - w$ shows that

$$|f(z) - f(w)| < \varepsilon \qquad \text{if} \quad |z - w| < \delta,$$

as desired.

Claim 2

For any real number $M > 0$, there is a real number $R > 0$ such that $|f(z)| > M$ for all complex numbers z such that $|z| > R$.

This claim shows that $|f(z)|$ grows large as $|z|$ grows large. To prove that $f(z)$ takes the value zero for some complex number z, we want to choose $|f(z)|$ to be as small as possible. Claim 2 shows we need only consider complex numbers z in a finite part of the plane when we minimize $|f(z)|$.

To prove Claim 2, we substitute a nonzero complex number z for x in (33) and factor out z^n. This shows that

$$|f(z)| = |z^n(a_n + a_{n-1}z^{-1} + \cdots + a_0(z^{-1})^n)|$$

(by (4), (6), (7), and (28))

$$= |z|^n \, |a_n + a_{n-1}z^{-1} + \cdots + a_0(z^{-1})^n|$$

(by (23))

$$\geq |z|^n(|a_n| - |a_{n-1}z^{-1} + \cdots + a_0(z^{-1})^n|)$$

(by inequality (18)). In other words, we have

$$|f(z)| \geq |z|^n(|a_n| - |g(z^{-1})|) \tag{37}$$

for

$$g(t) = a_{n-1}t + \cdots + a_0t^n. \tag{38}$$

We have $|a_n| > 0$ (by (12), (13), and the assumption after (33) that $a_n \neq 0$). We replace f with g, w with 0, and ε with $\frac{1}{2}|a_n|$ in Claim 1. Since $g(0) = 0$ (by (38)), there is a real number $\delta > 0$ such that

$$|g(t)| < \tfrac{1}{2}|a_n| \qquad \text{if} \quad |t| < \delta. \tag{39}$$

If we set $t = z^{-1}$ for $|z| > 1/\delta$, we have

$$|t| = |z^{-1}| = \frac{1}{|z|} < \delta$$

(by (27)), and (39) shows that

$$|g(z^{-1})| < \tfrac{1}{2}|a_n|.$$

Combining this inequality with inequality (37) gives

$$|f(z)| > |z|^n(|a_n| - \tfrac{1}{2}|a_n|) = \tfrac{1}{2}|a_n|\,|z|^n \tag{40}$$

for $|z| > 1/\delta$. If $|z|$ is also greater than

$$\left(\frac{2M}{|a_n|}\right)^{1/n} \tag{41}$$

(which makes sense because of the assumption after (33) that $n \geq 1$), inequality (40) shows that $|f(z)| > M$. In short, if we take R to be the larger of $1/\delta$ and the quantity in (41), we have $|f(z)| > M$ for all complex numbers z such that $|z| > R$, as desired.

Claim 3
There is a complex number w such that $|f(w)| \leq |f(z)|$ for all complex numbers z.

This claim shows that $|f(x)|$ has a minimum value $|f(w)|$; that is, there is a point w where $|f(x)|$ takes a value less than or equal to its value at every other point. We show in Claims 4 and 5 that $|f(w)|$ cannot be positive; then it must be zero, and the Fundamental Theorem holds.

Let T be a set of real numbers. A *lower bound* of T is a real number c such that $c \leq x$ for all numbers x in T. That is, a lower bound of a set T is a number less than or equal to every element of T. A *greatest lower bound* d of T is a lower bound of T such that $d \geq c$ for every lower bound c of T. In other words, a greatest lower bound of T is the largest possible lower bound of T. For example, the number $\pi = 3.14159\ldots$ is the greatest lower bound of the set of real numbers

$$\{4, 3.2, 3.15, 3.142, 3.1416, 3.14160, \ldots\}$$

obtained by terminating π after a finite number of digits and adding one to the last digit.

The following is a key property of the real numbers:

Completeness Property of the Real Numbers
If a nonempty set of real numbers has a lower bound, then it has a greatest lower bound.

Informally, the Completeness Property means that "there are no holes in the real number line." It holds because every infinite decimal represents a real number.

We use the greatest lower bound property to prove Claim 3. Let T be the set of all real numbers $|f(z)|$ as z varies over all complex numbers. T has zero as a lower bound (by inequality (12)), and so it has a greatest lower bound $d \geq 0$.

Let k be any positive integer. Since d is the greatest lower bound of T, $d + 1/k$ is not a lower bound of T. Thus, there is a complex number z_k such that

$$|f(z_k)| < d + \frac{1}{k}. \tag{42}$$

We consider the sequence $z_1, z_2, z_3, \ldots$ of complex numbers, which may include repetitions.

By Claim 2, there is a real number $R > 0$ such that

$$|f(z)| > d + 1 \qquad \text{if} \quad |z| > R. \tag{43}$$

Let S_1 be the square centered at the origin in the plane that has sides of length $2R$ (Figure 10.5). Every point z of the plane outside of S_1 lies more than R units from the origin, and so $|f(z)|$ is at least $d + 1$ (by (43)). Thus, all of the points $z_1, z_2, \ldots$ lie in S_1 (by (42)).

The coordinate axes divide S_1 into four squares of equal size, as in Figure 10.5. At least one of these four squares contains the points z_k for infinitely many values of k, and we let S_2 be such a square (Figure 10.6). We then subdivide S_2 into four squares of equal size. At least one of these squares contains the points z_k for infinitely many values of k, and we let S_3 be such a square. Continuing in this way, we obtain a sequence of squares

$$S_1 \supset S_2 \supset S_3 \supset \cdots \tag{44}$$

such that each square S_j contains the points z_k for infinitely many positive integers k.

The upper right corner of each square S_j is a complex number $p_j + q_j i$ for real numbers p_j and q_j. The nesting of the squares in (44) implies that the upper right corners $p_j + q_j i$ of the squares can only move down or to

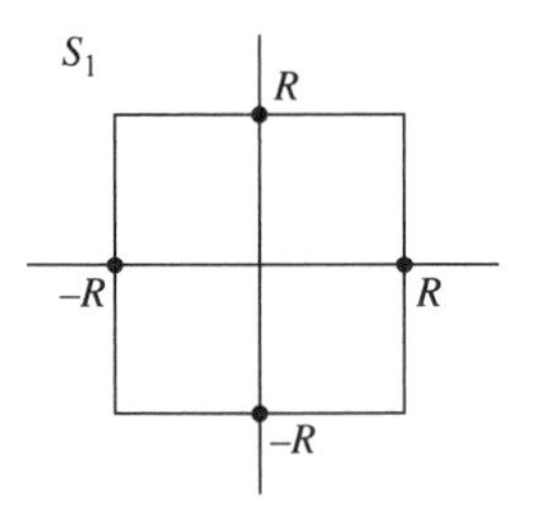

Figure 10.5

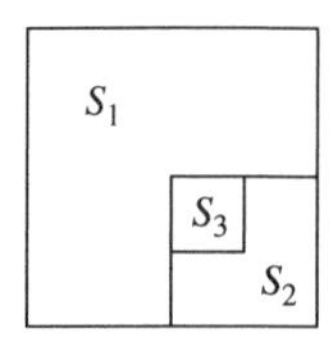

Figure 10.6

the left as j increases. Thus, we have

$$p_1 \geq p_2 \geq p_3 \geq \cdots \qquad \text{and} \qquad q_1 \geq q_2 \geq q_3 \geq \cdots. \tag{45}$$

The sets

$$\{p_1, p_2, p_3, \ldots\} \qquad \text{and} \qquad \{q_1, q_2, q_3, \ldots\} \tag{46}$$

are both bounded below by $-R$, since all of the squares S_j lie within S_1. Thus, we can let p and q be the greatest lower bounds of the two sets in (46). We set $w = p + qi$. We picture w as the complex number approached by the upper right corners of the squares S_j as j goes to infinity. We use inequality (42) to show that $|f(x)|$ takes its minimum value at w.

Because the sets in (46) are bounded below by $-R$, their greatest lower bounds p and q are greater than or equal to $-R$. On the other hand, p and q are both less than or equal to R, since the coordinates p_1 and q_1 of the upper right corner of S_1 equal R. The facts that p and q lie between $-R$ and R, inclusive, imply that $w = p + qi$ lies in the square S_1. It follows in the same way that w lies in all of the squares S_j, since the inequalities in (45) imply that p and q are the greatest lower bounds of the sets.

$$\{p_j, p_{j+1}, \ldots\} \qquad \text{and} \qquad \{q_j, q_{j+1}, \ldots\}$$

for every positive integer j.

Let ε be any positive real number. By Claim 1, there is a real number $\delta > 0$ such that

$$|f(z) - f(w)| < \varepsilon \qquad \text{if} \quad |z - w| < \delta. \tag{47}$$

Because each square in (44) is half as long as its predecessor, the lengths of the diagonals of the squares S_j shrink to zero as j increases. Accordingly, there is a positive integer u such that any two points of S_u are less than δ units apart. S_u contains w (by the previous paragraph), and it contains z_k for infinitely many positive integers k. For these values of k, we have

$$|z_k - w| < \delta,$$

which implies that

$$|f(z_k) - f(w)| < \varepsilon \tag{48}$$

(by (47)). It follows that

$$\begin{aligned} |f(w)| &= |f(w) - f(z_k) + f(z_k)| && \text{(by (5) and (10))} \\ &\leq |f(w) - f(z_k)| + |f(z_k)| && \text{(by (16))} \\ &< \varepsilon + d + \frac{1}{k} && \text{(by (15), (42), and (48)).} \end{aligned}$$

Because this inequality holds for infinitely many positive integers k, we have $|f(w)| \leq \varepsilon + d$. This holds for all positive real numbers ε, and so we have

$$|f(w)| \leq d. \tag{49}$$

On the other hand, since d is the greatest lower bound of the set of real numbers $|f(z)|$ as z varies over all complex numbers, we have $d \leq |f(z)|$ for all complex numbers z. Together with inequality (49), this shows that

$$|f(w)| \leq |f(z)|$$

for all complex numbers z, which completes the proof of Claim 3.

We have proved that $|f(x)|$ takes a minimum value at a point w. To complete the proof of the Fundamental Theorem of Algebra, we show that $f(w) = 0$. We eliminate the possibility that $|f(w)| > 0$ by showing that we could reduce $|f(x)|$ further in that case. We prove this first when w is the origin 0.

Claim 4
If $|f(0)| > 0$, *then there is a complex number* z *such that* $|f(z)| < |f(0)|$.

To prove the claim, we consider the expression for f in (33). We note that

$$|a_0| = |f(0)| > 0 \tag{50}$$

(by assumption). As stated before (33), we are assuming that f is not constant. Thus, we can let k be the smallest positive integer such that $a_k x^k$ is a nonzero term of f. We can write

$$\begin{aligned} |f(z)| &= |a_n z^n + \cdots + a_k z^k + a_0| \\ &\leq |a_n z^n + \cdots + a_{k+1} z^{k+1}| + |a_k z^k + a_0| \quad \text{(by (16))} \\ &= |z^k(a_n z^{n-k} + \cdots + a_{k+1} z)| + |a_k z^k + a_0| \end{aligned}$$

(by (4), (6), and (7))

$$= |z|^k |a_n z^{n-k} + \cdots + a_{k+1} z| + |a_k z^k + a_0| \tag{51}$$

(by (23)).

We consider first how small we can make the last quantity in (51). Let z, a_0, a_k lie at angles θ, α, β, respectively, measured counterclockwise from the positive x-axis. Then $a_k z^k$ lies at the angle $\beta + k\theta$ (by (22)). We choose the angle θ at which z lies so that

$$\beta + k\theta = \alpha + \pi$$

by setting

$$\theta = \frac{1}{k}(\alpha + \pi - \beta). \tag{52}$$

Since $a_k z^k$ has modulus $|a_k|\,|z|^k$ (by (23)), the two previous sentences and (19) show that

$$\begin{aligned} a_k z^k &= |a_k|\,|z|^k[\cos(\alpha+\pi) + i\sin(\alpha+\pi)] \\ &= -|a_k|\,|z|^k(\cos\alpha + i\sin\alpha) \end{aligned} \tag{53}$$

(by (20)). Since a_0 lies at angle α, we have

$$a_0 = |a_0|(\cos\alpha + i\sin\alpha).$$

Together with (53) and (7), this shows that

$$a_k z^k + a_0 = (|a_0| - |a_k|\,|z|^k)(\cos\alpha + i\sin\alpha). \tag{54}$$

Since $a_k \neq 0$, we have

$$|a_k| > 0 \tag{55}$$

(by (12) and (13)). Together with inequality (50), this shows that the first factor on the right-hand side of (54) is positive if

$$|z| < \left(\frac{|a_0|}{|a_k|}\right)^{1/k}. \tag{56}$$

In this case, (54) shows that

$$|a_k z^k + a_0| = |a_0| - |a_k|\,|z|^k. \tag{57}$$

The first term in (51) does not exist when n equals k, which occurs when f has only one term of positive degree. To analyze the first term in (51) when it does exist, we set

$$g(z) = a_n z^{n-k} + \cdots + a_{k+1} z.$$

By inequality (55), we can apply Claim 1 with f replaced by g, w replaced by 0, and ε replaced by $\frac{1}{2}|a_k|$. Since $g(0) = 0$, there is a real number $\delta > 0$ such that

$$|a_n z^{n-k} + \cdots + a_{k+1} z| < \tfrac{1}{2}|a_k| \tag{58}$$

if $|z| < \delta$.

The right-hand side of inequality (56) is positive (by (50) and (55)). Thus we can choose a complex number z that lies on the angle θ in (52) and whose modulus $|z|$ is a positive real number that is less than δ and also satisifes inequality (56). The two preceding paragraphs show that (57) and inequality (58) hold. Combining these relations with inequality (51) shows that

$$\begin{aligned} |f(z)| &< \tfrac{1}{2}|a_k|\,|z|^k + |a_0| - |a_k|\,|z|^k \\ &= |a_0| - \tfrac{1}{2}|a_k|\,|z|^k. \end{aligned}$$

This quantity is less than $|a_0|$, since $|a_k|$ and $|z|$ are both positive. Together with (50), this shows that $|f(z)|$ is less than $|f(0)|$, as claimed.

We can summarize the foregoing proof of Claim 4 as follows. We consider the term $a_k z^k$ of smallest positive degree appearing in f. Given that $|a_0| = |f(0)|$ is positive, we choose the angle at which z lies so that $a_k z^k$ is diametrically opposite to a_0. Then $a_k z^k + a_0$ has smaller modulus than a_0 when $|z|$ is not too large. We also choose $|z|$ to be small enough that the terms of f of degree greater than k are negligible compared to $a_k z^k$; this is based on the idea that higher powers of z go to zero faster than lower powers as z goes to zero. It follows that $|f(z)|$ is less than $|a_0| = |f(0)|$, as desired.

We have proved that $|f(x)|$ cannot have a minimum value at the origin 0 if $|f(0)| > 0$. We show next that $|f(x)|$ cannot have a minimum value at any point v such that $|f(v)| > 0$: we translate v to the origin and apply the previous result.

Claim 5

If v is a complex number such that $|f(v)| > 0$, then there is a complex number u such that $|f(u)| < |f(v)|$.

To prove the claim, we define a polynomial g with complex coefficients by setting

$$g(t) = f(t + v). \tag{59}$$

Since f is not constant, neither is g. Setting $t = 0$ shows that

$$|g(0)| = |f(v)| > 0.$$

By Claim 4, there is a complex number z such that

$$|g(z)| < |g(0)|.$$

Together with (59), this shows that

$$|f(z + v)| = |g(z)| < |g(0)| = |f(v)|,$$

as desired.

Claim 3 shows that $|f(x)|$ takes a minimum value at a point w. On the other hand, Claim 5 shows that $|f(x)|$ cannot take a minimum value at any point v such that $|f(v)| > 0$. Thus, we must have $|f(w)| = 0$ (by (12)), and so $f(w) = 0$ (by (13)). We have proved the following result:

Claim 6
Over the complex numbers, every polynomial in one variable of positive degree has a root.

This result leads directly to the Fundamental Theorem of Algebra. We call complex numbers *distinct* when no two of them are equal.

Theorem 10.1 (The Fundamental Theorem of Algebra)
Over the complex numbers, every polynomial $f(x)$ in one variable of positive degree n factors as

$$f(x) = r(x - w_1) \cdots (x - w_n)$$

for complex numbers $r, w_1, \ldots, w_n$, where $r \neq 0$ and the w_i are not necessarily distinct.

Proof
By Claim 6, there is a complex number w_1 such that $f(w_1) = 0$. Then the analogue of Theorem 1.10(ii) over the complex numbers shows that

$$f(x) = (x - w_1)f_1(x) \tag{60}$$

for a polynomial $f_1(x)$ of degree $n - 1$ over the complex numbers.

If $n = 1$, then $f_1(x)$ is a constant, and we are done. Otherwise, Claim 6 and the complex analogue of Theorem 1.10(ii) show that there is a complex number w_2 such that

$$f_1(x) = (x - w_2)f_2(x) \tag{61}$$

for a polynomial $f_2(x)$ of degree $n - 2$ over the complex numbers. Substituting (61) into (60) gives

$$f(x) = (x - w_1)(x - w_2)f_2(x).$$

We continue in this way until we have factored f completely. □

The Fundamental Theorem of Algebra shows that every polynomial in one variable of degree greater than 1 is reducible over the complex numbers. This is not true for polynomials in two or more variables. For example, $y^2 - x^3$ is irreducible over the complex as well as the real numbers: looking at powers of y shows that $y^2 - x^3$ could only factor as

$$(y - g(x))(y + g(x))$$

for some polynomial $g(x)$ such that $g(x)^2 = x^3$, but no such polynomial exists. The existence of irreducible polynomials of degree greater than 1 in two variables over the complex numbers ensures that the study of curves over the complex numbers is nontrivial. We pursue this study in the next section.

Exercises

10.1. Consider the complex numbers $z = 2 + 3i$, $w = 5 - 2i$, and $v = -1 + i$. Evaluate the following complex numbers, writing each one in the form $a + bi$ for real numbers a and b :
(a) $5iv + zw$. (b) $zv\bar{w}$.
(c) $v^4 - 3v^2 + 6$. (d) $z^3 - w$.
(e) $v^3 - iv + w$. (f) $i\bar{z} + 3w^2$.

10.2. (a) For any complex number z, prove that $z\bar{z} = |z|^2$.
(b) For any complex number $z \neq 0$, prove that $z^{-1} = |z|^{-2}\bar{z}$. (This makes it easy to find the inverse of a complex number written in the form $a + bi$ for real numbers a and b.)

10.3. This exercise is used in Exercises 11.13–11.15 and 12.24–12.27. Let

$$\omega = \frac{1}{2} + \frac{\sqrt{3}}{2}\,i. \tag{62}$$

(a) Write ω in polar form.
(b) Prove that the polynomial $x^3 + 1$ in one variable x factors over the complex numbers as

$$x^3 + 1 = (x + 1)(x - \omega)(x + \omega^2).$$

(c) Prove that there are exactly two complex numbers z such that $z^2 - z + 1 = 0$, namely, ω and $-\omega^2$.
(d) Prove that $\omega^{-1} = 1 - \omega$ and $(1 - \omega)^{-1} = \omega$.

10.4. Let x be an indeterminate, let n be a positive integer, and let w be a nonzero complex number given in polar form by (21). Prove that

$$x^n - w = (x - z_0)(x - z_1)\cdots(x - z_{n-1})$$

for

$$z_j = |w|^{1/n}\left[\cos\left(\frac{\psi + 2\pi j}{n}\right) + i\sin\left(\frac{\psi + 2\pi j}{n}\right)\right].$$

(This amplifies the discussion of nth roots after (25).)

10.5. Let x be an indeterminate, and let a, b, c be complex numbers such that $a \neq 0$. By the sentence after (25), there is a complex number z such that $z^2 = b^2 - 4ac$. Prove that

$$ax^2 + bx + c = a(x - v_1)(x - v_2)$$

for

$$v_j = (2a)^{-1}(-b + (-1)^j z).$$

(This extends the quadratic formula to the complex numbers. Like Exercises 10.3 and 10.4, this exercise gives a concrete illustration of the Fundamental Theorem of Algebra.)

We consider cubics over the real numbers in the remaining exercises, which extend the results of Section 9.

10.6. Prove the following result:

Theorem
Let C be a nonsingular, irreducible cubic. Let K be a conic that intersects C at six points E, F, G, H, W, X, listed by multiplicity. Let Y and Z be the third intersections of lines EG and FH. Then the third intersections of the lines WX and YZ are the same point.

(This theorem corresponds to Theorem 9.6 when the lines EF and GH are replaced by a conic K. One possible way to prove the theorem is to use Theorem 6.1 to "peel off" the conic K from the intersection of the cubic C and the cubic consisting of the three lines EG, FH, WX. Theorem 4.11 may help to prove that the latter cubic intersects K in the same six points, listed by multiplicity, as C does in cases where $E = G$, $F = H$, or $W = X$.)

10.7. Illustrate the theorem in Exercise 10.6 with a figure in each of the following cases. Restate the theorem as appropriate in each case in terms of tangents.

(a) The points E–H, W, X are all distinct.
(b) $E = G$, $F = H$, and $W = X$.
(c) $E = F$, $G = W$, and $H = X$.
(d) $E = G$ and $F = H$.
(e) $E = F$ and $G = H$.
(f) $E = F$ and $G = W$.
(g) $E = G$ and $F = W$.
(h) $E = G$.

10.8. Let C be a nonsingular, irreducible cubic with a flex O. Add points of C with respect to O as in Definition 9.3. Let P_1–P_6 be points of C that are not necessarily distinct.

(a) If a conic intersects C at P_1–P_6, listed by multiplicity, prove that $P_1 + \cdots + P_6 = O$ by combining the theorem in Exercise 10.6 with repeated applications of Exercise 9.2(a).
(b) If a curve of degree 2 intersects C at P_1–P_6, listed by multiplicity, prove that $P_1 + \cdots + P_6 = O$. (See part (a), Exercises 5.24 and 9.2(a), and Theorems 5.1 and 3.6(iii). Exercise 15.20 develops this exercise further.)

10.9. Let C be a nonsingular, irreducible cubic. Let P be a point of C at which the tangent contains a flex of C not equal to P. Prove that there is a conic that intersects C six times at P.

(Exercise 10.11 contains the converse of this result. One possible way to prove this result is to reduce to the case where P is the origin and C has equation $y^2 = x^3 + fx^2 + gx$ for $g \neq 0$. Show that $y^2 = fx^2 + gx$ is a conic that intersects C six times at the origin.)

10.10. Prove that a conic K intersects an irreducible cubic C at most twice at a flex of C.

(Hint: One possible approach is to use Theorems 3.4, 4.11, and 8.1(i) and Exercise 5.16 to reduce to the case where K is $y = x^2$, C is given by (5) of Section 8, and the flex is at infinity.)

10.11. Let C be a nonsingular, irreducible cubic, and let P be a point of C. We call P a *sextatic point* of C if there is a conic that intersects C six times at P. Prove that P is a sextatic point of C if and only if the tangent at P contains a flex of C not equal to P. (See Exercises 10.6 or 10.8, 10.9, and 10.10.)

10.12. Let C be a nonsingular, irreducible cubic with a flex O. Add points of C with respect to O as in Definition 9.3. Let P be a point of C, and define sextatic points as in Exercise 10.11.
(a) Prove that P is a sextatic point if and only if $6P = O$ and $3P \neq O$. (See Exercises 9.2 and 10.11.)
(b) Prove that P is a sextatic point of C if and only if P has order 2 or 6.

10.13. Let G and H be curves nonsingular at a point A, and let L be a line. Prove that two of the numbers $I_A(L,G)$, $I_A(L,H)$, $I_A(G,H)$ are equal and their common value is less than or equal to the third.

(This exercise analyzes the intersections in Theorem 9.4 in more detail when G and H are both nonsingular at A. The proof of that theorem can be adapted to do this exercise when $I_A(L,G)$ and $I_A(L,H)$ are finite and equal. Other cases can be handled by Theorems 9.4, 4.5, and 3.6. The line L is replaced with any curve nonsingular at A in Exercise 14.10.)

10.14. Let m and n be positive integers or ∞ such that $m \leq n$. In each part of this exercise, give an example of curves G and H that are nonsingular at the origin O and satisfy the given conditions.
(a) $I_O(G,H) = I_O(y,H) = m$ and $I_O(y,G) = n$.
(b) $I_O(y,G) = I_O(y,H) = m$ and $I_O(G,H) = n$.

(Cases where $n = \infty$ may need separate consideration. Exercise 10.13 and Theorem 3.6(iii) imply that parts (a) and (b) include all possible values for the three intersection multiplicities.)

10.15. For any positive integers m and n, find an example of curves G and H such that H is nonsingular at the origin O, y is not a factor of G, $I_O(y,H) = 1$, $I_O(y,G) = m$, and $I_O(G,H) = n$. (Thus, Theorems 9.4 and 9.5 are false without the assumption that G is nonsingular, even when H is nonsingular.)

10.16. Let C be a nonsingular, irreducible cubic. Let O be any point of C, not necessarily a flex. Define $P + Q$ for any points P and Q of C as in Definition 9.3. Let T be the third intersection of line OO. For any point P of C, prove that the third intersection of line PT is a point $-P$ such that $P + (-P) = O$.

(The proofs of Theorem 9.7 and (2) and (3) of Section 9 do not require O to be a flex. Thus, this exercise shows that C is an abelian group, as defined after the proof of Theorem 9.7, whose identity element is any point O of C, not necessarily a flex.)

10.17. Let C be a nonsingular, irreducible cubic, and let O and O' be two points of C. As in Exercise 10.16, use O to define $P+Q$ for any points P and Q of C, and use O' in place of O to define $P+'Q$.

(a) Prove that $(P+'Q)+O' = P+Q$ for any points P and Q of C.

(b) Set $f(X) = X+O'$ for any point X of C. Conclude from part (a) that

$$f(P)+'f(Q) = f(P+Q) \tag{63}$$

for any points P and Q of C.

(c) For any point R of C, prove that there is a unique point P of C such that $f(P) = R$.

(In general, an *isomorphism* between abelian groups $(C,+)$ and $(C',+')$ is a map f that matches up the elements of C and C' and satisfies (63) for all elements P and Q of C. There is an isomorphism between two abelian groups when they "look alike," differing only in the labeling of their elements. Parts (b) and (c) show that there is an isomorphism between any two of the abelian groups determined by a nonsingular, irreducible cubic for different choices of identity element.)

10.18. Let C be an irreducible cubic. Let P and Q be points of C at which C is nonsingular and which may or may not be distinct. Define the third intersection R of line PQ and C as after the proof of Theorem 9.1. Prove that C is nonsingular at R.

Exercises 10.19–10.21 *use the following terminology.* If C is an irreducible cubic that has a flex O, we let C_n be the set of nonsingular points of C. Define the sum of two points of C_n as in Definition 9.3. This sum is a point of C_n (by Exercise 10.18), and so C_n is closed under addition. The proofs of equations (2)–(4) of Section 9 show that addition is commutative, O is an identity element, and every element of C_n has an additive inverse. Addition on C_n is associative, since the proof of Theorem 9.6 requires only that C be nonsingular at the points in Figures 9.11 and 9.12, which, in Theorem 9.7, become the points in Figures 9.13 and 9.14. Thus, C_n is an abelian group. In the following exercises, C has the form of (25) of Section 9, except that the right-hand side of the equation now has repeated roots. Define addition on C_n by taking O to be the flex $(0,1,0)$ (as in Theorem 8.1(i)). Equations (26)–(39) of Section 9 describe addition on C_n.

10.19. Let C be $y^2 = x^3$ (Figure 8.5).

(a) Define a map f by setting $f(p) = (1/p^2, 1/p^3)$ for any nonzero real number p and setting $f(0) = (0,1,0)$. Prove that f matches up the real numbers with the points of C_n.

(b) For any real number p, prove that $f(-p) = -f(p)$ and that $f(-2p)$ lies on the tangent to C at $f(p)$.

(c) If p and q are real numbers such that the points $f(p)$, $f(q)$, and $f(-p-q)$ are distinct, prove that these points are collinear.

(d) Deduce that $f(p)+f(q)=f(p+q)$ for all real numbers p and q, whether or not they are distinct.

(Thus, addition on C_n looks like addition of real numbers. The map f is an isomorphism, as defined in Exercise 10.17.)

10.20. Let C be $y^2 = x^3 - x^2$ (Figure 8.7).

(a) Define a map f by setting

$$f(\cos\theta + i\sin\theta) = (d^2+1, d^3+d)$$

for $d = -\cot(\theta/2)$ when θ is a real number that is not an integral multiple of 2π and setting $f(1) = (0,1,0)$. Prove that f matches up the complex numbers of modulus 1 with the points of C_n. If $P = (\cos\theta, \sin\theta)$ is any point other than $(1,0)$ on the unit circle $x^2+y^2 = 1$, prove that f maps the complex number $\cos\theta + i\sin\theta$ associated with P to the unique point $Q = (d^2+1, d^3+d)$ of C_n such that the line through Q and the origin $(0,0)$ is parallel to the line through P and $(1,0)$ (Figure 10.7).

(b) For any complex number z of modulus 1, prove that $f(1/z) = -f(z)$ and that $f((z^2)^{-1})$ lies on the tangent at $f(z)$.

(c) If z and w are complex numbers of modulus 1 such that the points $f(z)$, $f(w)$, and $f((zw)^{-1})$ are distinct, prove that these points are collinear.

(d) Deduce that $f(z)+f(w) = f(zw)$ for all complex numbers z and w of modulus 1, whether or not they are distinct. (Thus, addition on C_n looks like multiplication of complex numbers of modulus 1: f is an isomorphism, as defined in Exercise 10.17.)

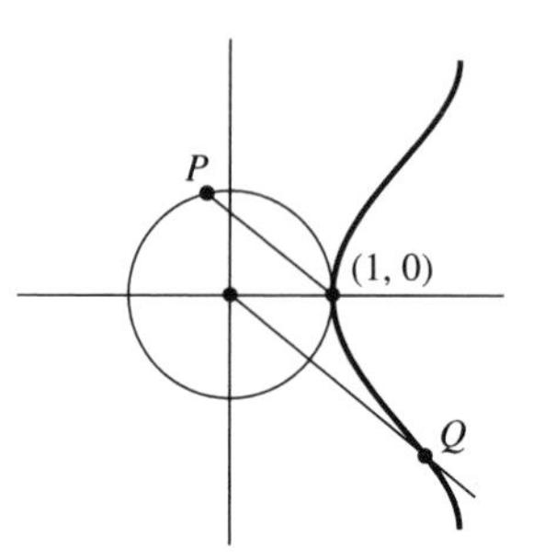

Figure 10.7

10.21. Let C be the cubic $y^2 = x^3 + x^2$ (Figure 8.6).

(a) Define a map f by setting $f(p) = (d^2-1, d^3-d)$ for

$$d = \frac{p+1}{p-1}$$

when p is a real number other than 0 and 1 and setting $f(1) = (0,1,0)$. Prove that f matches up the nonzero real numbers with the points of C_n.

(b) For any real number $p \neq 0$, prove that $f(1/p) = -f(p)$ and that $f(1/p^2)$ lies on the tangent at $f(p)$.
(c) If p and q are real numbers such that the points $f(p)$, $f(q)$, and $f(1/(pq))$ are distinct, prove that these points are collinear.
(d) Deduce that $f(p) + f(q) = f(pq)$ for all nonzero real numbers p and q, whether or not they are distinct. (Thus, addition on C_n looks like multiplication of nonzero real numbers. The map f is an isomorphism, as defined in Exercise 10.17.)

§11. Bezout's Theorem

Our goal is to prove that every irreducible cubic in the real projective plane has a flex or a singular point and is therefore classified by Theorems 8.3 and 8.4. We deduce this result in Section 12 from Bezout's Theorem, which we prove in this section.

We extend to the complex numbers the definitions and basic properties of the projective plane, curves, intersection multiplicities, and transformations. Bezout's Theorem states that curves of degrees m and n without common factors of positive degree intersect exactly mn times, counting multiplicities, over the complex numbers. The proofs of Theorems 4.4 and 4.5 on the intersections of lines and other curves extend to the complex numbers to give Bezout's Theorem when one of the curves is a line—that is, when m or n is 1. We deduce from this result that Bezout's Theorem holds for curves of all degrees by repeatedly reducing the highest exponent of y that appears in the equations of the curves.

If two homogeneous polynomials with real coefficients have a common factor of positive degree over the complex numbers, we prove that they also have a common factor of positive degree over the real numbers. We use this result in two ways. First, we use it in this section to deduce an analogue over the real numbers of Bezout's Theorem over the complex numbers: we prove that curves of degrees m and n without common factors of positive degree intersect at most mn times, counting multiplicities, in the real projective plane. Second, in the next chapter, we combine the result with Bezout's Theorem over the complex numbers to deduce that every nonsingular, irreducible cubic has a flex over the real numbers. It is common in mathematics to deduce a theorem over the real numbers from a theorem over the complex numbers.

We start by replacing the real numbers with the complex numbers in our basic definitions. The *real projective plane* is the standard name for the projective plane we have considered until now, which is defined using triples of real numbers. The *complex projective plane* consists of all triples (x, y, z) of complex numbers except $(0, 0, 0)$, where the triples (kx, ky, kz) all represent the same point as k varies over all nonzero complex numbers. If we think of triples of real numbers as triples of complex numbers, the real projective plane is contained in the complex projective plane, since each point of either plane equals exactly one of the triples in (1) of Section 2 over the real or complex numbers.

The Euclidean plane generalizes to the *complex affine plane*, which consists of all ordered pairs (x, y) of complex numbers. We identify each point (x, y) of the complex affine plane with the point $(x, y, 1)$ of the complex projective plane. Conversely, we can write any point (x, y, z) in the complex projective plane with $z \neq 0$ in exactly one way as $(x', y', 1)$ for complex numbers x' and y': we set $x' = z^{-1}x$ and $y' = z^{-1}y$. In this way, we match up the points of the complex affine plane with the points of the complex projective plane whose last coordinate is nonzero. We call the remaining points of the complex projective plane—those that have last coordinate zero—the *points at infinity*.

A *transformation* of the complex projective plane is a map

$$(x, y, z) \to (x', y', z')$$

from the complex projective plane to itself given by equations

$$\begin{aligned} x' &= ax + by + cz, \\ y' &= dx + ey + fz, \\ z' &= gx + hy + kz, \end{aligned}$$

where a–h, k are complex numbers such that these equations can be solved for x, y, z in terms of x', y', z'. As in Section 3, transformations are reversible, and a sequence of two transformations is again a transformation.

Let d be a nonnegative integer. A *homogeneous polynomial* of degree d over the complex numbers is a nonzero polynomial $F(x, y, z)$ with complex coefficients such that the exponents of x, y, and z in every term sum to d. We can write

$$F(x, y, z) = \sum e_{ij} x^i y^j z^{d-i-j} \tag{1}$$

for complex numbers e_{ij} that are not all zero. We also refer to a homogeneous polynomial over the complex numbers as a *complex curve*. We think of the complex curve $F(x, y, z)$ as the set of points (x, y, z) in the complex projective plane such that $F(x, y, z) = 0$. We refer to this complex curve as F or as the equation $F(x, y, z) = 0$ or its algebraic equivalents. We think of the homogeneous polynomials $kF(x, y, z)$ as the same complex curve for all complex numbers $k \neq 0$.

A nonzero polynomial

$$f(x, y) = \sum e_{ij} x^i y^j \tag{2}$$

with complex coefficients e_{ij} has degree d if d is the largest value of $i + j$

for which $e_{ij} \neq 0$. The *homogenization* of $f(x,y)$ is the homogeneous polynomial $F(x,y,z)$ in (1) that has the same degree d as $f(x,y)$. Setting $z = 1$ in the right-hand side of (1) gives the right-hand side of (2), and so the points $(a,b,1)$ of the complex projective plane that satisfy the equation $F(x,y,z) = 0$ are exactly the points (a,b) of the complex affine plane that satisfy the equation $f(x,y) = 0$. We refer to the complex curve F also as f or as the equation $f(x,y) = 0$ or its algebraic equivalents.

We must extend the basic properties of intersection multiplicities from the real to the complex projective plane. Let $O = (0,0)$ be the origin of the complex affine plane. We assume that the *intersection multiplicity* $I_O(f,g)$ of f and g at O is determined for all polynomials $f(x,y)$ and $g(x,y)$ with complex coefficients so that Properties 1.1–1.6 hold. We further assume that the *intersection multiplicity* $I_P(F,G)$ of F and G at P is determined for all homogeneous polynomials $F(x,y,z)$ and $G(x,y,z)$ with complex coefficients and all points P of the complex projective plane so that Properties 3.1 and 3.5 hold.

In Chapter IV, we determine intersection multiplicities for complex curves as above. We then define intersection multiplicities in the real projective plane to agree with those in the complex projective plane. That is, if the polynomials f, g, F, G has real coefficients, and if the point P has a triple of real numbers as homogeneous coordinates, we assign the intersection multiplicities $I_O(f,g)$ and $I_P(F,G)$ in the real projective plane the same values as in the complex projective plane. In other words, *curves with real coefficients intersect the same number of times at a point with real coordinates whether we think of the curves in the real or the complex projective plane.* Accordingly, once we show in Chapter IV that Properties 1.1–1.6, 3.1, and 3.5 hold for complex curves, it follows automatically that they hold for curves in the real projective plane.

A *line* in the complex projective plane is a complex curve of degree 1, namely,

$$px + qy + rz = 0$$

for complex numbers p, q, r that are not all zero. We claim that we can transform any line in the complex projective plane to $y = 0$. By interchanging variables, if necessary, we can assume that the coefficient q of y is nonzero. Using a transformation to multiply y by q^{-1} (as in (9) of Section 3) gives $px + y + rz = 0$ for real numbers p and r. This line is mapped to $y' = 0$ by the transformation

$$x' = x, \qquad y' = px + y + rz, \qquad z' = z,$$

as in (3) of Section 5.

Theorems 1.7–1.11, 3.4, 3.6, 3.7, and 4.2 follow directly from the intersection properties and, along with Definition 3.2 of the intersection multiplicities $I_{(a,b)}(f,g)$, they extend without change from the real to the complex numbers. Theorems 4.3 and 4.4 extend to the complex numbers with the change that the polynomial $r(x)$ in these theorems, which has no roots, is a nonzero constant, by the Fundamental Theorem of Algebra (Theorem 10.1). Since $r(x)$ has degree 0, the analogue of Theorem 4.4 over the complex numbers states that any homogeneous polynomial $G(x, y, z)$ of degree n that does not have y as a factor intersects the x-axis $y = 0$ exactly n times, counting multiplicities, in the complex projective plane.

Transformations preserve intersection multiplicities and factorizations of polynomials (by Property 3.5 and the discussion before Theorem 4.5). Together with the two previous paragraphs, this gives the following extension of Theorem 4.5 to the complex numbers:

Theorem 11.1
In the complex projective plane, let $L = 0$ be a line, and let $G = 0$ be a complex curve of degree n. If L is not a factor of G, then L and G intersect exactly n times, counting multiplicities, in the complex projective plane. □

The line L in Theorem 11.1 is replaced by a curve of any degree in Bezout's Theorem. The proof of Bezout's Theorem requires two preliminary theorems about multiplying homogeneous polynomials. Recall that homogeneous polynomials are nonzero, by definition.

Theorem 11.2
Let F, G, and H be homogeneous polynomials over the complex numbers.

(i) *Then FG is a homogeneous polynomial whose degree is the sum of the degrees of F and G.*
(ii) *If $FG = FH$, then $G = H$.*

Proof
(i) The degree of any term of FG is the sum of the degrees of terms of F and G. Part (i) follows, once we verify that FG is nonzero.

Among the terms of F that have the highest power of y, we choose the term that has the highest power of x, and we call this the leading term of F. We choose the leading term of G in the same way. The product of the leading terms of F and G is nonzero (by (24) of Section 10) and it has a higher power of y or x than the product of any other pair of terms of F and G. Thus, the product of the leading terms of F and G is a nonzero term of FG, and so FG is nonzero.

(ii) We can rewrite the equation $FG = FH$ as

$$F(G - H) = 0. \tag{3}$$

Part (i) and the equation $FG = FH$ imply that G and H have the same degree, and so $G - H$ is either a homogeneous polynomial or zero. If it were a homogeneous polynomial, its product with F would be nonzero (by part (i)), which would contradict (3). Thus $G-H$ is zero, and G equals H. $\square$

Let $H(x,z)$ be a homogeneous polynomial of positive degree that has positive coefficients and does not contain y. Setting $z=1$ gives a polynomial $H(x,1)$ in x alone. By the Fundamental Theorem of Algebra (Theorem 10.1), we can write

$$H(x,1) = r(x-w_1)^{s_1}\cdots(x-w_k)^{s_k}$$

for complex numbers $r \neq 0, w_1, \ldots, w_k$, and positive integers $s_1, \ldots, s_k$. (If $H(x,1)$ is a constant, there are no w_j.) Since H is homogeneous and does not contain y, it follows that

$$H(x,z) = r(x-w_1z)^{s_1}\cdots(x-w_kz)^{s_k}z^t \qquad (4)$$

for an integer $t \geq 0$. Since H has positive degree, we can move r into one of the factors on the right-hand side of (4) and write $H = L_1 \cdots L_m$ for lines L_i that need not be distinct. Thus, *any homogeneous polynomial in two variables that has positive degree factors over the complex numbers as a product of lines*. We use this observation to derive the second result about polynomial multiplication that we need.

Theorem 11.3

Over the complex numbers, let F, G, H be homogeneous polynomials such that H does not contain y or have a factor of positive degree in common with G. Then any common factor of HF and G is also a common factor of F and G.

Proof

Let R be a common factor of HF and G. Then R is a factor of G, and we can write $HF = RS$ for a homogeneous polynomial S. If H has degree 0, then it is a nonzero constant c, and the relation $F = c^{-1}RS$ shows that R is a common factor of F and G.

If H has degree 1, then it is a line. We can transform H to x, as discussed before Theorem 11.1. Since transformations preserve factorizations (as discussed before Theorem 4.5), we can assume that $H = x$. Since H has no factors of positive degree in common with G, x is not a factor of R. If x were not a factor of S either, the terms without x in R and S would form homogeneous polynomials R' and S' in y and z, and $R'S'$ would be nonzero (by Theorem 11.2(ii)); then RS would have nonzero terms without x, which would contradict the assumption that $xF = RS$. Thus, x is a factor of S, and we can write $S = xT$ for a homogeneous polynomial T. Substituting for S in $xF = RS$ gives $xF = xRT$. It follows that $F = RT$ (by Theorem 11.2(ii)), and so R is a common factor of F and G.

Finally, assume that H has degree $m > 1$. By the discussion before the theorem, we can write $H = LH'$ for a line L and a homogeneous polynomial H' of degree $m-1$. Since H has no factors of positive degree in common with G, neither do L and H'. Since R is a common factor of $HF = LH'F$ and G, it is a common factor of $H'F$ and G, by the previous paragraph. We continue to reduce the degree of H in this way until we are done. □

Let $I(F,G)$ be the total number of times, counting multiplicities, that complex curves F and G intersect in the complex projective plane. That is, $I(F,G)$ is the sum of the intersection multiplicities $I_P(F,G)$ for all points P in the complex projective plane. Since each of the numbers $I_P(F,G)$ is a nonnegative integer or ∞, so is $I(F,G)$.

If F, G, H are complex curves, the equation

$$I_P(F,GH) = I_P(F,G) + I_P(F,H)$$

holds at every point P (by Theorem 3.6(v)). Summing these relations for all points P in the complex projective plane shows that

$$I(F,GH) = I(F,G) + I(F,H). \tag{5}$$

Likewise, summing the relation in Theorem 3.6(iv) over all points P of the complex projective plane shows that

$$I(F,G) = I(F,G+FH) \tag{6}$$

if $G + FH$ is a homogeneous polynomial.

We say that *Bezout's Theorem holds* for complex curves F and G of degrees m and n if $I(F,G) = mn$. We want to prove that Bezout's Theorem holds whenever F and G have no common factors of positive degree.

Theorem 11.1 shows that Bezout's Theorem holds when F is a line that is not a factor of G. It follows that Bezout's Theorem holds when F does not contain y and has no factors of positive degree in common with G. To see this, suppose first that the degree m of F is positive. F is a product $L_1 \cdots L_m$ of lines L_i that need not be distinct, as discussed before Theorem 11.3. None of the lines L_i is factor of G, by assumption, and so we have

$$\begin{aligned} I(F,G) &= I(L_1 \cdots L_m, G) \\ &= I(L_1, G) + \cdots + I(L_m, G) \quad \text{(by (5))} \\ &= mn \end{aligned}$$

(by Theorem 11.1), as desired. On the other hand, if the degree m of F is zero, then F is a nonzero constant and there are no points on the curve $F = 0$. Bezout's Theorem holds in this case because

$$I(F,G) = 0 = 0n$$

(by Theorem 3.6(i) and (iii)).

Let F, G, H be homogeneous polynomials of respective degrees m, n, p, and assume that H does not contain y and has no factors of positive degree in common with G. The previous paragraph shows that $I(H, G) = pn$. Together with (5), this shows that

$$I(FH, G) = I(F, G) + pn.$$

It follows that $I(FH, G) = (m+p)n$ if and only if $I(F, G) = mn$. Since FH is a homogeneous polynomial of degree $m+p$ (by Theorem 11.2(i)), Bezout's Theorem holds for FH and G if and only if it holds for F and G. In effect, we can disregard factors in which y does not appear when we prove Bezout's Theorem.

The *degree in y* of a homogeneous polynomial F is the largest exponent of y that appears in the nonzero terms of F. We prove Bezout's Theorem by repeatedly reducing degrees in y, the same technique we used in Example 1.13 to compute intersection multiplicities. The next result formalizes this step.

Theorem 11.4

Let $F = 0$ and $G = 0$ be complex curves of respective degrees s and t in y. Assume that $s \geq t > 0$ and that F and G have no common factors of positive degree. Then there are complex curves $F_1 = 0$ and $G_1 = 0$ such that F_1 and G_1 have no common factors of positive degree, the degree of F_1 in y is less than s, the degree of G_1 in y is t, and Bezout's Theorem holds for F and G if and only if it holds for F_1 and G_1.

Proof

If we take G and factor out homogeneous polynomials of positive degree that do not contain y, we reduce the degree of G (by Theorem 11.2(i)). Thus, this process ends, and we can write $G = HG_1$, where $H(x, z)$ is a homogeneous polynomial that does not contain y, and G_1 has no factors of positive degree without y. G_1 has the same degree t in y as G. Since G has no factors of positive degree in common with F, neither do H and G_1.

Let $P(x, z)$ be the coefficient of y^s in F, and let $Q(x, z)$ be the coefficient of y^t in G_1. P and Q are homogeneous polynomials in x and z that do not contain y. QF and $Py^{s-t}G_1$ are both homogeneous polynomials of degree s in y in which y^s has coefficient PQ. Thus, QF and $Py^{s-t}G_1$ are homogeneous polynomials of the same degree, and so their difference

$$F_1 = QF - Py^{s-t}G_1, \tag{7}$$

is either zero or homogeneous of the same degree as QF. If F_1 is nonzero, its degree in y is less than s (by the second-to-last sentence). Since G_1 has no factors of positive degree in common with $Q(x, z)$ or F (by the previous paragraph), it has no factors of positive degree in common with QF (by Theorem 11.3). Since (7) implies that any common factor of F_1 and G_1 would also be a factor of QF, the last sentence shows that F_1 and G_1

have no common factors of positive degree. Since G_1 has positive degree (because it has degree $t > 0$ in y), it follows that F_1 is nonzero. Thus, the discussion accompanying (7) shows that F_1 is a homogeneous polynomial whose degree in y is less than s.

Bezout's Theorem holds for F and G if and only if it holds for F and G_1 if and only if it holds for QF and G_1 (by the second-to-last paragraph before the theorem). This occurs if and only if Bezout's Theorem holds for F_1 and G_1 (by (6) and (7)). We have seen that F_1 has degree less than s in y, G_1 has degree t in y, and F_1 and G_1 have no common factors of positive degree. □

We can now prove Bezout's Theorem which states that, if complex curves F and G have no common factors of positive degree, then the number of times they intersect, counting multiplicities, is the product of their degrees.

Theorem 11.5 (Bezout's Theorem)
Let $F = 0$ and $G = 0$ be complex curves of degrees m and n such that F and G have no common factors of positive degree. Then $F = 0$ and $G = 0$ intersect exactly mn times, counting multiplicities, in the complex projective plane.

Proof
If F and G both have positive degree in y, we can use Theorem 11.4 to reduce one of these degrees in y without changing the other. We repeat this process until one of the curves has degree 0 in y, and we are done by the third-to-last paragraph before Theorem 11.4. □

Let F and G be complex curves. Bezout's Theorem 11.5 shows that, if F and G have no common factors of positive degree, they intersect at only finitely many different points of the complex projective plane. On the other hand, F and G intersect at infinitely many different points of the complex projective plane if they have a common factor H of positive degree. This holds because the next result shows that H has infinitely many points, and these points lie on both F and G.

Theorem 11.6
Every complex curve of positive degree has infinitely many points.

Proof
Let $H(x, y, z) = 0$ be a complex curve of positive degree. By interchanging variables, if necessary, we can assume that H has a nonzero term of positive degree t in y. Setting $z = 1$ in $H(x, y, z)$ gives a polynomial $h(x, y)$ of positive degree t in y. By collecting the terms of h with the same powers of y, we can write

$$h(x, y) = \sum p_i(x) y^i,$$

where $p_t(x)$ is nonzero. We can write

$$p_t(x) = r(x - w_1) \cdots (x - w_k)$$

for complex numbers $r \neq 0$ and $w_1, \ldots, w_k$ that are not necessarily distinct (by the Fundamental Theorem of Algebra 10.1). For any complex number a other than $w_1, \ldots, w_k$, we have $p_t(a) \neq 0$ (by (24) of Section 10). Then $h(a, y)$ is a polynomial of positive degree in y, and so it has a root b in the complex numbers (by the Fundamental Theorem 10.1). As a varies over the complex numbers other than $w_1, \ldots, w_k$, this gives infinitely many points (a, b) of the complex affine plane on h. These correspond to infinitely many points $(a, b, 1)$ of the complex projective plane on H. □

Theorem 11.6 does not hold over the real numbers: Theorem 5.1 gives examples of curves of degree 2 that have no points or one point in the real projective plane.

To derive the analogue of Bezout's Theorem for the real numbers, we need to relate factorizations of homogeneous polynomials with real coefficients over the real and the complex numbers. If F is a homogeneous polynomial with complex coefficients, we define its *conjugate* $\bar{F}$ to be the homogeneous polynomial produced by conjugating the coefficients of F. For example, if F is

$$(2 + 3i)x^2yz + 7xz^3 - 8iy^4,$$

then $\bar{F}$ is

$$(2 - 3i)x^2yz + 7xz^3 + 8iy^4.$$

Theorem 11.7

Let F and G be homogeneous polynomials over the complex numbers, and set $H = FG$.

(i) *Then we have $\bar{H} = \bar{F} \cdot \bar{G}$.*
(ii) *If F and H have real coefficients, then so does G.*

Proof

(i) The coefficient of any term of H is a sum of products of coefficients of F and G. We obtain the corresponding coefficient of $\bar{H}$ in the same way from the coefficients of $\bar{F}$ and $\bar{G}$, since sums and products of complex numbers are preserved by conjugation (by (31) and (32) of Section 10). Thus, the relation $H = FG$ implies that $\bar{H} = \bar{F} \cdot \bar{G}$.

(ii) Since F and H have real coefficients, we have $\bar{F} = F$ and $\bar{H} = H$. Together with part (i), this shows that

$$FG = H = \bar{H} = \bar{F} \cdot \bar{G} = F\bar{G}.$$

Then G equals $\bar{G}$ (by Theorem 11.2(ii)), and so G has real coefficients.

□

The next result shows that a factor over the complex numbers of a homogeneous polynomial with real coefficients gives rise to a factor over the real numbers. Over the complex numbers, a homogeneous polynomial of positive degree is called *irreducible* if it does not equal the product of two homogeneous polynomials over the complex numbers of smaller degrees.

Theorem 11.8
Let F be a homogeneous polynomial with real coefficients. Let G be a homogeneous polynomial with complex coefficients that is irreducible and a factor of F over the complex numbers. Then either there is a nonzero complex number k such that kG has real coefficients or else $G\bar{G}$ is a homogeneous polynomial with real coefficients that is a factor of F over the real numbers.

Proof
Since G is irreducible over the complex numbers, so is

$$G_1 = a^{-1}G, \tag{8}$$

where a is the coefficient of a nonzero term of G. G_1 has a term with coefficient 1. If all the coefficients of G_1 are real, we are done by taking $k = a^{-1}$. Thus, we can assume that G_1 has a coefficient that is not real.

Since G is a factor of F, so is G_1, and we can write

$$F = G_1 S \tag{9}$$

for a homogeneous polynomial S with complex coefficients. Conjugating both sides of (9) gives

$$F = \overline{G_1} \cdot \bar{S} \tag{10}$$

(by Theorem 11.7(i)), where F equals $\bar{F}$ because it has real coefficients. Equations (5) and (10) show that

$$I(G_1, F) = I(G_1, \overline{G_1}) + I(G_1, \bar{S}). \tag{11}$$

Since G_1 is irreducible, it has positive degree (by definition), and so it contains infinitely many points of the complex projective plane (by Theorem 11.6). These points lie on F (since G_1 is a factor of F), and so we have

$$I(G_1, F) = \infty \tag{12}$$

(by Theorem 3.6). Since G_1 does not have real coefficients, it does not equal $\overline{G_1}$. It follows that G_1 and $\overline{G_1}$ are not scalar multiples of each other (since they have corresponding terms with coefficient 1), and so they have no common factors of positive degree (since G_1 is irreducible). Thus, we have

$$I(G_1, \overline{G_1}) < \infty \tag{13}$$

(by Bezout's Theorem 11.5).

Combining (11), (12), and (13) shows that $I(G_1, \bar{S}) = \infty$. It follows from Bezout's Theorem 11.5 that G_1 and $\bar{S}$ have a common factor of positive degree. Thus, since G_1 is irreducible, it is a factor of $\bar{S}$, and we can write $\bar{S} = G_1 T$ for a homogeneous polynomial T with complex coefficients. Substituting this expression for $\bar{S}$ into (10) gives

$$F = G_1 \overline{G_1} T. \tag{14}$$

Equation (8) implies that

$$\overline{G_1} = \overline{a^{-1}} \cdot \bar{G},$$

and combining this equation with (8) and (14) shows that

$$F = G\bar{G}U \tag{15}$$

for a homogeneous polynomial U with complex coefficients. Because conjugation interchanges G and $\bar{G}$ (by (30) of Section 10), it maps $G\bar{G}$ to itself (by Theorem 11.7(i)), and so $G\bar{G}$ has real coefficients. Since F has real coefficients, (15) and Theorem 11.7(ii) imply that $G\bar{G}$ is a factor of F over the real numbers. □

Theorem 11.8 has the following consequence, which lets us use Bezout's Theorem 11.5 to study the intersections of curves in the real projective plane that have no common factors of positive degree over the real numbers:

Theorem 11.9
Let F and G be homogeneous polynomial with real coefficients. If F and G have a common factor of positive degree over the complex numbers, then they have a common factor of positive degree over the real numbers.

Proof
Let H be a common factor of F and G over the complex numbers whose degree is positive and as small as possible. H is irreducible, since, otherwise, we could replace H with one of its factors. If there is a nonzero complex number k such kH has real coefficients, then the fact that kH is a factor of both F and G over the complex numbers implies that it is also a factor of both F and G over the real numbers (by Theorem 11.7(ii)). If there is no such complex number k, then $H\bar{H}$ is a homogeneous polynomial with real coefficients that is a factor of both F and G over the real numbers (by Theorem 11.8). □

We can now derive an analogue of Bezout's Theorem 11.5 for curves in the real projective plane.

Theorem 11.10
Over the real numbers, let $F=0$ and $G=0$ be curves of degrees m and n such that F and G have no common factors of positive degree over the real numbers. Then $F=0$ and $G=0$ intersect at most mn times, counting multiplicities, in the real projective plane.

Proof
Since F and G have no common factors of positive degree over the real numbers, the same holds over the complex numbers (by Theorem 11.9). Thus, F and G intersect exactly mn times, counting multiplicities, in the complex projective plane (by Bezout's Theorem 11.5). Theorem 11.10 follows, since F and G intersect the same number of times over the real or complex number at any point of the real projective plane, as discussed before Theorem 11.1. □

The curves F and G in Theorem 11.10 intersect fewer than mn times in the real projective plane when at least one of their points of intersection lies in the complex but not the real projective plane. Theorem 11.10 is illustrated by Theorems 4.4 and 4.5 and the discussion accompanying Figure 4.1 when one of the curves is a line and by Theorems 5.8 and 5.9 when one of the curves is a conic.

Exercises

11.1. Two polynomial equations with real coefficients are given in each part of this exercise. At what points of the complex projective plane do they intersect, and how many times do they intersect at each of these points? Use Bezout's Theorem 11.5 to check but not to obtain your answers. What is the total number of intersections, counting multiplicities, in the real projective plane?
(a) $y^3=x^2+1$, $y^3=-x^2-1$.
(b) $y=x^3$, $y^3=x^5$.
(c) $x^2+4y^2=1$, $x^2+4y^2=4$.
(d) $xy=1$, $y=x^2$.
(e) $x^3y=3x^3-1$, $y=x^3+1$.
(f) $x^2y=1$, $y=-x^2$.

11.2. Adapt the proof of Theorem 5.1 to prove that any two irreducible complex curves of degree 2 can be transformed into each other.

11.3. In the real projective plane, let $F=0$ and $G=0$ be curves, and let P be a point. Prove that $I_P(F,G)=\infty$ if and only if F and G have a common factor containing P. (This result is used in Exercises 14.9 and 14.10.)

11.4. A *circle* is a curve in the real projective plane given by an equation

$$(x-h)^2+(y-k)^2=r^2, \tag{16}$$

where h,k,r are real numbers with $r>0$.

(a) Prove that (16) gives a complex curve that contains exactly two points at infinity, namely $(1, i, 0)$ and $(1, -i, 0)$. (These two points are called *circular points*.)

(b) Prove that a curve in the real projective plane is a circle if and only if it has degree 2, it contains three noncollinear points in the Euclidean plane, and the complex curve with the same equation contains the points $(1, i, 0)$ and $(1, -i, 0)$.

(c) Deduce from part (b) and the extension of Theorem 5.10 to the complex projective plane that three noncollinear points in the Euclidean plane lie on a unique circle.

11.5. Let K_1 and K_2 be two circles that intersect at least once in the Euclidean plane. Prove that K_1 and K_2 intersect exactly twice, counting multiplicities, in the Euclidean plane by using Exercise 11.4, the extension of Theorem 5.8 to the complex projective plane, and the fact that any circle can be transformed in the real projective plane into $y = x^2$.

11.6. Consider the following result (Figure 11.1):

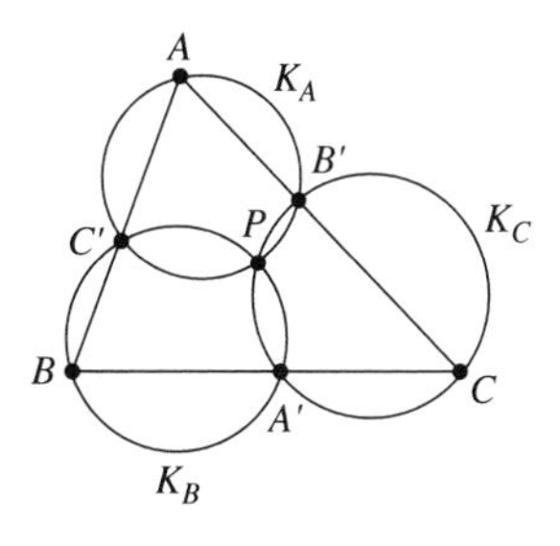

Figure 11.1

Miquel's Theorem
In the Euclidean plane, let A, B, C be three noncollinear points. Let A' be a point on line BC other than B and C, let B' be a point on line CA other than C and A, and let C' be a point on line AB other than A and B. Let K_A, K_B, K_C be the circles determined by the three triples of points A, B', C'; A', B, C'; A', B', C. Then the circles K_A, K_B, K_C have unique point P of the Euclidean plane in common.

Prove Miquel's Theorem as follows. Let F_B be the cubic formed by the circle K_B and the line CA, and let F_C be the cubic formed by the circle K_C and the line AB. Use Theorem 6.4 to "peel off" the line BC from the intersection of F_B and F_C, and deduce Miquel's Theorem from Exercises 11.4 and 11.5.

11.7 In the notation of Miquel's Theorem, use that result and its proof and Theorem 4.11 to prove that K_A contains A' if and only if K_B and K_C are tangent to the same line at A'. Illustrate this result with a figure.

11.8 In the notation of Miquel's Theorem, let K be the circle determined by the three points A, B, C. Prove as follows that A', B', C' are collinear if and only if P lies on K, and illustrate this result with a figure. If A', B', C' lie on a line L, prove that P lies on K by using Theorem 6.4 to "peel off" L from the intersection of the cubics F_B and F_C in Exercise 11.6. If P lies on K, use Exercise 11.4 and Theorem 6.1 to "peel off" K from the intersection of F_B and F_C.

11.9 In the Euclidean plane, let ABC be a triangle, and let Q be a point (Figure 11.2). Let A', B', C' be the feet of the perpendiculars from Q to BC, CA, AB, respectively. Let K be the circle determined by the three points A, B, C—the *circumcircle* of triangle ABC. Prove that A', B', C' are collinear if and only if Q lies on K. Use Exercises 11.6 and 11.8 and the following basic result from Euclidean geometry: If S, T, U are three points in the Euclidean plane, then T lies on the circle with diameter SU if and only if $\angle STU = 90°$ (Figure 11.3). Cases where any of the points A', B', C' equals A, B, or C require separate consideration.

(Because the points A', B', C' are not all equal, they determine a unique line when Q lies on K. This line is called the *Simson line* of the point Q and the triangle ABC.)

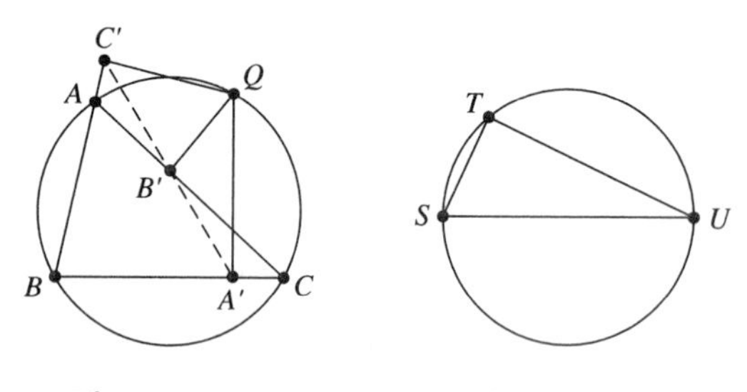

Figure 11.2 **Figure 11.3**

11.10 In the Euclidean plane, let K_1, K_2, K_3 be three circles such that any two of them intersect at two points (Figure 11.4). Let L_1, L_2, L_3 be the lines

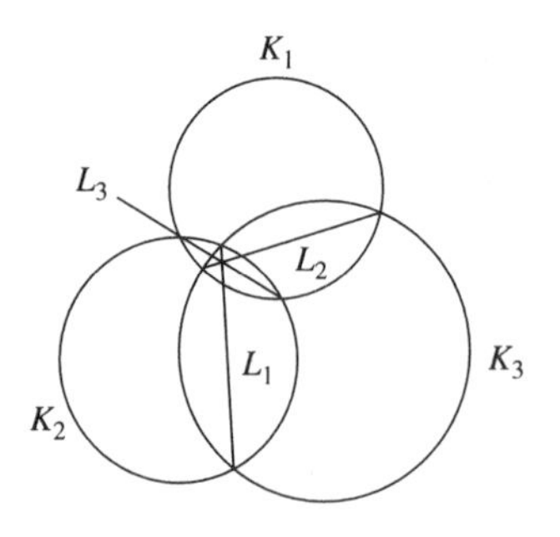

Figure 11.4

through the two intersection points of K_2 and K_3, K_3 and K_1, and K_1 and K_2, respectively. Prove that the lines L_1, L_2, L_3 are concurrent or parallel in the Euclidean plane by using Theorem 6.1 when K_1, K_2, K_3 do not have a point in common to "peel off" K_1 from the intersection of the cubic formed by K_2 and L_2 and the cubic formed by K_3 and L_3.

Exercises 11.11–11.20 *use the fact that Theorems* 4.6–4.8 *and the definitions of singular points, tangent lines, nonsingular curves, and flexes extend to the complex projective plane without change.* A *complex cubic* is a complex curve of degree 3. The results of Section 9 on addition of points from Theorem 9.1 through Theorem 9.7 extend without change to the complex numbers.

11.11. Over the real or complex numbers, let C be a nonsingular, irreducible cubic or complex cubic. Assume that C has three noncollinear flexes O, P, Q. Add points of C with respect to O as in Definition 9.3. Use Exercise 9.2 (which extends without change to the complex numbers) to prove that the nine points in Figure 11.5 are flexes of C and that no two of these points are equal.

O	Q	$2Q$
P	$P+Q$	$P+2Q$
$2P$	$2P+Q$	$2P+2Q$

Figure 11.5

11.12. In the notation of Exercise 11.11, use Exercise 9.2(a) (which extends without change to the complex numbers) to prove that the following twelve triples of points in Figure 11.5 are collinear:

(i) The three horizontal triples—O, Q, $2Q$; P, $P+Q$, $P+2Q$; $2P$, $2P+Q$, $2P+2Q$.

(ii) The three vertical triples—O, P, $2P$; Q, $P+Q$, $2P+Q$; $2Q$, $P+2Q$, $2P+2Q$.

(iii) The three triples of points on the lines in Figure 11.6—O, $P+Q$, $2P+2Q$; Q, $P+2Q$, $2P$; $2Q$, P, $2P+Q$.

(iv) The three triples of points on the lines in Figure 11.7—O, $2P+Q$, $P+2Q$; P, Q, $2P+2Q$; $2P$, $P+Q$, $2Q$.

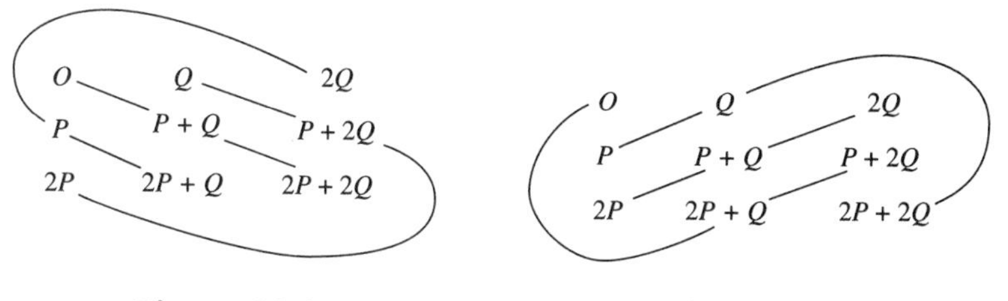

Figure 11.6 **Figure 11.7**

(The triples in (iii) and (iv) represent the six ways to choose one point from each row and column in Figure 11.5, in analogy with the definition of determinants in (22) of Section 12.)

11.13. In the notation of Exercise 11.11, assume that $P = (1, 0, 0)$, $2P = (0, 1, 0)$, $Q = (0, 0, 1)$, and $2Q = (1, 1, 1)$. Let ω be given by (62) of Section 10. Use Exercises 10.3 and 11.12 to prove that the nine points in Figure 11.5 are

$$\begin{array}{lll} (1,1,0), & (0,0,1), & (1,1,1), \\ (1,0,0), & (1,d,1), & (0,d,1), \\ (0,1,0), & (1,d,d), & (1,0,d), \end{array} \tag{17}$$

respectively, where d is either ω or $-\omega^2$.

11.14. Use Exercises 11.11 and 11.13 and Theorem 3.4 to prove that *every nonsingular, irreducible cubic in the real projective plane has at most three flexes.*

11.15. (a) Find the equation of a transformation that fixes $(1, 0, 0)$ and $(0, 1, 0)$ and interchanges $(0, 0, 1)$ and $(1, 1, 1)$.
(b) Prove that the transformation in (a) interchanges the nine points in (17) for $d = \omega$ with the nine points in (17) for $d = -\omega^2$.

11.16. (a) Can a reducible complex curve of positive degree be nonsingular in the complex projective plane? Justify your answer.
(b) Over the real numbers, can a reducible curve of positive degree be nonsingular in the real projective plane? Justify your answer.

11.17. Let w and v be complex numbers other than 0 and 1. In each part of this exercise, use Exercise 8.9 (which extends without change to the complex numbers) to determine when there is a transformation over the complex numbers that fixes $(0, 1, 0)$ and maps

$$y^2 = x(x-1)(x-w) \tag{18}$$

to

$$y^2 = x(x-1)(x-v). \tag{19}$$

(a) Prove that there is such a transformation fixing $(0, 0, 1)$ and $(1, 0, 1)$ if and only if $v = w$.
(b) Prove that there is such a transformation mapping $(1, 0, 1)$ to $(0, 0, 1)$ and $(0, 0, 1)$ to $(1, 0, 1)$ if and only if $v = 1 - w$.
(c) Prove that there is such a transformation fixing $(0, 0, 1)$ and mapping $(w, 0, 1)$ to $(1, 0, 1)$ if and only if $v = 1/w$.
(d) Prove that there is such a transformation mapping $(w, 0, 1)$ to $(0, 0, 1)$ and $(0, 0, 1)$ to $(1, 0, 1)$ if and only if $v = (w-1)/w$.
(e) Prove that there is such a transformation mapping $(1, 0, 1)$ to $(0, 0, 1)$ and $(w, 0, 1)$ to $(1, 0, 1)$ if and only if $v = 1/(1-w)$.
(f) Prove that there is such a transformation mapping $(w, 0, 1)$ to $(0, 0, 1)$ and fixing $(1, 0, 1)$ if and only if $v = w/(w-1)$.

11.18. Let w and v be complex numbers other than 0 and 1. If a complex cubic can be transformed into (18), prove that it can also be transformed into

(19) if and only if v is one of the numbers

$$w, \quad 1-w, \quad \frac{1}{w}, \quad \frac{w-1}{w}, \quad \frac{1}{1-w}, \quad \frac{w}{w-1}.$$

Use Theorem 8.1(i), Exercise 8.7, and either Theorem 8.2 or Exercise 8.9, which all extend without change to the complex numbers, and Exercise 11.17. (Exercise 12.29 shows that a complex cubic is nonsingular if and only if it can be transformed into (18) for a complex number w other than 0 and 1.)

11.19. (a) Adapt the proof of Theorem 8.4 to prove that a complex cubic is singular and irreducible if and only if it can be transformed into

$$y^2 = x^3 \qquad \text{or} \qquad y^2 = x^2(x+1). \tag{20}$$

(b) Prove that no complex cubic can be transformed into both of the equations in (20).

11.20. (a) For any integer $n \geq 4$, find a homogeneous polynomial F of degree n with real coefficients and a line L with real coefficients such that F is tangent to L in the complex projective plane but not the real projective plane.
(b) Can part (a) be done for any integer $n \leq 3$?

11.21. Let F be a homogeneous polynomial of odd degree with real coefficients. If F is irreducible over the real numbers, use Theorems 11.7 and 11.8 to prove that F is also irreducible over the complex numbers. (Exercise 14.3 sharpens this result.)

§12. Hessians

We finish characterizing nonsingular, irreducible cubics over the real numbers in this section. We prove that they all have flexes and so are determined by Theorem 8.3.

We consider only polynomials and curves with real coefficients in this section except where we explicitly state otherwise. We start by defining the first and second partial derivatives of polynomials in x and y. We use these to characterize the flexes and singular points of curves in the Euclidean plane. By translating these results into homogeneous coordinates, we associate each homogeneous polynomial F with a quantity H called the Hessian of F. H is either zero or a homogeneous polynomial, and the flexes and singular points of F are exactly the points of F that satisfy the equation $H = 0$.

When C is a nonsingular, irreducible cubic, we prove that its Hessian H is a cubic distinct from C. It follows from Bezout's Theorem 11.5 and Theorem 11.9 that C and H intersect nine times, counting multiplicities, in the complex projective plane. Because the number nine is odd, we deduce that C and H intersect in the real projective plane by considering the map of the complex projective plane that conjugates the homogeneous coordinates of every point. Thus, C has a flex in the real projective plane, as desired.

Let $f(x, y)$ be a polynomial with real coefficients. The *first partial derivative* f_x of f with respect to x is the derivative of f as a function of x when y is held constant. Likewise, the first partial derivative f_y of f with respect to y is the derivative of f as a function of y when x is held constant.

For example, suppose that

$$f(x, y) = x^4 + 7x^2y - 9xy^4 + 5y - 4.$$

Treating y as a constant and differentiating with respect to x gives

$$f_x = 4x^3 + 14xy - 9y^4. \tag{1}$$

Treating x as a constant and differentiating f with respect to y gives

$$f_y = 7x^2 - 36xy^3 + 5.$$

We obtain the values $f_x(a, b)$ and $f_y(a, b)$ of the partial derivatives at a point (a, b) of the Euclidean plane by setting $x = a$ and $y = b$ in the expressions for the partial derivatives. For instance, setting $x = 2$ and $y = -1$ in (1) gives

$$f_x(2, -1) = 4(2^3) + 14(2)(-1) - 9(-1)^4 = 32 - 28 - 9 = -5.$$

We can use partial derivatives to identify the singular points of a curve in the Euclidean plane.

Theorem 12.1

Let (a, b) be a point of the Euclidean plane on a curve $f(x, y) = 0$. Then f is nonsingular at (a, b) if and only if $f_x(a, b)$ and $f_y(a, b)$ are not both zero. In this case, the tangent to f at (a, b) is the line

$$f_x(a, b)(x - a) + f_y(a, b)(y - b) = 0.$$

Proof

By Theorem 4.10, we can write

$$f(x, y) = s(x - a) + t(y - b) + \sum e_{ij}(x - a)^i(y - b)^j,$$

where $i + j \geq 2$ for every term in the summation. Treating y as a constant and differentiating with respect to x gives

$$f_x = s + \sum ie_{ij}(x - a)^{i-1}(y - b)^j.$$

Because $(i - 1) + j \geq 1$ for every term in the summation, at least one of the exponents $i - 1$ or j is positive in each term. Thus, the summation has value zero when we set $x = a$ and $y = b$, and we have $f_x(a,b) = s$. Likewise, we have $f_y(a,b) = t$. Thus, the theorem follows from Theorem 4.10. □

If u and v each represent either x or y, we define the *second partial derivative* f_{uv} of a polynomial $f(x,y)$ with respect to u and v to be the result of differentiating f_u with respect to v. For example, differentiating the quantity in (1) with respect to x gives $f_{xx} = 12x^2 + 14y$, and differentiating (1) with respect to y gives $f_{xy} = 14x - 36y^3$.

Flexes are generalized inflection points. In single-variable calculus, the inflection points of a twice-differentiable function $y = f(x)$ occur only at points where $f''(x)$. The next result extends this to all algebraic curves in the Euclidean plane by using second partial derivatives to characterize flexes and singular points.

Theorem 12.2

Let (a,b) be a point of the Euclidean plane on a curve $f(x,y) = 0$. Then (a,b) is a flex or a singular point of f if and only if

$$f_{xx} f_y^2 + f_{yy} f_x^2 - 2 f_{xy} f_x f_y \tag{2}$$

takes the value zero at (a,b).

Proof

If (a,b) is a singular point of f, then f_x and f_y are both zero at (a,b) (by Theorem 12.1), and so the quantity in (2) is zero at (a,b). Thus, we can assume that f is nonsingular at (a,b). In this case, we must prove that (a,b) is a flex of f if and only if the quantity in (2) is zero at (a,b).

By Theorem 4.10, we can write

$$\begin{aligned} f(x,y) &= s(x-a) + t(y-b) + u(x-a)^2 + v(y-b)^2 \\ &\quad + w(x-a)(y-b) + \sum e_{ij}(x-a)^i (y-b)^j, \end{aligned} \tag{3}$$

where s and t are not both zero, f is tangent to the line

$$s(x-a) + t(y-b) = 0 \tag{4}$$

at (a,b), and $i + j \geq 3$ for every term in the sum. Because the quantity in (2) is symmetric in x and y, we can interchange x and y, if necessary, to ensure that $t \neq 0$.

Solving (4) for y gives

$$y = b - \frac{s}{t}(x-a).$$

Substituting the right-hand side of this equation for y in (3) is the same as substituting $-(s/t)(x-a)$ for $y - b$. This substitution takes the right-hand side of (3) to

$$\left(u+\frac{vs^2}{t^2}-\frac{ws}{t}\right)(x-a)^2+\sum e_{ij}\left(\frac{-s}{t}\right)^j(x-a)^{i+j}, \tag{5}$$

where $i+j \geq 3$ for every term in the summation.

Factoring $(x-a)^2$ out of (5) leaves $x-a$ as a factor of every term of the rightmost summation (since $i+j \geq 3$ in the summation). Thus, (5) has the form $(x-a)^2 h(x)$, where $h(x)$ is a polynomial such that

$$h(a)=u+\frac{vs^2}{t^2}-\frac{ws}{t}. \tag{6}$$

If $h(a) \neq 0$, then $f(x)$ intersects the tangent in (4) twice at (a,b) (by Theorem 4.2). If $h(a)=0$, but the polynomial in (5) is nonzero, then the largest power of $x-a$ we can factor out of (5) is at least three. Then f intersects its tangent at least three times at (a,b) (by Theorem 4.2). Finally, if the polynomial in (5) is zero, then the tangent at (a,b) is a factor of f (by Theorem 1.9(ii)), and f intersects its tangent infinitely many times at (a,b) (by Theorem 3.6(vi) and Definition 3.2). In short, f intersects its tangent at least three times at (a,b) if and only if $h(a)=0$. Thus, multiplying (6) by t^2 shows that (a,b) is a flex of f if and only if

$$ut^2+vs^2-wst=0. \tag{7}$$

The proof of Theorem 12.1 shows that

$$f_x(a,b)=s, \tag{8}$$

$$f_y(a,b)=t. \tag{9}$$

Differentiating (3) with respect to x gives

$$f_x(x,y)=s+2u(x-a)+w(y-b)+\sum ie_{ij}(x-a)^{i-1}(y-b)^j.$$

Differentiating this equation with respect to x and y gives

$$f_{xx}(x,y)=2u+\sum i(i-1)e_{ij}(x-a)^{i-2}(y-b)^j, \tag{10}$$

$$f_{xy}(x,y)=w+\sum ije_{ij}(x-a)^{i-1}(y-b)^{j-1}. \tag{11}$$

In each term of the rightmost summations in (10) and (11), either $x-a$ or $y-b$ has a positive exponent, since $i+j \geq 3$. Thus, substituting $x=a$ and $y=b$ in (10) and (11) makes the summations zero and shows that

$$f_{xx}(a,b)=2u, \tag{12}$$

$$f_{xy}(a,b)=w. \tag{13}$$

Interchanging x and y in (3) and (12) shows that

$$f_{yy}(a,b)=2v. \tag{14}$$

Equations (8), (9), and (12)–(14) show that (7) holds if and only if

$$\tfrac{1}{2}f_{xx}f_y^2 + \tfrac{1}{2}f_{yy}f_x^2 - f_{xy}f_xf_y$$

equals zero at (a, b). Multiplying this quantity by 2 gives the quantity in (2). Thus, the discussion before (7) shows that (a, b) is a flex of f if and only if the quantity in (2) is zero at (a, b). □

We translate Theorem 12.2 into homogeneous coordinates in order to extend it to the real projective plane. Let

$$F(x, y, z) = \sum e_{ijk}x^iy^jz^k \tag{15}$$

be a homogeneous polynomial of degree d with real coefficients. The homogeneity of F means that

$$i + j + k = d \tag{16}$$

for every term of F. The *partial derivatives* F_x, F_y, F_z of F are the results of differentiating F with respect to the given variable by holding the other two variables constant:

$$F_x = \sum ie_{ijk}x^{i-1}y^jz^k, \tag{17}$$

$$F_y = \sum je_{ijk}x^iy^{j-1}z^k, \tag{18}$$

$$F_z = \sum ke_{ijk}x^iy^jz^{k-1}. \tag{19}$$

Each partial derivative is either zero or a homogeneous polynomial of degree $d - 1$.

We define the partial derivatives of the zero polynomial to be zero. Equations (17)–(19) hold also in this case, where all the coefficients e_{ijk} are zero.

If F is given by (15), and if u and v are chosen from the variables x, y, z, the *second partial derivative* F_{uv} is the partial derivative of F_u with respect to v. For example, differentiating (17) with respect to y gives

$$F_{xy} = \sum ije_{ijk}x^{i-1}y^{j-1}z^k.$$

Differentiating (18) with respect to x gives the same result, and so we have $F_{xy} = F_{yx}$. By symmetry, we have the three equations

$$F_{xy} = F_{yx}, \qquad F_{xz} = F_{zx}, \qquad F_{yz} = F_{zy}. \tag{20}$$

We claim that the relation

$$xF_x + yF_y + zF_z = dF \tag{21}$$

holds for any homogeneous polynomial F of degree d with real coefficients. In fact, (17)–(19) show that the left-hand side of (21) equals

$$
\begin{aligned}
x \sum ie_{ijk}x^{i-1}y^jz^k &+ y \sum je_{ijk}x^iy^{j-1}z^k + z \sum ke_{ijk}x^iy^jz^{k-1} \\
&= \sum ie_{ijk}x^iy^jz^k + \sum je_{ijk}x^iy^jz^k + \sum ke_{ijk}x^iy^jz^k \\
&= \sum (i+j+k)e_{ijk}x^iy^jz^k = \sum de_{ijk}x^iy^jz^k \quad \text{(by (16))} \\
&= d \sum e_{ijk}x^iy^jz^k = dF \quad \text{(by (15))}.
\end{aligned}
$$

Equation (21) also holds when F is the zero polynomial—whose degree is undefined—and d is any integer, since both sides of the equation are zero.

We use determinants as a bookkeeping device to simplify algebra. We define the *determinant*

$$
\begin{vmatrix} a & b & c \\ d & e & f \\ g & h & i \end{vmatrix} = aei + bfg + cdh - afh - bdi - ceg, \tag{22}
$$

where a–i are real numbers or polynomials with real coefficients. We call a–i the *entries* of the determinant. The *rows* are the horizontal triples of entries on the left-hand side of (22): a, b, c is the first row, d, e, f is the second, and g, h, i is the third. The *columns* are the vertical triples of entries: a, d, g is the first column, b, e, h is the second, and c, f, i is the third. The right-hand side of (22) is more memorable if one notes that the first three terms are the products of the entries joined by lines in Figure 12.1, and the last three terms are the products of the entries joined by lines in Figure 12.2.

Theorem 12.3

(i) *For any values of* a–i, g'–i', *we have*

$$
\begin{vmatrix} a & b & c \\ d & e & f \\ g+g' & h+h' & i+i' \end{vmatrix} = \begin{vmatrix} a & b & c \\ d & e & f \\ g & h & i \end{vmatrix} + \begin{vmatrix} a & b & c \\ d & e & f \\ g' & h' & i' \end{vmatrix}. \tag{23}
$$

(ii) *If the third row of a determinant is a multiple of the first or second row, the value of the determinant is zero.*

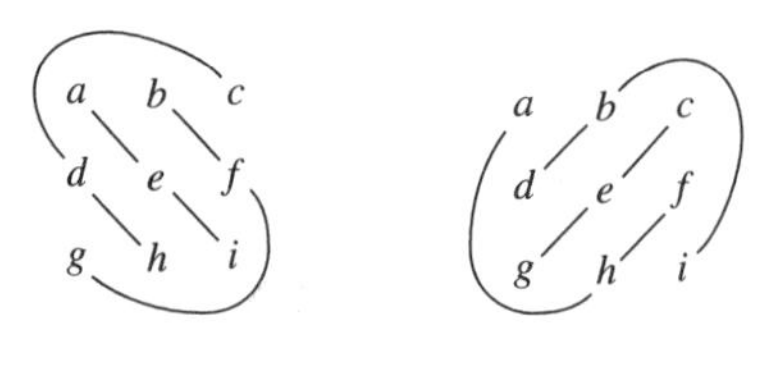

Figure 12.1

Figure 12.2

(iii) *The value of a determinant is unchanged by adding a multiple of the first or second row to the third.*

(iv) *When each entry in a determinant is multiplied by k, the value of the determinant is multiplied by k^3.*

Proof

(i) Equation (22) shows that the left-hand side of (23) equals

$$ae(i+i') + bf(g+g') + cd(h+h') - af(h+h') - bd(i+i') - ce(g+g').$$

Expanding each expression and collecting the terms without primed entries gives

$$aei + bfg + cdh - afh - bdi - ceg + aei' + bfg' + cdh' - afh' - bdi' - ceg'.$$

This is the right-hand side of (23).

(ii) Taking the third row in (22) to be k times the first gives

$$\begin{vmatrix} a & b & c \\ d & e & f \\ ka & kb & kc \end{vmatrix} = aekc + bfka + cdkb - afkb - bdkc - ceka,$$

and the terms on the right cancel to zero. Similarly, taking the third row in (22) to be k times the second gives

$$\begin{vmatrix} a & b & c \\ d & e & f \\ kd & ke & kf \end{vmatrix} = aekf + bfkd + cdke - afke - bdkf - cekd,$$

and again the terms on the right cancel to zero.

(iii) If we add k times the first row to the third, we obtain

$$\begin{vmatrix} a & b & c \\ d & e & f \\ g+ka & h+kb & i+kc \end{vmatrix} = \begin{vmatrix} a & b & c \\ d & e & f \\ g & h & i \end{vmatrix} + \begin{vmatrix} a & b & c \\ d & e & f \\ ka & kb & kc \end{vmatrix} = \begin{vmatrix} a & b & c \\ d & e & f \\ g & h & i \end{vmatrix},$$

by parts (i) and (ii). It follows in the same way that the determinant remains unchanged when we add a multiple of the second row to the third.

(iv) If we multiply each of the entries a–i in (22) by k, the right-hand side of the equation is multiplied by k^3. □

We claim that

$$\begin{vmatrix} a & b & c \\ d & e & f \\ g & h & i \end{vmatrix} = \begin{vmatrix} a & d & g \\ b & e & h \\ c & f & i \end{vmatrix} \tag{24}$$

for any values of a–i. Each of the rows $a, b, c; d, e, f; g, h, i$ on the left-hand side of (24) is a column of the right-hand side. Thus, (24) shows that *determinants are unaffected by interchanging rows and columns.* To verify (24), note that (22) shows that the right-hand side of (24) equals

$$aei + dhc + gbf - ahf - dbi - gec.$$

Since this equals the right-hand side of (22), equation (24) holds.

By Theorem 12.3(iii), the value of the left-hand side of (24) is unchanged if we add a multiple of the first or second row to the third. Looking then at the right-hand side of (24), we see that *the value of any determinant is unchanged if we add a multiple of the first or second column to the third.*

Let F be a homogeneous polynomial with real coefficients. We define the Hessian H of F by setting

$$H = \begin{vmatrix} F_{xx} & F_{xy} & F_{xz} \\ F_{xy} & F_{yy} & F_{yz} \\ F_{xz} & F_{yz} & F_{zz} \end{vmatrix}. \tag{25}$$

If F has degree d, each entry on the right-hand side of (25) is either zero or homogeneous of degree $d-2$. Thus, (22) and (25) imply that H is either zero or homogeneous of degree $3(d-2)$.

If we use (22) to expand (25), we obtain

$$H = F_{xx}F_{yy}F_{zz} + 2F_{xy}F_{yz}F_{xz} - F_{xx}F_{yz}^2 - F_{yy}F_{xz}^2 - F_{zz}F_{xy}^2. \tag{26}$$

Interchanging x and z on the right-hand side of this equation gives

$$F_{zz}F_{yy}F_{xx} + 2F_{zy}F_{yx}F_{zx} - F_{zz}F_{yx}^2 - F_{yy}F_{zx}^2 - F_{xx}F_{zy}.$$

This equals the right-hand side of (26) (by 20)). Likewise, interchanging y and z on the right-hand side of (26) gives

$$F_{xx}F_{zz}F_{yy} + 2F_{xz}F_{zy}F_{xy} - F_{xx}F_{zy}^2 - F_{zz}F_{xy}^2 - F_{yy}F_{xz}^2.$$

This also equals the right-hand side of (26) (by (20)). In short, *the Hessian remains unchanged when z is interchanged with x or y.*

We can now show that the Hessian is the analogue in homogeneous coordinates of the quantity in (2).

Theorem 12.4

Let P be a point on a curve F of degree greater than 1 in the real projective plane. Then P is a flex or a singular point of F if and only if P satisfies the equation $H = 0$.

Proof

At least one of the homogeneous coordinates of P is nonzero. We have seen that H remains unchanged if we interchange z with x or y. Such an interchange also preserves flexes and singular points (by Property 3.5). Thus, we can assume that the last coordinate of P is nonzero. Dividing by this coordinate, we can assume that $P = (a, b, 1)$ for real numbers a and b. Let F have degree $d > 1$.

To find the value of H at $(a, b, 1)$, we evaluate all the second partial derivatives on the right-hand side of (25) at $(a, b, 1)$. We add a times the first row and b times the second row to the third. This does not change the value of the determinant in (25), by Theorem 12.3(iii). The first entry in the third row becomes

$$aF_{xx} + bF_{xy} + F_{xz}$$

evaluated at $(a, b, 1)$. This quantity equals $(d-1)F_x$ at $(a, b, 1)$, as we see by replacing F with F_x and (x, y, z) with $(a, b, 1)$ in (21). (Note that we have replaced d in (21) with $d - 1$ because F_x has degree $d - 1$ or is 0.) Similarly, the second entry in the third row of the determinant in (25) becomes

$$\begin{aligned} aF_{xy} + bF_{yy} + F_{yz} &= aF_{yx} + bF_{yy} + F_{yz} \quad \text{(by (20))} \\ &= (d-1)F_y \quad \text{(by (21))} \end{aligned}$$

evaluated at $(a, b, 1)$. Likewise, the final entry in the third row of the determinant in (25) becomes

$$aF_{xz} + bF_{yz} + F_{zz} = aF_{zx} + bF_{zy} + F_{zz} = (d-1)F_z$$

evaluated at $(a, b, 1)$ (by (20) and (21)). In short, we have

$$H = \begin{vmatrix} F_{xx} & F_{xy} & F_{xz} \\ F_{xy} & F_{yy} & F_{yz} \\ (d-1)F_x & (d-1)F_y & (d-1)F_z \end{vmatrix} \tag{27}$$

at $(a, b, 1)$.

The value of this determinant is unchanged if we add a times the first column and b times the second column to the third (by the second paragraph after the proof of Theorem 12.3). As in the previous paragraph, the first two entries in the third column become $(d-1)F_x$ and $(d-1)F_y$ evaluated at $(a, b, 1)$. The last entry in the third column becomes

$$(d-1)(aF_x + bF_y + F_z) = (d-1)dF$$

evaluated at $(a, b, 1)$ (by (21)), and $F(a, b, 1)$ is zero (because P lies on F). Thus, we have

$$H = \begin{vmatrix} F_{xx} & F_{xy} & (d-1)F_x \\ F_{xy} & F_{yy} & (d-1)F_y \\ (d-1)F_x & (d-1)F_y & 0 \end{vmatrix}$$

at $(a, b, 1)$. Using (22) to evaluate this determinant gives

$$H = (d-1)^2(2F_{xy}F_xF_y - F_{xx}F_y^2 - F_{yy}F_x^2) \tag{28}$$

at $(a, b, 1)$. In short, we have used the algebraic properties of deteterminants to eliminate differentiation with respect to z from the Hessian so that we can interpret the Hessian in the Euclidean plane.

We set $f(x, y) = F(x, y, 1)$. We claim that setting $z = 1$ in F_x, F_y, F_{xx}, F_{xy}, F_{yy} gives f_x, f_y, f_{xx}, f_{xy}, f_{yy}. If so, the right-hand side of (28) is $-(d-1)^2$ times the quantity in (2). Since $d > 1$, this means that H equals 0 at $(a, b, 1)$ if and only if the quantity in (2) equals zero at (a, b).

To prove that

$$f_x(x, y) = F_x(x, y, 1), \tag{29}$$

we write

$$F(x, y, z) = \sum e_{ij}x^iy^jz^{d-i-j}. \tag{30}$$

Setting $z = 1$ gives

$$f(x, y) = \sum e_{ij}x^iy^j. \tag{31}$$

Differentiating (30) and (31) with respect to x gives

$$F_x(x, y, z) = \sum ie_{ij}x^{i-1}y^jz^{d-i-j} \tag{32}$$

and

$$f_x(x, y) = \sum ie_{ij}x^{i-1}y^j. \tag{33}$$

Setting $z = 1$ in the right-hand side of (32) gives the right-hand side of (33), and so (29) holds. Likewise, we have

$$f_y(x, y) = F_y(x, y, 1). \tag{34}$$

Similarly, setting $z = 1$ in

$$F_{xx} = \sum i(i-1)e_{ij}x^{i-2}y^jz^{d-i-j}$$

gives

$$\sum i(i-1)e_{ij}x^{i-2}y^j = f_{xx}.$$

By symmetry, setting $z = 1$ in F_{yy} gives f_{yy}. Finally, setting $z = 1$ in

$$F_{xy} = \sum ije_{ij}x^{i-1}y^{j-1}z^{d-i-j}$$

gives

$$\sum ije_{ij}x^{i-1}y^{j-1} = f_{xy}.$$

We have proved that $H = 0$ at $(a, b, 1)$ if and only if the quantity in (2) is zero at (a, b). This happens if and only if (a, b) is a flex or a singular point of f (by Theorem 12.2). We are done by the first paragraph of the proof. □

We show next that the Hessian H of a nonsingular, irreducible cubic C is not a scalar multiple of C over the real numbers. It follows from this, Bezout's Theorem 11.5, and Theorem 11.9 that C and H intersect nine times in the complex projective plane.

Theorem 12.5

Over the real numbers, if $C = 0$ is a nonsingular, irreducible cubic, its Hessian H is a homogeneous polynomial of degree 3 that is not a scalar multiple of C.

Proof

We claim first that $C = 0$ contains at least one point in the real projective plane. If it contains the point $(1, 0, 0)$, we are done. If not, x^3 has nonzero coefficient in C. Thus, setting y and z equal to 1, for example, gives a polynomial $C(x, 1, 1)$ of degree 3 in x. This polynomial has at least one root r in the real numbers (by the discussion accompanying (13) of Section 8), which gives a point $(r, 1, 1)$ of the real projective plane on the curve $C = 0$.

Second, we claim that C has a point that is not a flex. We have seen that C contains at least one point. If that point is not a flex, the claim holds. If it is a flex, then, by using Theorem 8.3 and replacing C with its image under a transformation, we can assume that C has the form

$$y^2 = x^3 + fx^2 + gx \tag{35}$$

for real numbers f and g such that $g \neq 0$. The y-axis $x = 0$ is tangent to C at the origin and contains the point $(0, 1, 0)$ that lies at infinity on C (by Theorem 8.2). It follows that C intersects its tangent at the origin exactly twice there (by Theorem 4.5 and Definition 4.9), and so the origin is a point of C that is not a flex.

Because it has a point that is not a flex, C is not a factor of its Hessian H (by Theorem 12.4). Taking $d = 3$ in the discussion after (25) shows that H is either zero or homogeneous of degree 3. Thus, H is homogeneous of degree 3 and is not a scalar multiple of C. □

The ordered triples (a, b, c) and (ta, tb, tc) represent the same point in the complex projective plane for every nonzero complex number t and every triple a, b, c of complex numbers not all zero. If we conjugate the coordinates, the triples still represent the same point, since the coordinates of $(\overline{ta}, \overline{tb}, \overline{tc})$ are the coordinates of $(\bar{a}, \bar{b}, \bar{c})$ multiplied by $\bar{t}$ (by (32) of Section 10). Moreover, the fact that a, b, c are not all zero implies that

$\bar{a}$, $\bar{b}$, $\bar{c}$ are not all zero. Thus, to any point $P = (a, b, c)$ in the complex projective plane, we can associate a point $\bar{P} = (\bar{a}, \bar{b}, \bar{c})$.

We defined the conjugate of a complex homogeneous polynomial before Theorem 11.7. Let $P = (a, b, c)$ be a point in the complex projective plane, and let F be a complex curve. The equations

$$F(a, b, c) = 0 \qquad \text{and} \qquad \bar{F}(\bar{a}, \bar{b}, \bar{c}) = 0$$

are equivalent because each is the conjugate of the other (by (30)–(32) of Section 10). Thus, P lies on F if and only if $\bar{P}$ lies on $\bar{F}$. This suggests our final intersection property, which states that conjugation preserves intersection multiplicities. We derive this property in Section 14.

Property 12.6
If F and G are complex curves and P is a point in the complex projective plane, then we have

$$I_P(F, G) = I_{\bar{P}}(\bar{F}, \bar{G}). \qquad \square$$

We can now prove that every nonsingular, irreducible cubic C has a flex in the real projective plane. Bezout's Theorem 11.5 and Theorem 11.9 imply that C intersects its Hessian nine times, counting multiplicities, in the complex projective plane. Because the nine intersections are interchanged in pairs by conjugation, the fact that nine is odd implies that at least one intersection is fixed by conjugation and lies in the real projective plane.

Theorem 12.7
Every nonsingular, irreducible cubic in the real projective plane has a flex.

Proof
Let C be a nonsingular, irreducible cubic with real coefficients, and let H be its Hessian. C and H have no common factors of positive degree over the real numbers, since C is irreducible and not a scalar multiple of H (by Theorem 12.5). Thus, C and H have no common factors of positive degree over the complex numbers (by Theorem 11.9). Therefore, since C and H are homogeneous polynomials of degree 3 (by Theorem 12.5), they intersect $3 \cdot 3 = 9$ times, counting multiplicities, in the complex projective plane (by Bezout's Theorem 11.5).

Assume that C and H intersect at a point Q in the complex projective plane such that $\bar{Q} \neq Q$. The map $P \to \bar{P}$ takes $\bar{Q}$ back to Q (by (30) of Section 10) and thereby pairs Q and $\bar{Q}$. Because C and H have real coefficients, they intersect the same number of times at Q and $\bar{Q}$ (by Property 12.6). Thus, the paired points Q and $\bar{Q}$ contribute an even number to the total number of times that C and H intersect in the complex projective plane.

The total number of times that C and H intersect in the complex projective plane is nine, an odd number (by the first paragraph of the proof). Together with the previous paragraph, this implies that C and H intersect at least once at a point R in the complex projective plane such that $\bar{R} = R$.

The homogeneous coordinates of R are complex numbers that are not all zero. By interchanging z with x or y, if necessary, we can assume that the last coordinate c of R is nonzero. Multiplying the coordinates of R by c^{-1} lets us write $R = (a, b, 1)$ for complex numbers a and b. The discussion after the proof of Theorem 12.5 shows that the relationship $\bar{R} = R$ holds for every choice of homogeneous coordinates for R. Thus, there is a complex number $t \neq 0$ such that multiplying the coordinates of $(a, b, 1)$ by t gives the coordinates of $(\bar{a}, \bar{b}, 1)$. Since both triples have last coordinate 1, we must have $t = 1$. Then we have $\bar{a} = a$ and $\bar{b} = b$, and so a and b are real numbers (by (29) of Section 10). Thus, $R = (a, b, 1)$ is a point of the real projective plane that lies on both C and H. Because C has no singular points, by assumption, R is a flex of C (by Theorem 12.4). □

We have now determined all irreducible cubics in the real projective plane. Combining Theorems 12.7 and 8.3 shows that *a cubic is nonsingular and irreducible if and only if it can be transformed into*

$$y^2 = x(x-1)(x-w) \qquad \textit{or} \qquad y^2 = x(x^2 + kx + 1)$$

for real numbers $w > 1$ *and* $-2 < k < 2$. Theorem 8.4 characterizes the singular, irreducible cubics.

Exercises

12.1. In each part of this exercise, we give a curve C with an unspecified constant term k, and we give a point R in the Euclidean plane. First, determine the value of k so that C contains R. Then use Theorem 12.1 to determine whether C is nonsingular at R and, if so, to write the tangent at R in one of the forms $y = mx + b$ or $x = a$.
(a) $x^3 + 6xy + y^2 = k$; $(4, -3)$.
(b) $2x^4 - x^2y^2 + y^3 = k$; $(2, 3)$.
(c) $36x - 3x^3y^2 - y^3 = k$; $(-1, 2)$.
(d) $x^4 - x^2y + 4y = k$; $(2, 1)$.
(e) $3x^4 + 6x^2y - y^3 = k$; $(1, -1)$.
(f) $x^2 - 6xy^2 - 36y = k$; $(3, -1)$.

12.2. Each part of Exercise 12.1 gives a curve with an unspecified constant term k. For what values of k does the curve have a singular point in the Euclidean plane?

12.3. Let F be a homogeneous polynomial of positive degree with real coefficients, and let (a, b, c) be a point in the real projective plane.

(a) Prove that (a, b, c) is a singular point of F if and only if F_x, F_y, and F_z are all zero at (a, b, c).

(b) If F is nonsingular at (a, b, c), prove that

$$F_x(a,b,c)x + F_y(a,b,c)y + F_z(a,b,c)z = 0$$

is the tangent at (a, b, c).

(See (21), (29), and (34) and Theorem 12.1.)

12.4. For any real number w, prove that the curve

$$(x + y + z)^3 = wxyz$$

in the real projective plane is nonsingular if and only if w is not equal to 0 or 27. (These cubics are discussed in Exercises 8.20, 8.31(c), and 12.9.)

12.5. For any real number t, prove that the curve

$$x^3 + y^3 + z^3 = txyz$$

in the real projective plane is nonsingular if and only if $t \neq 3$. (These cubics are discussed in Exercises 8.21–8.25, 8.31(d), 12.9, and 12.11.)

12.6. For any real number m, prove that the curve

$$x^2y + xy^2 + z^3 = mxyz$$

in the real projective plane is nonsingular if and only if $m \neq 3$. (These cubics are discussed in Exercises 8.26–8.28, 8.31(e), and 12.12.)

12.7. Let $f(x, y) = y^2 - q(x)$, where

$$q(x) = x^3 + ax^2 + bx + c \tag{36}$$

for real numbers a, b, c. Let $h(x, y)$ be the quantity in (2).

(a) Show that

$$h(x,y) = -4q''(x)q(x) + 2q'(x)^2 \tag{37}$$

at any point (x, y) on the graph of f in the Euclidean plane, where $q'(x)$ and $q''(x)$ are the first and second derivatives of q in the sense of single-variable calculus.

(b) Use (36) and ideas of single-variable calculus to prove that the right-hand side of (37) goes to $-\infty$ as x goes to $+\infty$.

(c) Let r be the largest root of $q(x)$. If $x - r$ is not a repeated factor of $q(x)$, prove that $h(r, 0) > 0$ and that $q(x) > 0$ for all $x > r$. (See (36) and (37) and Exercise 8.1.)

12.8. If the curve $f(x, y)$ in Exercise 12.7 is nonsingular, use parts (b) and (c) of that exercise and single-variable calculus to deduce that f has at least two flexes in the Euclidean plane. Conclude that every nonsingular, irreducible cubic in the real projective plane contains three collinear flexes.

(See Figures 8.3 and 8.4. Together with Exercise 11.14, this exercise shows that *every nonsingular irreducible cubic has exactly three flexes in the real projective plane.*)

12.9. (a) Prove that a cubic in the real projective plane is nonsingular and irreducible if and only if it can be transformed into one of the equations

$$x^3 + y^3 + z^3 = -6xyz,$$
$$(x + y + z)^3 = wxyz,$$

for a real number w not equal to 0 or 27. Use Exercises 8.18(d), 8.20(a), 8.25, 8.31(c) and (d), 12.4, 12.5, and 12.8. (See Figures 8.13, 8.14, and 8.17.)
(b) Prove that no cubic in the real projective plane can be transformed into more than one of the equations in part (a). Use Exercises 8.7, 8.20(b) and (c), 8.25, and 11.14.

12.10. Prove that a cubic in the real projective plane is nonsingular and irreducible if and only if it can be transformed into one of the equations

$$y(y - 3^{1/2}x)(y + 3^{1/2}x) = 1,$$
$$(y + 1)(y - 3^{1/2}x - 2)(y + 3^{1/2}x - 2) = u,$$

for a real number u not equal to 0 or 4. Prove that no cubic can be transformed into more than one of these equations. Use Exercises 8.19(b), 8.20(a), 8.25, and 12.9. (See Figures 8.10–8.12.)

12.11. Prove that a cubic in the real projective plane is nonsingular and irreducible if and only if it can be transformed into

$$x^3 + y^3 + z^3 = txyz$$

for a real number $t \neq 3$. Prove that no cubic can be transformed into more than one of these equations. Use Exercises 8.24 and 12.9. (See Figures 8.15–8.17.)

12.12. (a) Prove that a cubic in the real projective plane is nonsingular and irreducible if and only if it can be transformed into

$$x^2y + xy^2 + z^3 = mxyz$$

for a real number $m \neq 3$. Use Exercises 8.26(a), 8.27, 8.31(e), 12.6, 12.8, and Theorem 3.4. (See Figures 8.18 and 8.19.)
(b) Prove that no cubic can be transformed into more than one of the equations in part (a). Use Exercises 8.7, 8.26(a), 8.28, and 11.14.

12.13. In the notation of Exercise 12.7, set $p(x) = q(x)^{1/2}$. The graph of $y = p(x)$ is the top half of the graph of $y^2 = q(x)$, and $y = -p(x)$ is the bottom half (Figures 8.3–8.7).
(a) Deduce from Exercise 12.7(a) that the value of h at any point (x, y) on the graph of f is $-8p(x)^3p''(x)$.
(b) Conclude that the flexes of f in the Euclidean plane are exactly the points $(a, \pm p(a))$ such that $p''(a) = 0$.

(This shows that flexes generalize inflection points for cubics of the form $y^2 = q(x)$. Exercise 12.15 extends this result to all curves.)

12.14. Over the real numbers, let $f(x,y)$ be a polynomial.

(a) Let $r(x)$ be a differentiable function of x. Prove that

$$\frac{d}{dx}f(x,r(x)) = f_x(x,r(x)) + f_y(x,r(x))r'(x)$$

by writing $f(x,y) = \sum e_{ij}x^iy^j$ for real numbers e_{ij} and using single variable calculus.

(b) If $g(x)$ is a differentiable function such that the graph of $y = g(x)$ lies on the curve $f(x,y) = 0$, use part (a) to prove that

$$g'(x) = -\frac{f_x(x,g(x))}{f_y(x,g(x))} \tag{38}$$

for all values of x such that $f_y(x,g(x))$ is nonzero.

(c) Do part (b) by using Theorem 12.1 and the discussion after (12) of Section 4.

12.15. Over the real numbers, let $f(x,y)$ be a polynomial. Let $y = g(x)$ be a differentiable function whose graph lies on the curve $f(x,y) = 0$. Let a be a real number. Use Exercise 12.14 and Theorem 12.2 to prove that f has a flex at $(a,g(a))$ if and only if $g''(a) = 0$ and f is nonsingular at $(a,g(a))$. (This shows conclusively that flexes are generalizations of inflection points. One possible approach is to use Exercise 12.14(a) to differentiate both sides of (38) with respect to x. Use this result and Exercise 12.14(b) to prove that the quantity in (2) equals $-f_y^3g''(a)$ when all first and second partial derivatives of f are evaluated at $(a,g(a))$. When f is nonsingular at $(a,g(a))$, deduce from Exercise 12.14 and Theorem 12.1 that $g''(a)$ exists and $f_y(a,g(a))$ is nonzero. Why does the exercise follow?)

12.16. Over the real numbers, let $F = 0$ and $G = 0$ be curves of degrees m and n, and assume that F and G have no common factors of positive degree.

(a) Prove that the number of times, counting multiplicities, that F and G intersect in the real projective plane is $mn - 2k$ for an integer k with $0 \leq k \leq mn/2$.

(b) If F and G intersect at least $mn - 1$ times, counting multiplicities, in the real projective plane, prove that they intersect exactly mn times, counting multiplicities, in the real projective plane.

(c) If m and n are both odd, prove that F and G intersect at least once in the real projective plane.

(Part (b) is used in Exercises 15.20 and 15.23. This exercise can be done by adapting the proof of Theorem 12.7. Theorem 9.1 is a special case of part (b).)

12.17. Prove that a cubic in the real projective plane is irreducible if and only if it can be transformed into one of the equations

$$y^2 = x^3 + x + h, \qquad y^2 = x^3 - x + h,$$
$$y^2 = x^3 + 1, \qquad y^2 = x^3 - 1, \qquad y^2 = x^3,$$

as h varies over all real numbers. Prove that no cubic can be transformed into more than one of these equations. (See Exercises 8.7 and 8.9 and Theorems 8.1, 8.4, and 12.7.)

12.18. Prove that every singular, irreducible cubic in the real projective plane has one or three flexes. Use Exercise 12.7(a) and Theorems 8.4, 8.1(i), and 8.2. (See Figures 8.5–8.7.)

12.19. Over the real numbers, let

$$G = ax^2 + bxy + cy^2 + dxz + eyz + fz^2$$

be a homogeneous polynomial of degree 2. Let M be the determinant

$$\begin{vmatrix} 2a & b & d \\ b & 2c & e \\ d & e & 2f \end{vmatrix}.$$

(a) Use Theorems 12.4, 5.2, and 5.1 to prove that $M \neq 0$ if G is a conic and that $M = 0$ if G consists of two lines, one line doubled, or a single point.

(b) Deduce from part (a) and Theorem 5.1 that G is a conic if and only if $M \neq 0$ and G contains at least one point.

(c) Prove that $M \neq 0$ if G is the empty set. (*Hint*: One possible approach is to consider the graph of G in the complex projective plane.)

12.20. Does Theorem 12.4 remain true when F has degree 1? Justify your answer.

12.21. Let C be a nonsingular, irreducible cubic in the real projective plane. Define sextatic points as in Exercise 10.11. Prove that C has either three or nine sextatic points. (See Theorems 8.1–8.3 and 12.7 and Exercises 8.7, 10.11, 11.14, and 12.8.)

12.22. Let the notation be as in Exercise 12.21. Add points of C with respect to a flex O, as in Definition 9.3.

(a) Prove that C has a point of order 2 and a point of order 3 and that their sum has order 6. (See Exercise 9.2 and Theorems 8.2, 8.3, and 12.7.)

(b) If C has three sextatic points, prove that they are P, $3P$, $5P$, where P is point of C of order 6. Prove that the third intersections of the tangents at these points are the points $4P$, O, $2P$, respectively, and these are the three flexes of C. Illustrate this result with a figure. (See (a) and Exercises 9.2 and 10.12.)

(c) If C has nine sextatic points, prove that they are P, $3P$, $5P$, Q, and $kP + Q$ for $k = 1, \dots, 5$, where P has order 6, Q has order 2, and $Q \neq 3P$. (See (a), Exercises 10.12 and 11.14 and Theorems 8.2 and 8.3.)

Exercises 12.23–12.32 *use the discussion before Exercise* 11.11. The definitions of first and second partial derivatives in (17)–(19) and the subsequent discussion, the definition of Hessians in (25), and Theorems 12.1 and 12.4 all extend without change to the complex numbers.

12.23. Let $C(x, y, z) = y^2z - x^3 - fx^2z - gxz^2 - hz^3$ be the homogenization of (6) of Section 8 for complex numbers f, g, h.

(a) Use (25) to find the Hessian H of C and prove that it is nonzero.

(b) Prove that C and H are nonsingular and tangent to different lines at $(0, 1, 0)$.

(c) Use Theorems 4.11 and 8.1(i) and Exercise 8.7 (which all extend without change to the complex numbers) to prove that C and H intersect exactly once at every flex of C.

12.24. Prove that *every nonsingular complex cubic has exactly nine flexes in the complex projective plane, which lie by threes on twelve lines*. Prove that the flexes can be transformed into the points in (17) of Section 11 for $d = \omega$, where ω is given by (62) of Section 10. Use Theorems 3.4, 8.1, 12.4, and 12.5. (which all extend without change to the complex numbers), Bezout's Theorem 11.5, and Exercises 11.11–11.13, 11.15, and 12.23.

12.25. Let ω be given by (62) of Section 10. Prove that a complex cubic C contains the nine points

$$\begin{array}{ccc} (-1,1,0) & (1,0,-1) & (0,-1,1) \\ (\omega,1,0) & (1,0,\omega) & (0,\omega,1) \\ (-\omega^2,1,0) & (1,0,-\omega^2) & (0,-\omega^2,1) \end{array} \tag{39}$$

if and only if C is given by

$$ax^3 + ay^3 + az^3 + bxyz \tag{40}$$

for complex numbers a and b not both zero. (See Exercise 10.3.)

12.26. (a) Prove that the cubic in (40) has Hessian

$$rx^3 + ry^3 + rz^3 + sxyz$$

for $r = -6ab^2$ and $s = 216a^3 + 2b^3$.
(b) Prove that every complex cubic that contains the nine points in (39) has them as flexes. (See Theorems 12.1 and 12.4, part (a), and Exercise 12.25.)

12.27. (a) For any complex number t, prove that the complex cubic

$$x^3 + y^3 + z^3 = txyz \tag{41}$$

is nonsingular if and only if $t^3 \neq 27$.
(b) Conclude from part (a) and Exercises 12.24 and 12.26 that there is a transformation that maps the points in (39) to the points in (17) of Section 11 for $d = \omega$, where ω is given by (62) of Section 10.
(c) Use parts (a) and (b) and Exercises 12.24 and 12.25 to prove that a complex cubic is nonsingular if and only if it can be transformed into (41) for a complex number t such that $t^3 \neq 27$.

12.28. Prove that the four complex cubics given by (41) with $t^3 = 27$ and by the equation $xyz = 0$ are the four triples of lines containing the nine points of (39). (See Exercises 12.24–12.27.)

12.29. Prove that a complex cubic is nonsingular if and only if it can be transformed into

$$y^2 = x(x-1)(x-w)$$

for a complex number w other than 0 and 1. (Note that Theorems 8.1 and 8.2 extend without change to the complex numbers. Exercise 11.18 describes the latitude in the choice of w.)

12.30. (a) Let s, t, u, v be complex numbers such that $4s^3 + 27t^2$ and $4u^3 + 27v^2$ are both nonzero. Prove that we can transform the complex cubic

$$y^2 = x^3 + sx + t \tag{42}$$

into $y^2 = x^3 + ux + v$ if and only if

$$\frac{4s^3}{4s^3 + 27t^2} = \frac{4u^3}{4u^3 + 27v^2}.$$

(See Exercises 8.7 and 8.9 and Theorem 8.1(i), which all extend without change to the complex numbers.)

(b) Let C be a nonsingular complex cubic. Prove that we can associate a complex number j with C such that we can transform C into (42) for complex numbers s and t if and only if

$$\frac{4s^3}{4s^3 + 27t^2} = j.$$

(See Exercises 9.17 and 8.9, which extend without change to the complex numbers, Exercise 12.29, and part (a).)

12.31. Let C be a nonsingular complex cubic, and let j be the complex number associated with C in Exercise 12.30(b). For any complex numbers a, b, c, let

$$q(x) = x^3 + ax^2 + bx + c,$$

and define the *discriminant* Δ of C by (49) of Section 9. Why is $\Delta \neq 0$? Prove that C can be transformed into $y^2 = q(x)$ if and only if

$$\frac{4(3b - a^2)^3}{27\Delta} = j.$$

12.32. (a) For any integer $n \geq 4$, prove that there is a homogeneous polynomial of degree n that is irreducible over both the real and the complex numbers, that determines a nonsingular curve in the real projective plane, and that determines a complex curve that has a singular point in the complex projective plane.

(b) Can part (a) be done when $n \leq 3$? Justify your answer.

IV CHAPTER Intersection Properties

Introduction and History

Introduction

We proved many of the theorems in previous chapters by computing intersection multiplicities in two different ways and setting the results equal. Among the theorems we proved in this way are Pascal's Theorem 6.2 and its variant Theorem 6.3, Pappus' Theorem 6.5, and Theorem 9.7 on the associativity of addition on a cubic. The intersection properties guarantee that different ways of computing an intersection multiplicity give the same result. In Sections 13 and 14, we determine the multiplicity of every intersection, and we derive the intersection properties. This completes the proofs of the theorems in previous chapters.

We saw in Theorem 5.10 that a conic is uniquely determined by five points, no three of which are collinear. In Section 15, we ask what sets of points determine a unique cubic. The answer depends on systems of linear equations, which also arise in Section 14.

The general cubic has ten terms, as in (1) and (2) of Section 8. Because we can multiply all the coefficients by a nonzero number without changing the cubic, there are nine "degrees of freedom" in choosing the coefficients. This suggests that, in general, nine points determine a unique cubic. We must ensure, however, that there is no redundancy in the conditions imposed by the nine points. For instance, nine points lie on infinitely many cubics if every cubic through eight of the points also contains the ninth. We explore this possibility in Section 15.

History

Analytic geometers worked with multiple intersections informally until the late 1800s. Formal treatments of intersection multiplicities arose through work on Bezout's Theorem, singular points, and higher-dimensional algebraic geometry.

The most natural way to analyze the intersections of two curves $f(x, y) = 0$ and $g(x, y) = 0$ is the following technique called *elimination*. As in Example 1.13 and the proof of Bezout's Theorem 11.5, we eliminate the largest power of y in one of the polynomials f and g by adding suitable multiples of f and g together. We continue in this way until we eliminate all powers of y. This gives a polynomial $r(x)$ in x alone such that

$$r(x) = f(x, y)u(x, y) + g(x, y)v(x, y)$$

for polynomials u and v. If $r(x)$ has minimal degree, it is called the *resultant* of f and g. If f and g do not intersect at infinity or at two points in the complex affine plane with the same x-coordinate, the roots of $r(x)$ are the x-coordinates of the intersections of f and g, and the multiplicity of each root is the multiplicity of the coresponding intersection. In this case, Bezout's Theorem follows from the fact that the degree of $r(x)$ is the product of the degrees of f and g.

Elimination was discovered by Chinese mathematicians in the twelfth century. Newton claimed in 1665 that curves of degree m and n intersect in mn points when imaginary intersections are included. Colin Maclaurin explored this assertion and deduced in 1720 that an irreducible curve of degree n has at most $(n-1)(n-2)/2$ singular points. In 1764, Etienne Bezout and Leonhard Euler independently developed explicit elimination algorithms and deduced that the product of the degrees of two polynomials of appropriate form is the degree of their resultant. In 1840, James Sylvester developed the modern expression for the resultant as a determinant. Complete proofs of Bezout's Theorem appeared in the late 1800s, when resultants were combined with homogeneous coordinates.

A second approach to intersection multiplicities and Bezout's Theorem is based on "fractional power series." For any complex number r, a fractional power series about $x = r$ is an expression of the form

$$p((x-r)^{1/m}), \tag{1}$$

where m is a positive integer, $(x-r)^{1/m}$ is one of the m complex numbers whose mth power is $x - r$, and the quantity in (1) results from substituting $(x-r)^{1/m}$ for t in a power series

$$p(t) = c_0 + c_1 t + c_2 t^2 + \cdots$$

with complex coefficients c_i. If r is a complex number and $f(x, y)$ is a polynomial of degree d that has a nonzero y^d term, then $f(x,y)$ is a product of factors of the form

$$y - q(x), \tag{2}$$

where $q(x)$ is a fractional power series about $x = r$, as in (1). These factors are the sheets of the Riemann surface of f above values of x near r.

Newton introduced fractional power series to analyze the behavior or a curve near a singular point. He developed an algebraic algorithm for factoring a polynomial $f(x,y)$ into expressions of the form (2). In 1850, Victor Puiseux used complex analysis to prove that the points on a curve near a singular point are given by a finite number of expressions of the form (2). Georges Halphen showed in the 1870s how to use fractional power series to determine the intersection multiplicity of two curves at a point, and Bezout's Theorem follows from this.

We have so far described two ways to assign intersection multiplicities—resultants and fractional power series. We use a third method in Section 13. This method is based on abstract algebra, although we avoid abstract algebra in our presentation by working directly with polynomials. All three ways of assigning multiplicities give the same values.

Abstract algebra entered into the study of algebraic geometry in several ways. One was the widespread work on invariant theory in the late 1800s, which we mentioned at the end of the historical comments for Chapter I.

Abstract algebra also developed a role in algebraic geometry through the work of Richard Dedekind and Heinrich Weber in 1882. They sought to derive many of Riemann's results algebraically instead of analytically. Riemann had studied algebraic functions—the functions on a Riemann surface $f(x,y) = 0$ that are induced by rational functions of the coordinates x and y. Dedekind and Weber developed analogies between algebraic functions and algebraic numbers—roots of polynomials with rational coefficients. They took the ideas and structures of algebraic number theory and extended them to fields of algebraic functions.

Abstract algebra became linked to algebraic geometry in a third way through Riemann's introduction of "birational transformations." These are coordinate changes such that each new coordinate is a rational function of the old coordinates, and vice versa. Studying birational transformations leads again to studying algebraic function fields and their algebraic structures. Birational geometry was developed about 1870 by a school of geometers who sought to take Riemann's work, which was based on complex analysis, and reinterpret it in terms of the traditional study of algebraic curves via projective geometry. Among the most notable of these geometers were Alfred Clebsch, Max Noether (the father of Emmy Noether), and Luigi Cremona. Subsequent geometers have emphasized birational transformations instead of the (linear) transformations we introduced in Section 3 because birational transformations provide far more freedom. Since they alter curves substantially, birational transformations can be used to simplify, and thereby analyze, singular points. This makes it vital to find properties of curves that are preserved by birational transformations. One such property is the "genus" of an irreducible complex curve C: the genus is the nonnegative integer g such that C arises topologically from a sphere with g handles by identifying finitely many points together.

Max Noether proved a far-reaching generalization of Theorems 6.1 and 6.4 on "peeling off" conics and lines. Noether's "Fundamental Theorem" gives necessary and sufficient conditions in terms of the intersections of complex curves F, G, and H for there to exist homogeneous polynomials W and V such that

$$H = FW + GV.$$

By the intersection properties, this equation implies that

$$I_P(G, H) = I_P(G, F) + I_P(G, W)$$

for every point P in the complex projective plane, which means that we can "peel off" the intersections of G and F from the intersections of G and H. Exercise 14.12 shows that, if two curves G and H of degree n intersect a nonsingular, irreducible curve F of degree m in the same mn points, listed by multiplicity, then we can "peel off" these points from the intersections of G and H. This is essentially an early forerunner of Noether's Theorem due to Joseph-Diez Gergonne. Gergonne championed analytic over synthetic geometry in the 1820s, building upon the abridged notation introduced by Lamé and Bobillier.

Algebraic geometry became profoundly linked to both abstract algebra and algebraic topology in the late 1800s and early 1900s when analytic geometers sought to extend their studies to surfaces and spaces of all dimensions. They used tools from abstract algebra and algebraic topology to handle the increasingly general subject matter. Algebraic geometers strove to unify their work in the middle of the twentieth century by introducing new structures such as abstract varieties, sheaves, and schemes.

Colin Maclaurin raised the following issue in 1720. On the one hand, requiring a curve to contain a particular point imposes a linear condition on the coefficients of the curve. A general curve of degree n has $\binom{n+2}{2}$ coefficients. Because we can multiply the coefficients by a nonzero number without changing the curve, a curve of degree n has

$$\binom{n+2}{2} - 1 = \frac{(n+2)(n+1)}{2} - 1 = \frac{n(n+3)}{2}$$

"degrees of freedom." Accordingly, we expect that a curve of degree n

is uniquely determined by $n(n+3)/2$ of its points. On the other hand, Bezout's Theorem shows that two complex curves of degree n without multiple intersections intersect at n^2 points. The last two sentences appear to conflict because

$$n^2 \geq n(n+3)/2$$

for $n \geq 3$.

This apparent difficulty was explored by Leonhard Euler in 1748 and by Gabriel Cramer in 1750, and it is now known as "Cramer's paradox." Euler and Cramer suggested that there may be redundancies among the conditions that points impose on curves. We examine this idea for cubics in Section 15. Taking $n = 3$ in the first part of the previous paragraph shows that a cubic is uniquely determined by $n(n+3)/2 = 9$ conditions, provided that the conditions are not redundant. In fact, the $n^2 = 9$ points where two cubics intersect impose redundant conditions—otherwise, the points would not lie on two cubics. If two cubics intersect in nine points, any cubic through eight of the points necessarily contains the ninth, as Exercise 15.7 shows. If nine points are to determine a unique cubic, we show in Section 15 that we can choose eight of the points quite generally, but we must ensure that the ninth point does not lie on every cubic through the first eight.

§13. Independence and Intersections

We have not yet established the Intersection Properties 1.1–1.6, 3.1, 3.5, and 12.6. We do so in this section and the next.

We begin the section by introducing the idea of polynomials being "independent" with respect to two curves f and g at the origin O. We use this idea to determine the value of $I_O(f,g)$, the number of times that f and g intersect at O. Properties 1.1–1.5 follows almost immediately.

Property 1.6 states that

$$I_O(f,gh) = I_O(f,g) + I_O(f,h) \tag{1}$$

for any polynomials f, g, h. We devote much of this section to proving that the left-hand side of (1) is greater than or equal to the right-hand side. The proof requires the following fact: if a polynomial $p(x,y)$ is a factor of the product of polynomials $u(x,y)$ and $v(x,y)$, and if p has no factors of positive degree in common with v, then it is a factor of u. We deduce this fact from the result that two complex curves without a common factor of positive degree intersect at only finitely many points.

We end this section by using the idea of independence to prove a result about the algebra of polynomials that we need to complete the proof of (1) in the next section.

We work over the complex numbers throughout this section and the next. In particular, we work over the complex numbers in determining $I_O(f,g)$ and $I_P(F,G)$ even when the polynomials f, g, F, G have real coefficients and the point P has real coordinates. As we observed before Theorem 11.1, this ensures that polynomials with real coefficients intersect the same number of times at points with real coordinates regardless of whether we think of the polynomials as curves in the real of the complex projective plane. Once we derive the Intersection Properties 1.1–1.6, 3.1, and 3.5 over the complex numbers, they hold automatically over the real numbers.

Throughout this section and the next, we let $f(x,y), g(x,y)$, and $h(x,y)$ be polynomials, and we let O be the origin $(0,0)$. Our method of assigning intersection multiplicities is based on the idea of independent polynomials.

Definition 13.1
Let n be a positive integer, and let $q_1(x,y), \ldots, q_n(x,y)$ be polynomials. We call $q_1, \ldots, q_n$ *dependent* with respect to f and g at O if there are polynomials $r(x,y), s(x,y), t(x,y)$, and complex numbers $b_1, \ldots, b_n$ such that

$$r(b_1 q_1 + \cdots + b_n q_n) = sf + tg, \tag{2}$$

r is nonzero at the origin, and $b_1, \ldots, b_n$ are not all zero. □

We can think of (2) as saying that $b_1 q_1 + \cdots + b_n q_n$ becomes zero under multiplication by a polynomial r nonzero at the origin and subtraction of polynomial multiples of f and g. Definition 13.1 says that $q_1, \ldots, q_n$ are dependent with respect to f and g at O if there is a sum of scalar multiples of the q's that becomes zero in this way and has scalars that are not all zero.

We call $q_1, \ldots, q_n$ *independent* with respect to f and g at O if they are not dependent. In other words, $q_1, \ldots, q_n$ are independent if they do not satisfy any equation of the form (2) where r is nonzero at the origin and $b_1, \ldots, b_n$ are not all zero.

We agree, by convention, that the empty set of polynomials is independent with respect to f and g at O. This ensures that that there are n polynomials independent with respect to f and g at O when $n = 0$.

We use independent polynomials to determine intersection multiplicities at the origin.

Definition 13.2
We define $I_O(f,g)$, the *intersection multiplicity* of f and g at the origin O, as follows. We set $I_O(f,g) = \infty$ if, for every positive integer c, there are at least c polynomials independent with respect to f and g at O. Otherwise, we set $I_O(f,g) = e$, where e is the largest integer such that there are e

polynomials independent with respect to the intersection of f and g at O. □

Definition 13.2 states roughly that $I_O(f,g)$ is the largest number of polynomials independent with respect to f and g at O.

Let $q_1,\ldots,q_n$ be dependent with respect to f and g at O, and let $p_1,\ldots,p_m$ be additional polynomials in x and y. We claim that the polynomials

$$q_1,\ldots,q_n, \quad p_1,\ldots,p_m, \tag{3}$$

are also dependent with respect to f and g at O. In fact, since $q_1,\ldots,q_n$ are dependent, there are polynomials r,s,t and complex numbers $b_1,\ldots,b_n$ such that (2) holds, r is nonzero at the origin, and $b_1,\ldots,b_n$ are not all zero. Adding in zero times each of the polynomials $p_1,\ldots,p_m$ gives

$$r(b_1q_1+\cdots+b_nq_n+0p_1+\cdots+0p_m)=sf+tg.$$

By Definition 13.1, this shows that the polynomials in (3) are dependent, as claimed. Thus, *any finite enlargement of a set of dependent polynomials is also dependent.*

In particular, suppose that there is a positive integer v such that every set of v polynomials is dependent with respect to f and g at O. The same holds for every finite set of more than v polynomials, by the previous paragraph, since any such set contains a subset of v polynomials. Thus, $I_O(f,g)$ is less than v (by Definition 13.2).

Why should we think that Definition 13.2 gives a reasonable value for $I_O(f,g)$? The easiest answer comes from observing how naturally Definition 13.2 leads to the Intersection Properties 1.1–1.5.

There are always n polynomials independent with respect to f and g at O for $n=0$. Accordingly, Definition 13.2 shows that $I_O(f,g)$ is a nonnegative integer or ∞. This gives Property 1.1. Since f and g play symmetric roles in Definitions 13.1 and 13.2, the value of $I_O(f,g)$ is unaffected by interchanging f and g, and so Property 1.2 holds.

We claim that $I_O(f,g)$ is zero if f or g is nonzero at the origin. By symmetry, we can asume that $f(0,0)\neq 0$. For any polynomial q, we can write

$$fq=qf+0g, \tag{4}$$

where 0 is the zero polynomial. This shows that (2) holds with $r=f, n=1, b_1=1, s=q$, and $t=0$. Since f is nonzero at the origin, this shows that every polynomial q is dependent with respect to f and g at O. Taking $v=1$ in the third-to-last paragraph shows that $I_O(f,g)$ is zero, as claimed.

On the other hand, assume that f and g are both zero at the origin. We claim that the constant polynomial 1 is independent with respect to f

and g at O. In fact, suppose that the equation

$$r(b1) = sf + tg \tag{5}$$

holds for polynomials r, s, t and a complex number b. Since the right-hand side of this equation is zero at the origin (because both f and g have this property), so is the left. This means that either r is zero at the origin or else $b = 0$ (by (24) of Section 10). Thus, the constant polynomial 1 is independent with respect to f and g at O (by Definition 13.1), as claimed. Accordingly, $I_O(f,g)$ is at least one (by Definition 13.2).

The two preceding paragraphs show that $I_O(f,g) \geq 1$ if and only if f and g are both zero at the origin. This establishes Property 1.3.

We can now see one reason why the factor of r is included in (2): it ensures that $I_O(f,g) = 0$ if f or g does not contain the origin (by the discussion accompanying (4)). The requirement that $r(0,0) \neq 0$ ensures that $I_O(f,g) \geq 1$ if f and g both contain the origin (by the discussion accompanying (5)).

We turn next to Property 1.4, which states that $I_O(x,y) = 1$. Let q_1 and q_2 be polynomials in x and y, and let c_1 and c_2 be their respective constant terms. The polynomial $c_2q_1 - c_1q_2$ has constant term zero, and so we can factor x or y out of every term of this polynomial. Thus, we can write

$$c_2q_1 - c_1q_2 = sx + ty \tag{6}$$

for polynomials s and t. If c_1 and c_2 are not both zero, (6) shows that q_1 and q_2 are dependent with respect to x and y at the origin (by taking $r = 1$ in Definition 13.1). On the other hand, if the constant term c_1 of q_1 is zero, we can factor x or y out of every term of q_1 and write

$$1q_1 + 0q_2 = q_1 = sx + ty$$

for polynomials s and t, which shows in this case as well that q_1 and q_2 are dependent. In short, any two polynomials q_1 and q_2 are dependent with respect to x and y at the origin. It follows, by taking $v = 2$ in the third paragraph after Definition 13.2, that $I_O(x,y) \leq 1$. Since x and y are both zero at the origin, we have $I_O(x,y) \geq 1$ (by Property 1.3, which we have already established). Thus, we have $I_O(x,y) = 1$, and Property 1.4 holds.

We turn next to Property 1.5, which states that

$$I_O(f,g) = I_O(f,g+fh). \tag{7}$$

We show first that

$$I_O(f,g) \geq I_O(f,g+fh). \tag{8}$$

Since this inequality holds automatically when the left-hand side is infinite, we can assume that $I_O(f,g)$ is finite. Let n be any integer greater than $I_O(f,g)$, and let $q_1, \ldots, q_n$ be any n polynomials in x and y. By

Definition 13.2, $q_1, \ldots, q_n$ are dependent with respect to f and g at O. By Definition 13.1, there are polynomials r, s, t, and complex numbers $b_1, \ldots, b_n$ such that the quantities

$$r(b_1 q_1 + \cdots + b_n q_n) \tag{9}$$

and

$$sf + tg \tag{10}$$

are equal, r is nonzero at the origin, and $b_1, \ldots, b_n$ are not all zero. Because we can rewrite the quantity in (10) as

$$(s - th)f + t(g + fh), \tag{11}$$

the equality of the quantities in (9) and (11) shows that $q_1, \ldots, q_n$ are also dependent with respect to f and $g + fh$ at the origin. Since this holds for any n polynomials $q_1, \ldots, q_n$, where n is any positive integer greater than $I_O(f,g)$, inequality (8) follows from Definition 13.2. If we replace g with $g + fh$ and h with $-h$ in inequality (8), we see that

$$I_O(f, g + fh) \geq I_O(f, (g + fh) + f(-h)) = I_O(f, g).$$

Together with inequality (8), this establishes (7) and proves Property 1.5.

Our next goal is to prove that the left-hand side of (1) is greater than or equal to the right-hand side. We need a result about factoring polynomials that follows from the fact that two complex curves without a common factor of positive degree intersect at only finitely many different points. The latter fact follows from Bezout's Theorem 11.5, but we cannot deduce it in that way because we are now deriving the intersection properties we used to prove Bezout's Theorem. Instead, we adapt the proof of Bezout's Theorem by counting distinct points of intersection instead of intersection multiplicities. We start with the following analogue of Theorem 11.1 that does not depend on the intersection properties:

Theorem 13.3

In the complex projective plane, let $L = 0$ be a line, and let $G = 0$ be a complex curve. If L is not a factor of G, then $L = 0$ intersects $G = 0$ at only finitely many different points in the complex projective plane.

Proof

We can transform the line $L = 0$ to the x-axis $y = 0$ (by the third paragraph before Theorem 11.1). Since transformations preserve factorizations (as noted before Theorem 4.5), we can assume that $L = 0$ is the line $y = 0$. Since G does not have y as a factor, it has nonzero terms in which y does not appear. By collecting these terms into a homogeneous polynomial $H(x, z)$ in x and z and factoring y out of the other terms, we can write

$$G(x, y, z) = yQ(x, y, z) + H(x, z) \tag{12}$$

for a homogeneous polynomial Q. Setting $z = 1$ in $H(x, z)$ gives a nonzero polynomial $H(x, 1)$ in x, which factors as

$$H(x, 1) = r(x - a_1)^{s_1} \cdots (x - a_k)^{s_k} \tag{13}$$

for complex numbers $r \neq 0, a_1, \ldots, a_k$, and positive integers $s_1, \ldots, s_k$, by the Fundamental Theorem of Algebra 10.1. Equation (12) implies that the points of the complex affine plane that lie on both $y = 0$ and $G = 0$ are the points $(b, 0, 1)$ for complex numbers b such that $H(b, 1) = 0$. This means that b is one of the numbers $a_1, \ldots, a_k$ (by (13) above and (24) of Section 10). Thus, $y = 0$ and $G = 0$ intersect at only finitely many different points of the complex affine plane. Since $y = 0$ has a unique point $(1, 0, 0)$ at infinity, $y = 0$ and $G = 0$ intersect at only finitely many different points in the complex projective plane. □

Let $F = 0, G = 0$, and $H = 0$ be complex curves. We repeatedly use the observation that *$F = 0$ intersects $GH = 0$ at only finitely many different points if and only if it intersects both $G = 0$ and $H = 0$ at only finitely many different points.* This follows from the fact that a point of the complex projective plane satisfies the equation $GH = 0$ if and only if it satisfies $G = 0$ or $H = 0$ (by (24) of Section 10).

Let $H(x, z)$ be a homogeneous polynomial in x and z that has no factor of positive degree in common with a homogeneous polynomial $F(x, y, z)$. Because $H(x, z)$ either is a nonzero constant or factors as a product of lines (as discussed before Theorem 11.3), Theorem 13.3 and the previous paragraph imply that F and H intersect at only finitely many different points of the complex projective plane. Applying the previous paragraph again shows that, for any homogeneous polynomial $G(x, y, z)$, F and GH intersect at finitely many different points of the complex projective plane if and only if F and G intersect at finitely many different points.

We can use Theorems 11.2 and 11.3 here because we proved them without referring to intersection multiplicities. We define the degree in y of a homogeneous polynomial as before Theorem 11.4. We have the following analogue of Theorem 11.4 that does not depend on the intersection properties:

Theorem 13.4
Let $F = 0$ and $G = 0$ be complex curves of respective degrees s and t in y. Assume that $s \geq t > 0$ and that F and G have no common factors of positive degree. Then there are complex curves $F_1 = 0$ and $G_1 = 0$ such that F_1 and G_1 have no common factors of positive degree, the degree of F_1 in y is less than s, the degree of G_1 in y is t, and F and G intersect at only finitely many different points of the complex projective plane if and only if F_1 and G_1 do so.

Proof
As in the first paragraph of the proof of Theorem 11.4, we can write $G = HG_1$ for a homogeneous polynomial $H(x, z)$ in which y does not appear and a homogeneous polynomial $G_1(x, y, z)$ of degree t in y that has no factors of positive degree without y. Let $P(x, z)$ be the coefficient of y^s in F and let $Q(x, z)$ be the coefficient of y^t in G_1. The second paragraph of the proof of Theorem 11.4 shows that

$$F_1 = QF - Py^{s-t}G_1 \tag{14}$$

is a homogeneous polynomial that has degree less than s in y and has no factors of positive degree in common with G_1. F and G intersect at only finitely many different points of the complex projective plane if and only if F and G_1 do so, and this occurs if and only if QF and G_1 do so (by the discussion before the theorem). The theorem follows because (14) implies that QF and G_1 intersect at the same points as do F_1 and G_1. □

We can now prove the following weak form of Bezout's Theorem:

Theorem 13.5
Two complex curves that have no common factors of positive degree intersect at only finitely many points of the complex projective plane.

Proof
Let F and G be the complex curves. If they both have positive degree in y, we can use Theorem 13.4 to reduce one of these degrees. We repeat this step until one of the complex curves has degree 0 in y, and we are done by the second-to-last paragraph before Theorem 13.4. □

We use the previous theorem to deduce the following result on multiplication of polynomials, which lets us prove that the left-hand side of (1) is greater than or equal to the right:

Theorem 13.6
Let P, U, V be homogeneous polynomials in x, y, z. If P is a factor of UV and has no factors of positive degree in common with V, then P is a factor of U.

Proof
If P is a nonzero constant, it is automatically a factor of U. Thus, we can assume that P has positive degree (since homogeneous polynomials are nonzero, by definition). Then the complex curve P contains infinitely many different points (by Theorem 11.6, whose proof does not depend on the intersection properties), and so P intersects UV at infinitely many different points of the complex projective plane (since P is a factor of UV). P intersects V at only finitely many different points of the complex projective plane (by Theorem 13.5 and the assumption that P and V have

no common factors of positive degree). Combining the last two sentences with the paragraph after the proof of Theorem 13.3 shows that P and U intersect at infinitely many different points of the complex projective plane. Thus, P and U have a common factor R of positive degree (by Theorem 13.5). We write

$$P = RP^* \qquad \text{and} \qquad U = RU^* \tag{15}$$

for homogeneous polynomials P^* and U^*.

Because P is a factor of UV, we can write

$$UV = PQ \tag{16}$$

for a homogeneous polynomial Q. Substituting the expressions in (15) for P and U in (16) gives

$$RU^*V = RP^*Q.$$

We can cancel the factors of R in this equation (by Theorem 11.2(ii)), and so we have

$$U^*V = P^*Q. \tag{17}$$

Equation (17) shows that P^* is a factor of U^*V. Since P has no factors of positive degree in common with V, neither does P^*. If P^* is a factor of U^*, then P is a factor of U (by (15)). Thus, we can replace P and U with P^* and U^*. This reduces the degree of P (by (15) and Theorem 11.2(i), since R has positive degree). We repeat this process of reducing the degree of P until P becomes constant, and we are done by the first sentence of the proof. □

Theorems 11.2(ii) and 13.6 on homogeneous polynomials in x, y, z imply the following result on polynomials in x and y:

Theorem 13.7
Let f, g, h, p, u, v be polynomials in x and y.

(i) *If $fg = fh$ and f is nonzero, then g equals h.*
(ii) *If p is a factor of uv and has no factors of positive degree in common with v, then p is a factor of u.*

Proof
(i) The assumption that $fg = fh$ implies that $f(g-h) = 0$. If $g - h$ were a nonzero polynomial, the homogenizations of f and $g - h$ would have product zero, which would contradict Theorem 11.2(i). Thus, $g - h$ is zero, and so g equals h.

(ii) If p is zero, then so is u (since p is a factor of uv but has no factors of positive degree in common with v). If u is zero, it automatically has p as a factor. If v is zero, then p is a nonzero constant (since it has no factors of positive degree in common with v), and so it is a factor of u. Thus, we can assume that p, u, v are all nonzero.

Let P, U, V be the homogenizations of p, u, v. P is a factor of UV, and it has no factors of positive degree in common with V (by the corresponding assumptions about p, u, v). Then P is factor of U (by Theorem 13.6), and so p is a factor of u. □

Readers familiar with polynomial algebra can confirm that Theorem 13.7(ii) is essentially the result that every polynomial of positive degree in two variables factors in a unique way as a product of irreducible polynomials. Instead of proving this algebraically, we have deduced it from the geometric Theorems 11.6 and 13.5.

One of the ways we use Theorem 13.7 is to deduce that two complex curves intersect infinitely many times at the origin if they have a common factor containing the origin. Over the complex numbers, a polynomial in x and y of positive degree is called *irreducible* if it does not factor as a product of two polynomials of smaller degrees.

Theorem 13.8

If f and g are complex curves that have a common factor containing the origin O, then $I_O(f,g)$ is ∞.

Proof

Let $w(x,y)$ be the common factor of f and g. We claim that w has an irreducible factor $p(x,y)$ that contains the origin. If w is zero, we can take p to be x. If w is nonzero, then, since it contains the origin, it is not a constant, and we can write it as a product of irreducible polynomials by factoring it as far as possible. One of these irreducible factors contains the origin (by (24) of Section 10), and the claim holds.

Since p has positive degree, we can assume that x appears in p (by interchanging x and y, if necessary). We claim that the polynomials $y, y^2, \ldots, y^n$ are independent with respect to f and g at O for every positive integer n. To see this, let r, s, t be polynomials in x and y, and let $b_1, \ldots, b_n$ be complex numbers such that

$$r(b_1y + b_2y^2 + \cdots + b_ny^n) = sf + tg \tag{18}$$

and $b_1, \ldots, b_n$ are not all zero. Since w is a common factor of f and g, so is p. Then p is a factor of the right-hand side of (18), and so it is also a factor of the left-hand side. Because x appears in p, and because the polynomial $b_1y + \cdots + b_ny^n$ is nonzero, p is not a factor of $b_1y + \cdots + b_ny^n$. Then p has no factors of positive degree in common with $b_1y + \cdots + b_ny^n$, since p is irreducible. It follows from Theorem 13.7(ii) and the fact that p is factor of the left-hand side of (18) that p is a factor of r. Since p contains the origin, so does r. In short, (18) implies that r contains the origin when $b_1, \ldots, b_n$ are not all zero. By Definition 13.1, this shows that $y, \ldots, y^n$ are independent, as claimed.

We have shown that there are n polynomials independent with respect to f and g at O for every positive integer n. It follows that $I_O(f,g) = \infty$ (by Definition 13.2). □

We can now prove that the left-hand side of (1) is greater than or equal to the right. This is half of Property 1.6. We complete the proof of Property 1.6 and derive the remaining intersection properties in the next section.

Theorem 13.9

Let f, g, h be polynomials in x and y, and let O be the origin. Then we have

$$I_O(f, gh) \geq I_O(f,g) + I_O(f,h). \tag{19}$$

Proof

We can assume that f and g have no common factor that contains the origin: if they did, then so would f and gh, and inequality (19) would hold because $I_O(f, gh)$ is ∞ (by Theorem 13.8). We can also assume that f contains the origin; if not, inequality (19) would hold because both sides would be zero (by Properties 1.1 and 1.3, which we have already proved). These two assumptions imply that g is nonzero: if g were zero, then f itself would be a common factor of f and g that contains the origin.

Let $p_1, \ldots, p_m$ and $q_1, \ldots, q_n$ be polynomials in x and y such that $p_1, \ldots, p_m$ are independent with respect to f and g at O and $q_1, \ldots, q_n$ are independent with respect to f and h at O. We claim that the $m + n$ polynomials

$$p_1, \ldots, p_m, \quad q_1 g, \ldots, q_n g, \tag{20}$$

are independent with respect to f and gh at O.

Assume that there are polynomials $r(x, y)$, $s(x, y)$, $t(x, y)$ and complex numbers $a_1, \ldots, a_m$, $b_1, \ldots, b_n$ such that

$$r(a_1 p_1 + \cdots + a_m p_m + b_1 q_1 g + \cdots + b_n q_n g) = sf + tgh \tag{21}$$

and

$$r(0,0) \neq 0. \tag{22}$$

Moving all the terms in (21) with a factor of g to the right-hand side of the equation shows that

$$r(a_1 p_1 + \cdots + a_m p_m) = sf + (th - rb_1 q_1 - \cdots - rb_n q_n)g. \tag{23}$$

Since $p_1, \ldots, p_m$ are independent with respect to f and g at O, equation (23) and inequality (22) imply that $a_1, \ldots, a_m$ are all zero (by Definition 13.1). Then we can factor g out of the left-hand side of (21) and obtain the relation

$$r(b_1q_1 + \cdots + b_nq_n)g = sf + tgh. \tag{24}$$

If f and g have a common factor d, we can write

$$f = df^* \qquad \text{and} \qquad g = dg^* \tag{25}$$

for polynomials f^* and g^*. In fact, we can always do so by taking $d = 1$, $f = f^*$, and $g = g^*$. If f^* and g^* have a common factor e of positive degree, then we can write $f^* = ef^{**}$ and $g^* = eg^{**}$ for polynomials f^{**} and g^{**}, and we can increase the degree of d by replacing d with de, f^* with f^{**}, and g^* with g^{**}. We can repeat this step only a finite number of times because g is nonzero (by the first paragraph of the proof). In the end, we have written f and g in the form of (25), where f^* and g^* have no common factors of positive degree.

Substituting the expressions for f and g from (25) into (24) gives

$$r(b_1q_1 + \cdots + b_nq_n)\, dg^* = s\, df^* + t\, dg^*h. \tag{26}$$

Since g is nonzero, so is its factor d. By Theorem 13.7(i), we can cancel the factor of d on both sides of (26) and obtain the relation

$$r(b_1q_1 + \cdots + b_nq_n)g^* = sf^* + tg^*h. \tag{27}$$

This equation shows that g^* is a factor of sf^*, since it is a a factor of the other terms in the equation. Because f^* and g^* have no common factors of positive degree, g^* is a factor of s (by Theorem 13.7(ii)). Writing $s = g^*w$ for a polynomial w and substituting this expression for s in (27) gives

$$r(b_1q_1 + \cdots + b_nq_n)g^* = g^*wf^* + tg^*h. \tag{28}$$

Since g is nonzero, so is its factor g^*. By Theorem 13.7(i), we can cancel the factor of g^* from both sides of (28) and obtain

$$r(b_1q_1 + \cdots + b_nq_n) = wf^* + th. \tag{29}$$

In effect, we have cancelled a factor of g from every term of (24) by cancelling first d and then g^*.

Multiplying both sides of (29) by d and using the first equation in (25) gives

$$rd(b_1q_1 + \cdots + b_nq_n) = wf + t\, dh. \tag{30}$$

Since d is a common factor of f and g, it is nonzero at the origin (by the first sentence of the proof). Then rd is also nonzero at the origin (by inequality (22) above and (24) of Section 10). Thus, since $q_1, \ldots, q_n$ are independent with respect to f and h at the origin, (30) shows that $b_1, \ldots, b_n$ are all zero.

We have proved that $a_1, \ldots, a_m$ and $b_1, \ldots, b_n$ are all zero whenever (21) and inequality (22) hold. By Definition 13.1, this shows that the

$m+n$ polynomials in (20) are independent with respect to the intersection of f and gh at the origin, as we claimed.

Inequality (19) follows easily. Suppose first that $I_O(f,g)$ and $I_O(f,h)$ are both finite. By Definition 13.2, there are $m = I_O(f,g)$ polynomials independent with respect to the intersection of f and g at O, and there are $n = I_O(f,h)$ polynomials independent with respect to the intersection of f and h at O. The claim shows that there are $m+n$ polynomials independent with respect to f and gh at O, and so we have

$$I_O(f,gh) \geq m+n = I_O(f,g) + I_O(f,h)$$

(by Definition 13.2), as desired.

On the other hand, assume that $I_O(f,g)$ or $I_O(f,h)$ is ∞. By Definition 13.2, we can find m polynomials independent with respect to f and g at O and n polynomials independent with respect to f and h at O, where we can make m or n greater than any given positive integer. Accordingly, we can make $m+n$ greater than any given positive integer, where (by the claim) there are $m+n$ polynomials independent with respect to f and gh at O. It follows from Definition 13.2 that $I_O(f,gh) = \infty$, and so inequality (19) holds in this case as well. □

We end this section by using the idea of polynomials being dependent to derive a result on the algebra of polynomials. We require this result in the proof of Theorem 14.5 in order to bring additional terms into the parentheses on the left-hand side of (2).

Theorem 13.10

Assume that $I_O(f,g)$ is finite, and let $r(x,y)$ be a polynomial nonzero at the origin. Then there are polynomials c, d, s, t in x and y such that

$$c(rd-1) = sf + tg \tag{31}$$

and c and d are nonzero at the origin.

Proof

$I_O(f,g)$ is a nonnegative integer e (by assumption). The $e+1$ polynomials $1, x, \ldots, x^e$ are dependent with respect to f and g at the origin (by Definition 13.2). By Definition 13.1, there are polynomials u, v, w in x and y and complex numbers $b_0, \ldots, b_e$ such that

$$u(b_0 + b_1x + \cdots + b_ex^e) = vf + wg, \tag{32}$$

u is nonzero at the origin, and $b_0, \ldots, b_e$ are not all zero. Let m be the smallest integer such that b_m is nonzero. We can factor x^m out of the quantity in parentheses and rewrite (32) as

$$px^m = vf + wg, \tag{33}$$

where $p(x,y)$ is a polynomial that takes the nonzero value $u(0,0)b_m$ at the origin. Likewise, there are polynomials p^*, v^*, w^* and there is a nonnegative integer n such that

$$p^* y^n = v^* f + w^* g \tag{34}$$

and p^* is nonzero at the origin.

Since $r(x, y)$ is nonzero at the origin, its constant term a is nonzero. We can write

$$a^{-1} r = 1 + q, \tag{35}$$

where $q(x, y)$ is a polynomial without a constant term. We define a polynomial $d(x, y)$ by setting

$$d = a^{-1}(1 - q + q^2 - \cdots + (-1)^{m+n} q^{m+n}). \tag{36}$$

Since $q(x, y)$ has no constant term, $d(x, y)$ takes the nonzero value a^{-1} at the origin. Equations (35) and (36) show that

$$rd - 1 = (1 + q)(1 - q + q^2 - \cdots + (-1)^{m+n} q^{m+n}) - 1.$$

Multiplying out the right-hand side of this equation and cancelling terms leaves

$$rd - 1 = (-1)^{m+n} q^{m+n+1}. \tag{37}$$

Since every term of q has x or y as a factor (because q has no constant term), every term of q^{m+n+1} has x^m or y^n as a factor. It follows from (33) and (34) that

$$pp^* q^{m+n+1}$$

is a sum of multiples of f and g. Together with (37), this shows that (31) holds for $c = pp^*$ and polynomials s and t. Since p and p^* are nonzero at the origin, so is c (by (24) of Section 10). □

Exercises

13.1. Over the real numbers, let f and g be polynomials in x and y that are both singular at the origin O. Let

$$ax^2 + bxy + cy^2 \qquad \text{and} \qquad a'x^2 + b'xy + c'y^2$$

be the sums of the terms of degree 2 in f and g, respectively. The points of Euclidean space $(0, 0, 0)$, (a, b, c), (a', b', c') lie in a plane, and there is a point (u, v, w) of Euclidean space that does not lie in this plane. Let $h(x, y)$ • be the polynomial

$$ux^2 + vxy + wy^2.$$

Prove that $1, x, y, h(x, y)$ are independent with respect to f and g at O, and deduce that $I_O(f, g) \geq 4$.

13.2. Let f, g, h be polynomials in x and y such that g is nonzero at the origin O. Use Definitions 13.1 and 13.2 to prove that

$$I_O(f, gh) = I_O(f, h).$$

Do not use Theorem 13.9 or Property 1.6. (This gives an alternate proof of Theorem 1.8.)

We work over the real numbers in Exercises 13.3–13.21 *except where indicated in Exercises* 13.19 *and* 13.21. *We use the following terminology.* The *polar* of a point $P = (s, t, u)$ with respect to a homogeneous polynomial $F(x, y, z)$ of degree d is the quantity

$$sF_x + tF_y + uF_z, \tag{38}$$

which is either zero or a homogeneous polynomial of degree $d - 1$.

13.3. Let Q be a point of a curve F, and let P be a point of the real projective plane that need not be distinct from Q. Prove that the polar of P with respect to F contains Q if and only if either F is singular at Q or else F is nonsingular at Q and the tangent at Q contains P. (See Exercise 12.3.)

13.4. Prove that the polar of any point with respect to a conic is a line.

(*Hint*: Why do Exercise 13.3 and the discussion after Theorem 7.5 imply that the polar is nonzero?)

13.5. Let P and Q be points that are not necessarily distinct, and consider their polars with respect to a conic. Prove that the polar of P contains Q if and only if the polar of Q contains P.

(*Hint*: One possible approach is to use (38) and direct computation to evaluate polars with respect to a conic given by (1) of Section 5.)

13.6. Define harmonic conjugates as in Exercise 5.18. Consider the following result (Figure 13.1):

Theorem

Let P be a point that does not lie on a conic K. Then there is a line l that has the following properties:

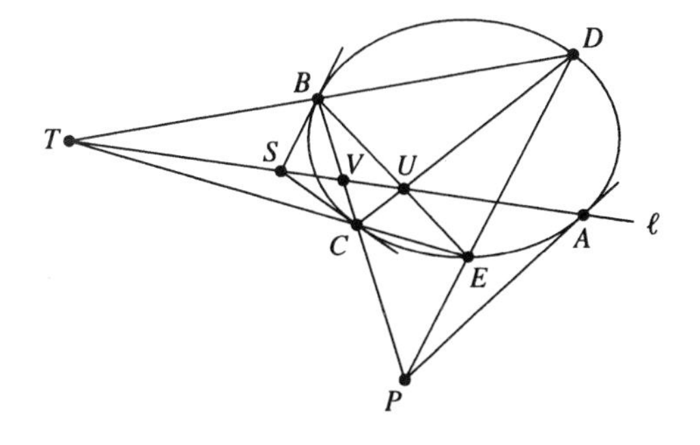

Figure 13.1

(i) *A point A of K lies on l if and only if the tangent at A contains P.*
(ii) *For every pair of points B and C of K collinear with P, l contains the point S where the tangents at B and C intersect.*
(iii) *For any two pairs B, C and D, E of points of K collinear with P, l contains the points $T = BD \cap CE$ and $U = BE \cap CD$.*
(iv) *For every pair of points B and C of K collinear with P, l contains the harmonic conjugate V of P with respect to B and C.*

Prove this theorem by taking l to be the polar of P with respect to K and using Exercises 13.3–13.5, 5.5, and 5.18 and the discussion after Theorem 7.5.

13.7. State the version of the theorem in Exercise 13.6 that holds in the Euclidean plane when P lies at infinity. State this version in common Euclidean terms, using Exercise 5.19 to eliminate the term "harmonic conjugate." Illustrate the result you state with a figure.

13.8. Deduce the theorems in Exercise 7.2 from Exercise 13.6.

13.9. Define harmonic conjugates as in Exercise 5.18. Let A, B, C, D be four points on a conic K. Let E, F, G be the points where AB intersects $\tan C$, $\tan D$, CD, respectively.

(a) If $E \neq F$, prove that G has the same harmonic conjugate with respect to A and B as with respect to E and F.

(*Hint*: One approach is to deduce from Exercise 3.12 that $\tan C$ and $\tan D$ intersect at a point collinear with the harmonic conjugates of G with respect to E and F and with respect to C and D. Conclude from Exercise 13.6 that the polar of G with respect to K intersects line AB at a unique point that is the harmonic conjugate of G both with respect to A and B and with respect to E and F.)

(b) If $E = F$, deduce from Exercise 13.6 that G is the harmonic conjugate of E with respect to A and B. Illustrate this result.

13.10. In the Euclidean plane, let A and B be two points on a hyperbola such that line AB does not contain the point of intersection of the asymptotes. Deduce from Exercises 13.9 and 5.19 that line AB intersects the asymptotes at two points E and F that have the same midpoint M as A and B (Figure 13.2).

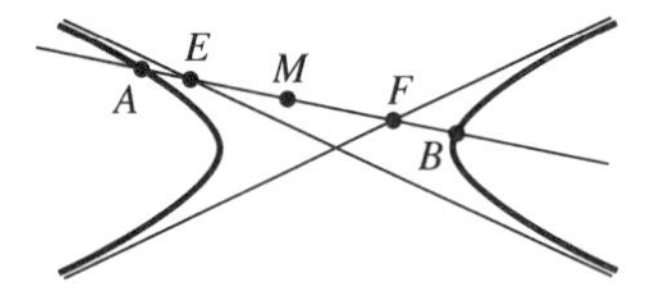

Figure 13.2

13.11. Define harmonic conjugates as in Exercise 5.18. Prove the following result:

Butterfly Theorem

Let A, B, C, D be four points on a conic K. Set $P = AB \cap CD$, and let n be a line through P that does not contain any of the points A–D. Assume that n intersects K at two points G and H. Set $E = AC \cap n$ and $F = BD \cap n$.

(i) *If $E \neq F$, then P has the same harmonic conjugate with respect to G and H as with respect to E and F.*

(ii) *If $E = F$, then E is the harmonic conjugate of P with respect to G and H.*

(Figure 13.3 illustrates part (i). We can prove part (i) by deducing from Exercises 3.12, 3.14, and 5.22 that $AC \cap BD$ is collinear with $AD \cap BC$ and the harmonic conjugate of P with respect to E and F. It follows from Exercises 3.14 and 13.6 that the polar of P intersects n at a unique point Q that is the harmonic conjugate of P both with respect to G and H and with respect to E and F. Part (ii) follows directly from Exercises 3.14 and 13.6.)

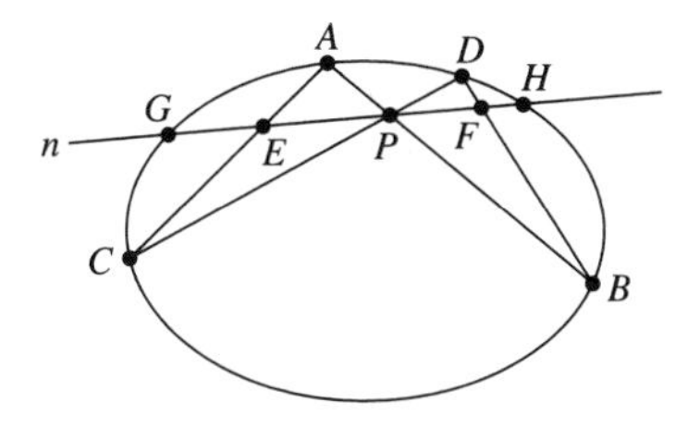

Figure 13.3

13.12. State the version of the Butterfly Theorem in Exercise 13.11 that holds in the Euclidean plane in the following cases. Use Exercise 5.19 to state your results in terms of midpoints instead of harmonic conjugates. Illustrate parts (i) and (ii) of each version with figures in the Euclidean plane.

(a) P is the only point at infinity named.

(b) CD is the line at infinity.

13.13. Consider the polars of points with respect to a conic K. Prove that the polar of a point of K is the tangent at that point.

13.14. Consider the polars of points with respect to a conic. Prove that every line of the real projective plane is the polar of exactly one point. (See Exercises 13.4 and 13.5.)

13.15. Let $F(x, y, z)$ be a homogeneous polynomial of degree d, and set $f(x, y) = F(x, y, 1)$. Let (a, b) be a point of the Euclidean plane, and write

$$f(x, y) = \sum e_{ij}(x - a)^i (y - b)^j$$

(as before Theorem 4.10). Prove that the polar of $(a, b, 1)$ with respect to F is

$$\sum (d - i - j) e_{ij} (x - az)^i (y - bz)^j z^{d-i-j-1}.$$

13.16. Let F be a curve in the real projective plane that has degree greater than 1 and is nonsingular at a point P. Prove that the polar of P with respect to F is a curve of positive degree that is also nonsingular at P and is tangent to the same line there as F. (See Exercise 13.15.)

13.17. Let C be a nonsingular, irreducible cubic. Let P be a point of C that lies on the tangents at four points R, S, T, U other than P. Prove that no three of the five points P, R, S, T, U are collinear and that the three points $RS \cap TU$, $RT \cap SU$, and $RU \cap ST$ lie on C. Illustrate this result with a figure.

(*Hint*: One possible approach is to use (38), Exercises 13.3 and 13.16, and Theorems 4.5, 4.11, 5.1, and 5.9 to show that the polar of P with respect to C is a conic that intersects C once at each of the points R, S, T, U and twice at P. We can then apply the theorem in Exercise 10.6 with E, F, G, H, W, X equal to R, S, P, P, T, U, respectively. Exercises 8.5 and 9.9 contain related results.)

13.18. In the notation of Exercise 13.17, prove that the tangents at P, $RS \cap TU$, $RT \cap SU$, and $RU \cap ST$ have the same point as their third point of intersection with C. Illustrate this result with a figure.

(*Hint*: To prove that the tangents at P and $RS \cap TU$ have the same point as their third points of intersection with C, one possible approach is to take the points E, F, G, H in Theorem 9.6 to be R, R, S, S, respectively.)

13.19. Consider the polar of a point P with respect to a curve F. Prove that the polar of P is the zero polynomial if and only if F either is constant or factors over the complex numbers as a product of lines through P. (See Exercise 13.15.)

13.20. Consider the polar of a point P with respect to a curve F of positive degree. Prove that the polar contains P if and only if P lies on F. (See (21) of Section 12.)

13.21. Consider the following result, which complements Exercises 8.5 and 9.9 and the Hint to Exercise 13.17:

Theorem

Let P be a flex of a nonsingular, irreducible cubic C. Then the polar of P with respect to C is a homogeneous polynomial of degree 2 that consists of two lines—the tangent to C at P and a line that intersects C at exactly the points of C other than P whose tangents contain P.

Prove the theorem by considering the polar of P over the complex numbers and using Exercises 13.3 and 13.16 and Theorem 8.2, which all extend without change to the complex numbers, and Exercise 12.29.

§14. Spanning and Homogeneous Coordinates

We complete the proofs of the intersection properties in this section. In the first half of the section, we extend Theorem 13.9 to a proof of Property 1.6 by proving that

$$I_O(f, gh) \le I_O(f, g) + I_O(f, h)$$

for any polynomials f, g, h. The key is to characterize the largest possible collections of polynomials independent with respect to two curves at the origin. We do so by introducing the idea of polynomials "spanning" the intersection of two curves at the origin. In the second half of the section, we determine intersection multiplicities at every point of the complex projective plane, not just the origin. This enables us to deduce the remaining Intersection Properties 3.1, 3.5, and 12.6.

We continue to work over the complex numbers in this section. As in the last section, this ensures that polynomials with real coefficients intersect the same number of times in both the real and the complex projective planes at points with real coordinates.

A *homogeneous linear equation* is an equation of the form

$$a_1x_1 + \cdots + a_nx_n = 0, \tag{1}$$

where the coefficients $a_1, \ldots, a_n$ are complex numbers and $x_1, \ldots, x_n$ are variables. The term "homogeneous" indicates that the right-hand side of the equation is zero rather than a nonzero number. A *system* of homogeneous linear equations consists of a finite number of homogeneous linear equations to be satisfied simultaneously, such as

$$\begin{aligned} x_1 - 2x_2 + x_3 + 4x_4 + x_5 &= 0, \\ 2x_1 - 6x_2 + x_3 - 2x_4 + 3x_5 &= 0, \\ -x_1 + 5x_2 + 4x_3 + 3x_4 + 2x_5 &= 0. \end{aligned} \tag{2}$$

We can satisfy a system of homogeneous linear equations in a trivial way by setting all the variables equal to zero. A *nontrivial solution* of the system is a choice of values for the variables so that all the equations hold and the value of at least one variable is nonzero.

One of the fundamental results of elementary linear algebra is this: If a system of homogeneous linear equations has more variables than equations, then it has a nontrivial solution. This is true because we can eliminate one equation and one variable at a time until all the equations have been eliminated. Because we started with more variables than equations, there is at least one variable left when all the equations have been eliminated, and this variable can take any value.

For example, consider the system (2) of three equations in five unknowns. We can eliminate x_1 from the last two equations in (2) by subtracting twice the first equation from the second and by adding the first equation to the third. This gives

$$\begin{aligned} -2x_2 - x_3 - 10x_4 + x_5 &= 0, \\ 3x_2 + 5x_3 + 7x_4 + 3x_5 &= 0. \end{aligned} \tag{3}$$

Any solution of the system in (2) gives a solution of the system in (3). Conversely, any values of x_2–x_5 that satisfy the system in (3) correspond to a solution of the system in (2) when the first equation in (2) is used to determine the value of x_1. We have reduced (2) to the system (3) by eliminating one equation and one variable.

Similarly, we can eliminate x_2 from the second equation in (3) by adding $\frac{3}{2}$ times the first equation to the second. This gives

$$\tfrac{7}{2}x_3 - 8x_4 + \tfrac{9}{2}x_5 = 0. \tag{4}$$

The systems of equations in (3) and (4) have corresponding solutions when the value of x_2 is determined by the first equation in (3).

For any values of x_4 and x_5, (4) determines the value of x_3, and the first equations in (3) and (2) determine the values of x_2 and x_1. If we choose values for x_4 and x_5 that are not both zero, we obtain a nontrivial solution of system (2).

For example, suppose we take

$$x_4 = 1 \qquad \text{and} \qquad x_5 = 2. \tag{5}$$

Equation (4) becomes $\frac{7}{2}x_3 - 8(1) + \frac{9}{2}(2) = 0$, which gives

$$x_3 = -\tfrac{2}{7}. \tag{6}$$

The first equation in (3) becomes $-2x_2 - (-\frac{2}{7}) - 10(1) + 2 = 0$, which gives

$$x_2 = -\tfrac{27}{7}. \tag{7}$$

The first equation in (2) becomes $x_1 - 2(-\frac{27}{7}) + (-\frac{2}{7}) + 4(1) + 2 = 0$, which gives

$$x_1 = -\tfrac{94}{7}. \tag{8}$$

Equations (5)–(8) give a nontrivial solution to the system of equations in (2), as desired.

The same procedure works whenever we start with more variables than equations. We call a homogeneous linear equation "zero" if it has the form $0x_1 + \cdots + 0x_n = 0$, where all variables have coefficient zero. Such an equation holds for all values of $x_1, \cdots, x_n$ because it reduces to $0 = 0$.

Theorem 14.1

A system of homogeneous linear equations that has more variables than equations has a nontrivial solution.

Proof

Let Equation E be the first nonzero equation in the system, if there are any. It contains a variable x_t with nonzero coefficient. We eliminate

x_t from the other equations by adding multiples of Equation E. Setting aside Equation E leaves a system that has one equation fewer and does not contain x_t. Equation E determines the value of x_t.

We have eliminated one nonzero equation and one variable. We continue in this way until no nonzero equations remain. Because we started with more variables than equations and have eliminated the same number of variables as equations, at least one variable remains at the end. Because any remaining equations are zero and hold automatically, we can choose any nonzero values for the remaining variables. This gives a nontrivial solution of the original system. □

We now introduce the idea of polynomials spanning intersections. *As in the last section, we assume that f, g, h are polynomials in x and y and that O is the origin.*

Definition 14.2

Let n be a nonnegative integer, and let $q_1, \ldots, q_n$ be polynomials in x and y. We say that $q_1, \ldots, q_n$ *span* the intersection of f and g at O if, for every polynomial $e(x, y)$, there are polynomials $r(x, y)$, $s(x, y)$, $t(x, y)$, and complex numbers $b_1, \ldots, b_n$ such that

$$r(e + b_1 q_1 + \cdots + b_n q_n) = sf + tg \tag{9}$$

and r is nonzero at the origin. □

If we take $n = 0$ in Definition 14.2, there are no q's and b's, and the definition says that the empty set of polynomials spans the intersection of f and g at O if, for every polynomial $e(x, y)$, we can write

$$re = sf + tg$$

for polynomials $r(x, y)$, $s(x, y)$, $t(x, y)$ such that $r(0, 0) \neq 0$.

In general, we might think of (9) as saying that the quantity in parentheses becomes zero under multiplication by r and addition of the terms $-sf$ and $-tg$. That is, every polynomial $e(x, y)$ must equal a sum $-b_1 q_1 - \cdots - b_n q_n$ of constant multiples of $q_1, \ldots, q_n$ under multiplication by a polynomial nonzero at the origin and addition of polynomial multiples of f and g.

The next two theorems relate spanning and independence with respect to two curves at the origin. Theorem 14.3 shows that a set of independent polynomials that does not span the intersection of the curves at the origin is contained in a larger set of independent polynomials. Thus, a set of independent polynomials that is as large as possible must necessarily span the intersection.

Theorem 14.3

Let n be a nonnegative integer, and let $q_1, \ldots, q_n$ be polynomials in x and y that are independent with respect to f and g at O. If $q_1, \ldots, q_n$ do not span the intersection of f and g at O, then there is a polynomial $u(x, y)$ such that $q_1, \ldots, q_n$, u are independent with respect to f and g at O.

Proof

Since $q_1, \ldots, q_n$ do not span the intersection of f and g at O, Definition 14.2 implies that there is a polynomial $u(x, y)$ that does not satisfy any equation of the form

$$r(u + b_1 q_1 + \cdots + b_n q_n) = sf + tg$$

for polynomials r, s, t and complex numbers $b_1, \ldots, b_n$, where r is nonzero at the origin. We claim that $q_1, \ldots, q_n$, u are independent with respect to f and g at O.

Assume that there are polynomials $r(x, y)$, $s(x, y)$, $t(x, y)$ and complex numbers $b_1, \ldots, b_n$, c such that $r(0, 0)$ is nonzero and the relation

$$r(b_1 q_1 + \cdots + b_n q_n + cu) = sf + tg \tag{10}$$

holds. If c were nonzero, multiplying (10) by c^{-1} would give

$$r(u + c^{-1} b_1 q_1 + \cdots + c^{-1} b_n q_n) = c^{-1} sf + c^{-1} tg,$$

which would contradict the first sentence of the proof. Thus, c is zero, and (10) becomes

$$r(b_1 q_1 + \cdots + b_n q_n) = sf + tg. \tag{11}$$

Since $q_1, \ldots, q_n$ are independent with respect to f and g at O, equation (11) implies that $b_1, \ldots, b_n$ are all zero (by Definition 13.1). In short, (10) holds only when $b_1, \ldots, b_n$, c are all zero. Thus, $q_1, \ldots, q_n$, u are independent with respect to f and g at O (by Definition 13.1), as claimed. □

The next theorem, like the previous one, prepares us to use spanning to recognize maximal sets of polynomials independent with respect to two curves at the origin. The proof uses Theorem 14.1 to show that any set of polynomials spanning an intersection at the origin has at least as many elements as any set of independent polynomials.

Theorem 14.4

Let m be a nonnegative integer. If the intersection of f and g at the origin is spanned by m polynomials, then any set of polynomials independent with respect to f and g at the origin contains at most m polynomials.

Proof

We must show that any set of more than m polynomials is dependent. Let n be an integer greater than m, and let $q_1, \ldots, q_n$ be n polynomials in

x and y. We prove that $q_1, \ldots, q_n$ are dependent. By assumption, there are m polynomials $p_1, \ldots, p_m$ in x and y that span the intersection.

Because $p_1, \ldots, p_m$ span the intersection, for each polynomial q_j there are polynomials r_j, s_j, t_j in x and y and there are complex numbers $b_{1j}, \ldots, b_{mj}$ such that r_j is nonzero at the origin and the equation

$$r_j(q_j + b_{1j}p_1 + \cdots + b_{mj}p_m) = s_jf + t_jg$$

holds (by Definition 14.2). Multiplying this equation by

$$r_1 \cdots r_{j-1}r_{j+1} \cdots r_n$$

gives the relation

$$r(q_j + b_{1j}p_1 + \cdots + b_{mj}p_m) = u_jf + v_jg \tag{12}$$

for polynomials r, u_j, v_j in x and y, where

$$r = r_1 \cdots r_n$$

does not depend on j and is nonzero at the origin (by (24) of Section 10).

We multiply (12) by a complex number c_j to be determined. This gives

$$r(c_jq_j + b_{1j}c_jp_1 + \cdots + b_{mj}c_jp_m) = c_ju_jf + c_jv_jg.$$

Summing these equations as j runs from 1 through n and collecting terms gives the equation

$$\begin{aligned} r[(c_1q_1 + \cdots + c_nq_n) + (b_{11}c_1 + \cdots + b_{1n}c_n)p_1 + \cdots \\ + (b_{m1}c_1 + \cdots + b_{mn}c_n)p_m] = uf + vg \end{aligned} \tag{13}$$

for polynomials

$$u = c_1u_1 + \cdots + c_nu_n \qquad \text{and} \qquad v = c_1v_1 + \cdots + c_nv_n.$$

We think of $c_1, \ldots, c_n$ as unknowns to be determined so that the coefficients of $p_1, \ldots, p_m$ in (13) are all zero. This gives a system of homogeneous linear equations

$$\begin{aligned} b_{11}c_1 + \cdots + b_{1n}c_n &= 0, \\ &\vdots \\ b_{m1}c_1 + \cdots + b_{mn}c_n &= 0. \end{aligned}$$

This system has more variables than equations, since it has n variables and m equations, where n is greater than m. Thus, the system has a nontrivial solution (by Theorem 14.1). Substituting these values for $c_1, \ldots, c_n$ in (13) gives the relation

$$r(c_1q_1 + \cdots + c_nq_n) = uf + vg,$$

where r is nonzero at the origin and $c_1, \ldots, c_n$ are complex numbers that are not all zero. Hence, $q_1, \ldots, q_n$ are dependent (by Definition 13.1), as desired. □

At last we can complete the proof of Property 1.6, which determines the intersections of products.

Theorem 14.5
Let f, g, h be polynomials, and let O be the origin. Then we have

$$I_O(f, gh) = I_O(f, g) + I_O(f, h).$$

Proof
We must prove that

$$I_O(f, gh) \leq I_O(f, g) + I_O(f, h), \tag{14}$$

since Theorem 13.9 gives the reverse inequality. Because inequality (14) holds automatically when either term on the right is ∞, we can assume that $I_O(f, g)$ and $I_O(f, h)$ are nonnegative integers m and n, respectively.

Since $I_O(f, g) = m$, Definition 13.2 shows that there are m polynomials $p_1, \ldots, p_m$ independent with respect to f and g at O and that no collection of more than m polynomials is independent. Theorem 14.3 implies that $p_1, \ldots, p_m$ span the intersection of f and g at O. Likewise, there are n polynomials $q_1, \ldots, q_n$ that span the intersection of f and h at O.

We claim that the $m + n$ polynomials

$$p_1, \ldots, p_m, \quad q_1 g, \ldots, q_n g \tag{15}$$

span the intersection of f and gh at O. If so, every set of polynomials independent with respect to f and gh at O has at most $m + n$ elements (by Theorem 14.4). It follows from Definition 13.2 that

$$I_O(f, gh) \leq m + n = I_O(f, g) + I_O(f, h),$$

as desired.

To prove that the polynomials in (15) span the intersection of f and gh at O, we let $e(x, y)$ be any polynomial. Because $p_1, \ldots, p_m$ span the intersection of f and g at O, there are polynomials $r(x, y), s(x, y), t(x, y)$ and complex numbers $a_1, \ldots, a_m$ such that

$$r(e + a_1 p_1 + \cdots + a_m p_m) = sf + tg, \tag{16}$$

and $r(0, 0)$ is nonzero (by Definition 14.2).

We use Theorem 13.10 to move the tg term from the right-hand side of (16) into the parentheses on the left. Since $I_O(f, h)$ is finite and r is nonzero at the origin, Theorem 13.10 shows that there are polynomials c, d, k, l such that

$$c(rd - 1) = kf + lh$$

and c and d are nonzero at the origin. Distributing c in the last equation and multiplying both sides of the equation by $-tg$ gives

$$-crdtg + ctg = -tkgf - tlgh. \tag{17}$$

If we multiply (16) by c, add (17), and cancel the terms ctg on both sides of the result, we obtain

$$cr(e + a_1p_1 + \cdots + a_mp_m - dtg) = (cs - tkg)f - tlgh. \tag{18}$$

In effect, the tg term in (16) has become the $-dtg$ term inside the parentheses in (18).

Since $q_1, \ldots, q_n$ span the intersection of f and h at O, there are polynomials u, v, w and complex numbers $b_1, \ldots, b_n$ such that u is nonzero at the origin and

$$u(dt + b_1q_1 + \cdots + b_nq_n) = vf + wh. \tag{19}$$

If we multiply (18) by u, multiply (19) by crg, and add the results, the terms $\pm$ $crudtg$ on the left cancel, and we obtain the relation

$$\begin{aligned} &cru(e + a_1p_1 + \cdots + a_mp_m + b_1q_1g + \cdots + b_nq_ng) \\ &\quad = (csu - tkgu + crgv)f + (-tlu + crw)gh. \end{aligned} \tag{20}$$

The product cru is nonzero at the origin because all of its factors are (by (24) of Section 10). We have shown that an equation of the form (20) holds for every polynomial $e(x, y)$, and so the polynomials in (15) span the intersection of f and gh at O (by Definition 14.2). This completes the proof, as observed after (15). □

Definitions 13.1 and 14.2 characterize polynomials that are "dependent" or "spanning" with respect to curves f and g at the origin. Readers familiar with linear algebra might note that these definitions are analogous to the definitions of linearly dependent and spanning vectors in a vector space. In fact, Definitions 13.1 and 14.2 are exactly the usual definitions of linearly dependent and spanning vectors in a specially constructed vector space. (In the language of commutative algebra, we construct this vector space by taking the quotient algebra of polynomials in x and y modulo the ideal generated by f and g and localizing at the multiplicitively closed set of polynomials nonzero at the origin.) Accordingly, Theorem 14.3 and 14.4 are special cases of standard results in elementary linear algebra. We proved Theorems 14.3 and 14.4 by translating the linear algebra proofs into the present context. The language and results of linear and abstract algebra would also have simplified the proofs of Theorems 13.8–13.10 and 14.5.

We have derived the Intersection Properties 1.1–1.6. We must still determine intersection multiplicities at every point in the complex rejective plane—not only at the origin—and we must derive the Intersection Properties 3.1, 3.5, and 12.6. We start by extending Definition 13.1 of dependent polynomials to homogeneous coordinates.

Definition 14.6

Let n be a positive integer, and let $F, G, Q_1, \ldots, Q_n$ be homogeneous polynomials in x, y, z. Let P be a point in the complex projective plane, and let $L = 0$ be a line that does not contain P. Let $k_1, \ldots, k_n$ be the degrees of $Q_1, \ldots, Q_n$, and let k be the largest of these integers. We say that $Q_1, \ldots, Q_n$ are *dependent* with respect to F and G at P using L if there are homogeneous or zero polynomials R, S, T in x, y, z and complex numbers $b_1, \ldots, b_n$ such that the equation

$$R(b_1 L^{k-k_1} Q_1 + \cdots + b_n L^{k-k_n} Q_n) = SF + TG \tag{21}$$

holds, R is nonzero and does not contain P, and $b_1, \ldots, b_n$ are not all zero. □

The factors of L in (21) ensure that each nonzero term in parenthesis has the same degree k and so the left-hand side of the equation is homogeneous or zero. Of course, we say that $Q_1, \ldots, Q_n$ are *independent* with respect to F and G at P using L if they are not dependent. We consider the empty set of polynomials to be independent, and so there are always n independent polynomials for $n = 0$.

We generalize Definition 13.2 to homogeneous coordinates as follows:

Definition 14.7

Let F and G be homogeneous polynomials in x, y, z, let P be a point in the complex projective plane, and let $L = 0$ be a line that does not contain P. We define $I_P^L(F, G)$, the *intersection multiplicity* of F and G at P using L, as follows. We set $I_P^L(F, G) = \infty$ if, for every positive integer c, there are at least c homogeneous polynomials independent with respect to F and G at P using L. Otherwise, we set $I_P^L(F, G) = e$, where e is the largest integer such that there are e polynomials independent with respect to F and G at P using L. □

The next result confirms that Definition 14.7 reduces to Definition 13.2 when P is the origin O and $L = 0$ is the line at infinity $z = 0$.

Theorem 14.8

Let F and G be homogeneous polynomials in x, y, z, let O be the origin, and let $z = 0$ be the line at infinity. Let f and g be the polynomials in x and y obtained by setting $z = 1$ in F and G. Then we have

$$I_O^z(F, G) = I_O(f, g). \tag{22}$$

Proof

We show first that

$$I_O^z(F, G) \leq I_O(f, g). \tag{23}$$

Since this inequality holds automatically when the right-hand side is infinite, we can assume that $I_O(f, g)$ is a nonnegative integer. Let n be any integer greater than $I_O(f, g)$. Let $Q_1, \ldots, Q_n$ be any n homogeneous polynomials in x, y, z. Let $k_1, \ldots, k_n$ be their degrees, and let k be the largest of these integers. Let $q_1, \ldots, q_n$ be the polynomials in x and y obtained by setting $z = 1$ in the Q_i. Since n is greater than $I_O(f,g)$, $q_1, \ldots, q_n$ are dependent with respect to f and g at O (by Definition 13.2). Thus, there are polynomials r, s, t in x and y and complex numbers $b_1, \ldots, b_n$ such that

$$r(b_1 q_1 + \cdots + b_n q_n) = sf + tg, \tag{24}$$

$$r(0,0) \neq 0, \tag{25}$$

and $b_1, \ldots, b_n$ are not all zero (by Definition 13.1). Multiplying the terms of the polynomials in (24) by appropriate powers of z gives homogeneous or zero polynomials R, S, T in x, y, z such that

$$R(b_1 z^{k-k_1} Q_1 + \cdots + b_n z^{k-k_n} Q_n) = SF + TG, \tag{26}$$

and

$$R(0,0,1) \neq 0. \tag{27}$$

Thus, $Q_1, \ldots, Q_n$ are dependent with respect to F and G at O using the line $z = 0$ (by Definition 14.6). Since this holds for any n homogeneous polynomials $Q_1, \ldots, Q_n$, where n is any integer greater than $I_O(f,g)$, inequality (23) follows from Definition 14.7.

We must show the reverse inequality

$$I_O(f,g) \leq I_O^z(F,G). \tag{28}$$

Since this inequality holds automatically if the right-hand side is ∞, we can assume that $I_O^z(F,G)$ is a nonnegative integer. Let n be any integer greater than $I_O^z(F,G)$, and let $q_1, \ldots, q_n$ be any n polynomials in x and y. Assume first that $q_1, \ldots, q_n$ are all nonzero. Let $Q_1, \ldots, Q_n$ be their homogenizations, let $k_1, \ldots, k_n$ be the degrees of the Q_i, and let k be the largest of the integers k_i. Since n is greater than $I_O^z(F,G)$, $Q_1, \ldots, Q_n$ are dependent with respect to F and G at O using the line $z = 0$ (by Definition 14.7). Thus, there are homogeneous or zero polynomials R, S, T in x, y, z and complex numbers $b_1, \ldots, b_n$ such that (26) and (27) hold and $b_1, \ldots, b_n$ are not all zero. Setting $z = 1$ in R, S, T gives polynomials r, s, t in x and y. Setting $z = 1$ in (26) gives (24), and inequality (27) gives inequality (25). Thus, since $b_1, \ldots, b_n$ are not all zero, $q_1, \ldots, q_n$ are dependent with respect to the intersection of f and g at O (by Definition 13.1). This also holds when one of the q's, say q_w, is zero: the conditions of Definition 13.1 are satisfied by taking $b_w = 1, b_j = 0$ for $j \neq w$, $r = 1, s = 0$, and $t = 0$. In short, any n polynomials $q_1, \ldots, q_n$ in x and y are dependent with respect to f and g at O for any integer n greater than $I_O^z(F,G)$. Inequality (28) follows (by Definition 13.2).

Combining inequalities (23) and (28) gives equation (22). □

Substituting for variables preserves algebraic relations among polynomials, such as those in (21). It follows that transformations preserve the quantities $I_P^L(F,G)$.

Theorem 14.9

Let F and G be homogeneous polynomials in x,y,z. Let P be a point in the complex projective plane, and let $L=0$ be a line that does not contain P. If F',G',P',L' are the images of F,G,P,L under a transformation, then we have

$$I_P^L(F,G) = I_{P'}^{L'}(F',G'). \tag{29}$$

Proof

Since L does not contain P, L' does not contain P' (by (16) of Section 3). Thus, $I_{P'}^{L'}(F',G')$ is defined.

We show first that

$$I_P^L(F,G) \le I_{P'}^{L'}(F',G'). \tag{30}$$

Because this inequality holds automatically when the right-hand side is ∞, we can assume that $I_{P'}^{L'}(F',G')$ is a nonnegative integer. Let n be an integer greater than $I_{P'}^{L'}(F',G')$, and let $Q_1,\ldots,Q_n$ be n homogeneous polynomials in x,y,z. Let $k_1,\ldots,k_n$ be the degrees of the Q_i, and let k be the largest of these integers. Let the given transformation map (x,y,z) to (x',y',z') as in Definition 3.3, and let $Q_1',\ldots,Q_n'$ be the images of $Q_1,\ldots,Q_n$ under the transformation. Since n is greater than $I_{P'}^{L'}(F',G')$, $Q_1',\ldots,Q_n'$ are dependent with respect to F' and G' at P' using L' (by Definition 14.7). By Definition 14.6, there are homogeneous or zero polynomials R',S',T' in x',y',z' and complex numbers $b_1,\ldots,b_n$ such that the equation

$$R'(b_1L'^{k-k_1}Q_1' + \cdots + b_nL'^{k-k_n}Q_n') = S'F' + T'G' \tag{31}$$

holds, R' does not contain P', and $b_1,\ldots,b_n$ are not all zero. The transformation is given by expressions for x',y',z' in terms of x,y,z, as in (5) of Section 3; substituting these expressions for x',y',z' takes R',S',T' to homogeneous or zero polynomials R,S,T in x,y,z and takes (31) to

$$R(b_1L^{k-k_1}Q_1 + \cdots + b_nL^{k-k_n}Q_n) = SF + TG.$$

R does not contain P (by (16) of Section 3). Thus, $Q_1,\ldots,Q_n$ are dependent with respect to F and G at P using L (by Definition 14.6). Because this holds for any n homogeneous polynomials $Q_1,\ldots,Q_n$ where n is any integer greater than $I_{P'}^{L'}(F',G')$, inequality (30) follows from Definition 14.7.

Because transformations are reversible, we have the analogue of inequality (30) for the transformation taking (x',y',z') to (x,y,z). This shows that

$$I_{P'}^{L'}(F', G') \leq I_P^L(F, G).$$

Together with inequality (30), this gives (29). □

The last two theorems enable us to prove that the value of $I_P^L(F,G)$ does not depend on the choice of the line L. This lets us define the intersection multiplicity of F and G at P as the common value of $I_P^L(F,G)$ for all lines L that do not contain P.

Theorem 14.10

Let F and G be homogeneous polynomials in x, y, z, and let P be a point in the complex projective plane. Let $L = 0$ and $M = 0$ be lines that do not contain P. Then we have

$$I_P^L(F, G) = I_P^M(F, G). \tag{32}$$

Proof

We show first that

$$I_P^L(F, G) \leq I_P^M(F, G). \tag{33}$$

Let D and E be two points of M, and let D' and E' be two points on the line at infinity $z = 0$. P is not collinear with D and E (since P does not lie on M), and the origin O is not collinear with D' and E' (since O does not lie on the line at infinity). Thus, there is a transformation that maps P to O, D to D', and E to E' (by Theorem 3.4, which extends without change to the complex numbers). This transformation maps line $DE = M$ to the line $D'E'$ at infinity. We can replace P, L, F, G, M in inequality (33) with their images under the transformation (by Theorem 14.9). Thus, it suffices to prove that

$$I_O^L(F, G) \leq I_O^z(F, G)$$

for any line L that does not contain the origin O. By Theorem 14.8, this is equivalent to showing that

$$I_O^L(F, G) \leq I_O(f, g), \tag{34}$$

where f and g are the polynomials in x and y obtained by setting $z = 1$ in F and G.

Since inequality (34) holds automatically when $I_O(f,g) = \infty$, we can assume that $I_O(f,g)$ is a nonnegative integer. Let n be any integer greater than $I_O(f,g)$, and let $Q_1, \ldots, Q_n$ be n homogeneous polynomials in x, y, z. If we prove that $Q_1, \ldots, Q_n$ are dependent with respect to F and G at O using L, inequality (34) follows from Definition 14.7.

Let $k_1, \ldots, k_n$ be the degrees of $Q_1, \ldots, Q_n$, and let k be the largest of these integers. Let $l, q_1, \ldots, q_n$ be the polynomials in x and y obtained by setting $z = 1$ in $L, Q_1, \ldots, Q_n$. Because n is greater than $I_O(f,g)$, the n polynomials

$$l^{k-k_1}q_1, \ldots, l^{k-k_n}q_n$$

are dependent with respect to f and g at O (by Definition 13.2). By Definition 13.1, there are polynomials r, s, t in x and y and complex numbers $b_1, \ldots, b_n$ such that the equation

$$r(b_1 l^{k-k_1}q_1 + \cdots + b_n l^{k-k_n}q_n) = sf + tg \tag{35}$$

holds, r is nonzero at the origin, and $b_1, \ldots, b_n$ are not all zero. Multiplying the terms of the polynomials in (35) by appropriate powers of z gives homogeneous or zero polynomials R, S, T in x, y, z such that the equation

$$R(b_1 L^{k-k_1}Q_1 + \cdots + b_n L^{k-k_n}Q_n) = SF + TG$$

holds and R does not contain O. Since $b_1, \ldots, b_n$ are not all zero, this shows that $Q_1, \ldots, Q_n$ are dependent with respect to F and G at O using L (by Definition 14.6).

We have established inequality (34). By the first paragraph of the proof, it follows that inequality (33) holds for any lines L and M that do not contain a point P. Interchanging the roles of L and M shows that

$$I_P^M(F, G) \leq I_P^L(F, G).$$

Together with inequality (33), this establishes (32). □

Now that we have proved that the value of $I_P^L(F, G)$ does not depend on the choice of the line L, we can define intersection multiplicities as follows:

Definition 14.11
Let F and G be homogeneous polynomials in x, y, z and let P be a point in the complex projective plane. The *intersection multiplicity* of F and G at P, written $I_P(F, G)$, is the common value of the quantities $I_P^L(F, G)$ for all lines L that do not contain P. □

Definition 14.11 makes sense because Theorem 14.10 shows that the quantities $I_P^L(F, G)$ are equal for all lines L that do not contain P.

We can use Theorem 14.8 and Definition 14.11 to give a quick proof of Property 3.1. Let F and G be homogeneous polynomials in x, y, z, and let f and g be the polynomials in x and y obtained by setting $z = 1$ in F and G. Property 3.1 states that

$$I_O(F, G) = I_O(f, g), \tag{36}$$

which means that Definitions 14.11 and 13.2 give the same values for intersection multiplicities at the origin O. In fact, Definition 14.11 shows that the left-hand side of (36) equals $I_O^z(F,G)$, and Theorem 14.8 shows that the right-hand side does, too. Thus, (36) holds, and we have established Property 3.1.

Theorem 14.9 and Definition 14.11 provide a quick proof of Property 3.5, which shows that transformations preserve intersection multiplicities. Let F and G be homogeneous polynomials in x, y, z, and let P be a point in the complex projective plane. Let F', G', P' be the images of F, G, P under a transformation. Property 3.5 states that

$$I_P(F,G) = I_{P'}(F',G'). \tag{37}$$

Let L be a line that does not contain P, and let L' be the image of L under the transformation. Theorem 14.9 shows that

$$I_P^L(F,G) = I_{P'}^{L'}(F',G').$$

Each side of this equation equals the corresponding side of (37) (by Definition 14.11). Thus, (37) holds, and we have established Property 3.5.

The only intersection property that remains to be proved is Property 12.6, which states that conjugation preserves intersection multiplicities. This follows from the fact that conjugating polynomials preserves algebraic relations such as (21).

Conjugation preserves products of polynomials (by Theorem 11.7(i)). We claim that it also preserves sums of polynomials. In fact, the relation

$$\overline{F+G} = \bar{F} + \bar{G} \tag{38}$$

holds for all homogeneous polynomials

$$F = \sum e_{ijk}x^iy^jz^k \qquad \text{and} \qquad G = \sum d_{ijk}x^iy^jz^k$$

because (31) of Section 10 shows that

$$\sum \overline{(e_{ijk} + d_{ijk})}x^iy^jz^k = \sum \overline{e_{ijk}}x^iy^jz^k + \sum \overline{d_{ijk}}x^iy^jz^k.$$

In the notation of the previous paragraph, we also have the relation

$$\bar{F} = G \quad \text{if and only if} \quad F = \bar{G}, \tag{39}$$

since (30) of Section 10 shows that

$$\overline{e_{ijk}} = d_{ijk} \quad \text{if and only if} \quad e_{ijk} = \overline{d_{ijk}}.$$

We can now prove Property 12.6, which states that

$$I_P(F,G) = I_{\bar{P}}(\bar{F},\bar{G}) \tag{40}$$

for any complex curves F and G and any point P in the complex projective plane. We show first that

$$I_P(F,G) \leq I_{\bar{P}}(\bar{F},\bar{G}). \tag{41}$$

Since this holds automatically when the right-hand side is infinite, we can assume that $I_{\bar{P}}(\bar{F},\bar{G})$ is a nonnegative integer. Let n be any integer greater than $I_{\bar{P}}(\bar{F},\bar{G})$. Let $Q_1,\ldots,Q_n$ be n homogeneous polynomials in x, y, z. Let their degrees be $k_1,\ldots,k_n$, and let k be the largest of these integers. Let $L=0$ be a line that does not contain P. Then $\bar{L}=0$ is a line that does not contain $\bar{P}$ (as discussed before Property 12.6). Since n is greater than $I_{\bar{P}}(\bar{F},\bar{G})$, the n polynomials $\overline{Q_1},\ldots,\overline{Q_n}$ are dependent with respect to $\bar{F}$ and $\bar{G}$ at $\bar{P}$ using $\bar{L}$ (by Definitions 14.11 and 14.7). By Definition 14.6, there are homogeneous or zero polynomials R,S,T in x,y,z and complex numbers $b_1,\ldots,b_n$ such that the equation

$$R(b_1\bar{L}^{k-k_1}\overline{Q_1}+\cdots+b_n\bar{L}^{k-k_n}\overline{Q_n}) = S\bar{F}+T\bar{G} \tag{42}$$

holds, R does not contain P, and $b_1,\ldots,b_n$ are not all zero. Conjugating both sides of (42) shows that

$$\bar{R}(\overline{b_1}L^{k-k_1}Q_1+\cdots+\overline{b_n}L^{k-k_n}Q_n) = \bar{S}F+\bar{T}G \tag{43}$$

(by Theorem 11.7(i) and (38) and (39)). Since R does not contain P and $b_1,\ldots,b_n$ are not all zero, $\bar{R}$ does not contain $\bar{P}$ (as discussed before Property 12.6) and $\overline{b_1},\ldots,\overline{b_n}$ are not all zero. Thus, (43) shows that $Q_1,\ldots,Q_n$ are dependent with respect to F and G at P using L (by Definition 14.6). Since this holds for any n homogeneous polynomials $Q_1,\ldots,Q_n$, where n is any integer greater than $I_{\bar{P}}(\bar{F},\bar{G})$, inequality (41) follows from Definitions 14.7 and 14.11.

If we replace F,G,P with $\bar{F},\bar{G},\bar{P}$ in inequality (41), we also replace $\bar{F},\bar{G},\bar{P}$ with F,G,P (by (39) above and (30) of Section 10). Then inequality (41) becomes the relation

$$I_{\bar{P}}(\bar{F},\bar{G}) \le I_P(F,G).$$

Together with inequality (41), this gives (40), and so Property 12.6 holds.

We have proved all of the intersection properties at last. Readers should congratulate themselves for their perseverance.

Exercises

14.1. Let f and g be polynomials in x and y, and let O be the origin. If the intersection of f and g at O is not spanned by any finite set of polynomials in x and y, prove that $I_O(f,g)$ is ∞. Otherwise, prove that $I_O(f,g)$ is the least integer $n \ge 0$ such that the intersection of f and g at O is spanned by n polynomials. (See Theorems 14.3 and 14.4.)

14.2. Let f and g be polynomials in x and y, and let O be the origin. Let n be a nonnegative integer, and let $q_1,\ldots,q_n$ be polynomials in x and y that span the intersection of f and g at O. Prove that $I_O(f,g) = n$ if and only if $q_1, \ldots, q_n$ are independent with respect to f and g at O.

(*Hint*: If $q_1, \ldots q_n$ are dependent with respect to f and g at O, we can show that the intersection of f and g at O is spanned by $n-1$ of the polynomials $q_1, \ldots, q_n$.)

14.3. Let $F(x,y,z)$ be a homogeneous polynomial of positive degree with real coefficients. Use Theorems 11.8, 11.7, and 13.6 to prove that the following conditions are equivalent, i.e., that (i) holds if and only if (ii) does:
(i) F is irreducible over the real but not the complex numbers.
(ii) We can write $F = \pm G\bar{G}$, where $G(x,y,z)$ is a homogeneous polynomial with complex coefficients that is irreducible over the complex numbers and is not a scalar multiple of a homogeneous polynomial with real coefficients.

14.4. Let k be a positive integer, and set

$$G(x,y,z) = y^{2k} - i(x^2+z^2)(x^2+4z^2)\cdots(x^2+k^2z^2).$$

(a) Prove that the complex curve G is nonsingular in the terminology introduced before Exercise 11.11. Why does it follow that G is irreducible?
(b) Prove that $F = G\bar{G}$ is a homogeneous polynomial of degree $4k$ with real coefficients that is irreducible over the real numbers and contains no points in the real projective plane. (The fact that the curve $F = 0$ in the real projective plane is empty means that it is nonsingular (by the definition before Theorem 8.3)).

14.5. Let $F(x,y,z)$ be a homogeneous polynomial that has real coefficients and is irreducible over the real but not the complex numbers. If $F = 0$ is nonsingular in the real projective plane, prove that there are no points on this curve in the real projective plane and that the degree of F is a multiple of four.

(Exercise 14.4 shows that such curves exist. One possible way to prove that the degree of any such curve is a multiple of four is to take the map $P \to \bar{P}$ introduced before Property 12.6 and consider its effect on the intersections of the complex curves G and $\bar{G}$ in Exercise 14.3, as determined by Bezout's Theorem 11.5.)

We work over the real numbers in Exercises 14.6–14.19.

14.6. Let F and G be homogeneous polynomials in x,y,z, and let P be a point.
(a) Prove that $I_P(F,G) \geq 4$ if F and G are both singular at P. Use Exercise 13.1 and Properties 3.1 and 3.5.
(b) Prove that $I_P(F,G) = 1$ if and only if F and G both contain P, are nonsingular at P, and are tangent to different lines there. Use part (a) and Theorems 3.6 and 4.11.

(Exercises 4.26 and 4.27 give another proof of these results.)

14.7. Let $f(x,y)$ and $g(x,y)$ be polynomials. Prove that $I_O(f,g)$ is finite if only if there is a polynomial $r(x,y)$ nonzero at the origin and there is an integer $e \geq 0$ such that any polynomial $h(x,y)$ in which every term has degree at least e satisfies an equation of the form

$$rh = fs + gt$$

for polynomials $s(x,y)$ and $t(x,y)$. (See Theorem 14.4 and the proof of Theorem 13.10.)

14.8. Let $f(x,y)$ and $g(x,y)$ be polynomials such that f is tangent to $y = 0$ at the origin O.

(a) For any positive integer d, prove that we can write

$$g = q + fm + p$$

for a polynomial $q(x)$ in x alone and polynomials $m(x,y)$ and $p(x,y)$ such that every term of p has degree at least d.

(b) If $I_O(f,g) = n < \infty$, prove that there are polynomials $u(x,y), v(x,y)$, and $w(x,y)$ such that

$$ug = x^n v + fw \tag{44}$$

and u and v are nonzero at the origin.

(Hint: One possible approach is to take $d = e + 1$ in part (a), where e is the integer in Exercise 14.7.)

14.9. Let F, G, H be curves, and let P be a point of F at which F is nonsingular.

(a) Prove that $I_P(G,H)$ is greater than or equal to the smaller of $I_P(F,G)$ and $I_P(F,H)$.

(*Hint*: One possible approach is is to use Exercise 11.3 when $I_P(F,G)$ or $I_P(F,H)$ is infinite. If both of these quantities are finite, we can reduce to the case where P is the origin and F is tangent to $y = 0$ at the origin and then apply Exercise 14.8(b) to f and g and to f and h, where $f(x,y), g(x,y), h(x,y)$ are the restrictions of F, G, H to the Euclidian plane. Exercise 10.15 shows that part (a) requires the assumption that F is nonsingular at P.)

(b) If G is also nonsingular at P and $I_P(F,G) > I_P(F,H)$, deduce from part (a) that $I_P(G,H) = I_P(F,H)$. (This shows that we can replace the line L in Theorem 9.4 with any curve F nonsingular at the point being considered.)

14.10. Let F, G, H be curves that are all nonsingular at a point P. Prove that two of the numbers $I_P(F,G), I_P(F,H), I_P(G,H)$ are equal and that their common value is less than or equal to the value of the third. (See Exercise 14.9(b). Exercise 10.15 shows the need to assume that F, G, H are all nonsingular. Exercise 10.14 shows that all values of intersection multiplicities allowed by this exercise actually occur.)

14.11. Over the real numbers, let F, G, H be homogeneous polynomials such that G and H have the same degree and are not scalar multiples of each other. Let P be a point of the real projective plane such that F is nonsingular at P and intersects G and H the same finite number k of times at P. Prove that there is a unique real number s such that F intersects $sG + H$ more than k times at P. For any real number r other than s, prove that F intersects $rG + H$ exactly k times at P. (See Exercise 14.8(b).)

14.12. In the real projective plane, let $F=0$ be a nonsingular, irreducible curve of degree m, Let $G=0$ and $H=0$ be distinct curves that have the same degree n and intersect F in the same mn points, listed by multiplicity, in the real projective plane. Prove that $m \leq n$ and that there is a homogeneous polynomial W of degree $n-m$ such that

$$I_P(G,H) = I_P(G,F) + I_P(G,W) = I_P(H,F) + I_P(H,W)$$

for every point P in the real projective plane.

(This extends Theorems 6.1 and 6.4 on "peeling off" conics and lines to "peeling off" nonsingular, irreducible curves of all degrees. The assumption that the curves $G=0$ and $H=0$ are distinct means that the homogeneous polynomials G and H are not scalar multiples of each other (as discussed after Theorem 3.6). One approach to this exercise is to deduce from Exercise 14.11 that there is a real number $s \neq 0$ such that $sG+H$ intersects F more than mn times, counting multiplicities.)

14.13. Let F be the irreducible cubic $y^2 = x^2(x+1)$ in (23) of Section 8. Find two different lines that intersect G three times at the origin. (This shows that we cannot omit the assumption in Exercise 14.12 that F is nonsingular, since $m=3$ is greater than $n=1$ in this case.)

14.14. Let $S(x,y,z)$ be a homogeneous polynomial in x, y, z. Let $(x,y,z) \to (x',y',z')$ be a transformation, and let x,y,z be expressed in terms of x',y',z' and constants A–I as in (6) of Section 3. Let $S'(x',y',z')$ be the result of substituting the expressions for x,y,z in $S(x,y,z)$. Prove that $S'_{x'}$ is the result of substituting the expressions for x,y,z in $AS_x + DS_y + GS_z$. State analogous results for $S'_{y'}$ and $S'_{z'}$.

(*Hint*: We can write $S(x,y,z) = \sum e_{ijk} x^i y^j z^k$ for numbers e_{ijk} and find $S'(x',y',z')$ by substituting the expressions for x,y,z.)

14.15. Let a transformation map a curve S to a curve S' as in Exercise 14.14. Let the transformation map a point P to a point P'. Deduce from Exercise 14.14 that the transformation maps the polar of P with respect to S to the polar of P' with respect to S'.

14.16. Do Exercise 13.14 by using Exercise 14.15 to transform any given conic into the unit circle $x^2+y^2=z^2$.

14.17. Prove the theorem in Exercise 13.21 without going up to the complex numbers by using Exercise 14.15 to transform the given cubic C into one of the forms in Theorem 8.3.

14.18. Consider the following result:

Theorem
Let $F'(x',y',z')=0$ be the image of a curve $F(x,y,z)=0$ under a transformation taking (x,y,z) to (x',y',z'). Let $H(x,y,z)$ be the Hessian of $F(x,y,z)$, and let $H'(x',y',z')$ be the Hessian of $F'(x',y',z')$ with respect to the variables x',y',z'. Then the transformation takes $H(x,y,z)=0$ to $H'(x',y',z')=0$.

Prove the theorem for transformations of the following forms:

(a) $x' = x + rz, y' = y + sz, z' = z$ for real numbers r and s.

(*Hint*: One possible approach is to deduce from Exercise 14.14 that $F'_{x'z'}$ is the image under the transformation of $-rF_{xx} - sF_{xy} + F_{xz}$. Use this and similar expressions for other second partial derivatives to compute $H'(x', y', z')$. Theorem 12.3 and the discussion after its proof may be helpful.)

(b) $x' = rx, y' = sy, z' = tz$ for nonzero real numbers r, s, t.

(c) $x' = z, y' = y, z' = x$.

(d) $x' = x, y' = z, z' = y$.

14.19. Prove the theorem in Exercise 14.18 by using that exercise and by combining Exercise 3.18 with the proof of Theorem 3.4 to show that every transformation is a product of the transformations in parts (a)–(d) of Exercise 14.18.

14.20. This exercise uses the discussions before Exercises 11.11 and 12.23. Let C be a nonsingular complex cubic. C has nine flexes in the complex projective plane (by Exercise 12.24). Let H be the Hessian of C. Prove that the complex cubics other than C that include the nine flexes of C among their own flexes are exactly the complex cubics $rC + H$ for all complex numbers r. Prove that all but four of the cubics $rC + H$ are nonsingular and the remaining four consist of the four triples of lines in Exercise 11.12. (See the theorem in Exercise 14.18, which extends without change to the complex numbers, and Exercises 12.24–12.28.)

§15. Determining Cubics

Five points, no three of which are collinear, lie on a unique conic (by Theorem 5.10). We derive an analogous result for cubics in this section: Starting with eight points such that no four are collinear and no seven lie on a conic and adding any other point except at most one gives nine points that lie on a unique cubic.

We work over the real numbers in this section and its exercises except where we state otherwise in the proof of Theorem 15.6.

The general equation of a cubic is

$$ax^3 + bx^2y + cxy^2 + dy^3 + ex^2 + fxy + gy^2 + hx + iy + j = 0 \tag{1}$$

in rectangular coordinates and

$$ax^3 + bx^2y + cxy^2 + dy^3 + ex^2z + fxyz + gy^2z + hxz^2 + iyz^2 + jz^3 = 0 \tag{2}$$

in homogeneous coordinates. To specify that a cubic contains a certain

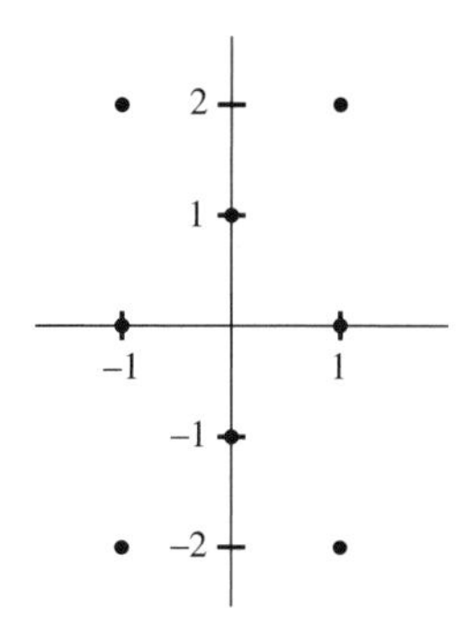

Figure 15.1

point, we substitute the coordinates of the point for the variables in (1) or (2) and obtain a homogeneous linear equation in the ten coefficients a–j. In the next example, requiring a cubic to contain the eight points in Figure 15.1 gives eight homogeneous linear equations in the ten coefficients a–j. We use these equations to eliminate all but two of the coefficients.

EXAMPLE 15.1
Determine all cubics through the eight points $(-1,2)$, $(-1,0)$, $(-1,-2)$, $(0,1)$, $(0,-1)$, $(1,2)$, $(1,0)$, $(1,-2)$ (Figure 15.1)

Solution
Because the points are given in rectangular, instead of homogeneous, coordinates, we use (1) instead of (2). Substituting the coordinates of the points for x and y in (1) gives a system of eight homogeneous linear equations in the ten coefficients a–j. We use each equation to eliminate one of the coefficients.

Substituting $(1,0)$ and $(-1,0)$ in (1) gives

$$a+e+h+j=0 \qquad \text{and} \qquad -a+e-h+j=0.$$

Adding and subtracting these equations and dividing the results by 2 gives the equivalent equations

$$e+j=0 \qquad \text{and} \qquad a+h=0.$$

By rewriting these equations as

$$j=-e \qquad \text{and} \qquad h=-a,$$

we can eliminate h and j from (1) and get

$$\begin{aligned} &ax^3+bx^2y+cxy^2+dy^3+ex^2+fxy+gy^2 \\ &\quad -ax+iy-e=0. \end{aligned} \tag{3}$$

Similarly, substituting $(0,1)$ and $(0,-1)$ in (3) gives

$$d+g+i-e=0 \qquad \text{and} \qquad -d+g-i-e=0.$$

Adding and subtracting these equations and dividing the results by 2 gives the equivalent equations

$$g-e=0 \qquad \text{and} \qquad d+i=0.$$

By rewriting these equations as

$$g=e \qquad \text{and} \qquad i=-d,$$

we can eliminate g and i from (3) and get

$$\begin{aligned} &ax^3+bx^2y+cxy^2+dy^3+ex^2+fxy+ey^2 \\ &\quad -ax-dy-e=0. \end{aligned} \tag{4}$$

Substituting $(1,2)$ and $(-1,2)$ in (4) and collecting terms gives

$$2b+4c+6d+4e+2f=0 \qquad \text{and} \qquad 2b-4c+6d+4e-2f=0.$$

Adding and subtracting these equations and dividing the results by 4 gives the equivalent equations

$$b+3d+2e=0 \qquad \text{and} \qquad 2c+f=0.$$

By rewriting these equations as

$$b=-3d-2e \qquad \text{and} \qquad f=-2c,$$

we can eliminate b and f from (4) and get

$$\begin{aligned} &ax^3+(-3d-2e)x^2y+cxy^2+dy^3+ex^2-2cxy+ey^2 \\ &\quad -ax-dy-e=0. \end{aligned} \tag{5}$$

Finally, substituting $(1,-2)$ and $(-1,-2)$ in (5) and collecting terms gives

$$8c+8e=0 \qquad \text{and} \qquad -8c+8e=0.$$

Adding and subtracting these equations and dividing by 16 gives the equivalent equations $c=0$ and $e=0$. Thus, (5) becomes

$$ax^3-3dx^2y+dy^3-ax-dy=0. \tag{6}$$

We have used the eight given points to eliminate the eight coefficients b, c, e–j from (1), leaving the two coefficients a and d. The only restriction on a and d is that they are not both zero (so that (6) is a cubic).

If we collect multiples of a and d, (6) becomes

$$aC+dD=0, \tag{7}$$

where C is x^3-x and D is y^3-3x^2y-y. As a and d vary over all pairs of

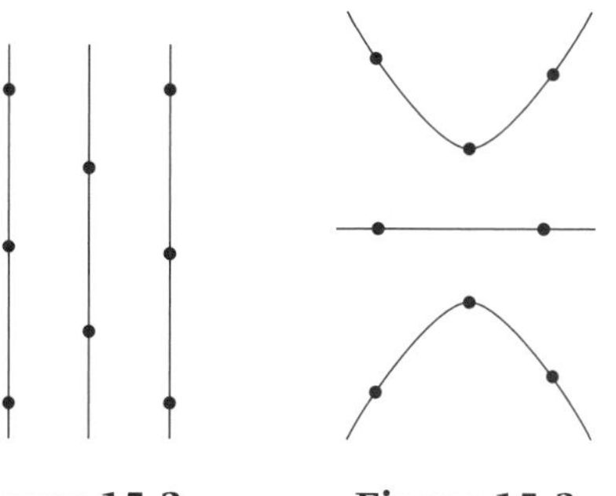

Figure 15.2 **Figure 15.3**

numbers that are not both zero, (7) gives all cubics through the eight specified points. Factoring

$$x^3 - x = x(x+1)(x-1)$$

shows that the cubic $C = 0$ consists of the three vertical lines $x = 0$, $x = -1$, and $x = 1$. (See Figure 15.2, which shows the eight points in Figure 15.1). Factoring

$$y^3 - 3x^2y - y = y(y^2 - 3x^2 - 1)$$

shows that the cubic $D = 0$ consists of the x-axis $y = 0$ and the hyperbola $y^2 - 3x^2 = 1$ (Figure 15.3).

If $d \neq 0$, dividing (7) by d gives

$$rC + D = 0 \tag{8}$$

for $r = a/d$. If $d = 0$, we have $a \neq 0$ (since a and d are not both zero), and dividing (7) by a gives

$$C = 0. \tag{9}$$

Conversely, for any real number r, (8) has the form of (7), and so does (9). In short, the cubics through the eight given points are given by (8) for all real numbers r and (9). □

Figures 15.4–15.12 show (8) for various values of r. Each figure is based on a computer plot of (6) where a is the chosen value of r and d is 1. Since D corresponds to $r = 0$, we can picture Figure 15.3 between Figures 15.8 and 15.9. By thinking of C as $rC + D$ for $r = \infty$ (where ∞ stands for both $+\infty$ and $-\infty$), we can picture Figure 15.2 as the transition from Figure 15.12 back to Figure 15.4.

The general cubic in (1) and (2) has ten coefficients a–j. When we specified eight points on the cubic in Example 15.1, we obtained a system of eight homogeneous linear equations in the coefficients. We solved the system by expressing eight of the coefficients in terms of the other

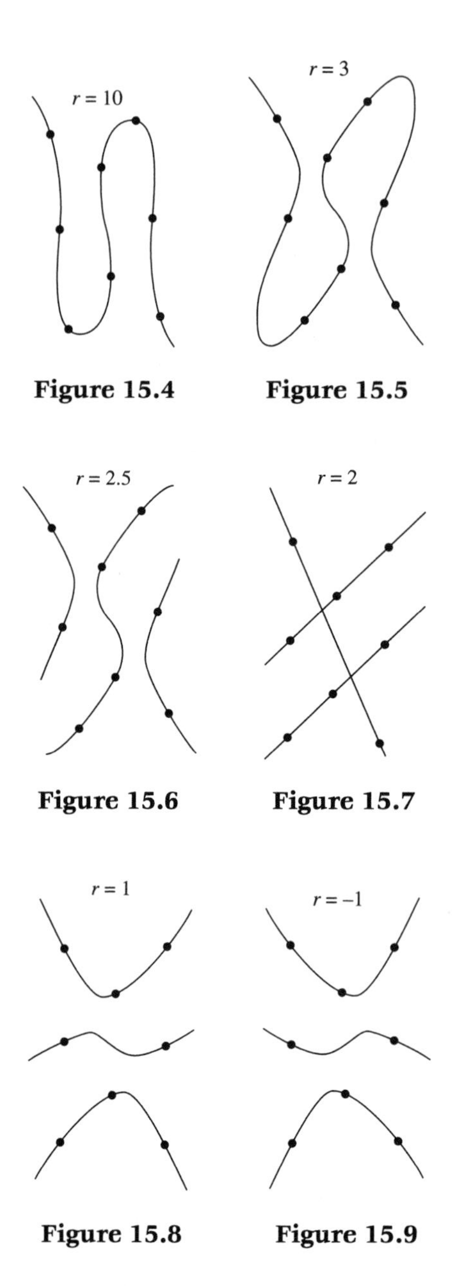

Figure 15.4 **Figure 15.5**

Figure 15.6 **Figure 15.7**

Figure 15.8 **Figure 15.9**

two. This gave the family of cubics in (7) with two parameters a and d. Dividing by a or d gave the family of cubics in (8) and (9) with one parameter r (where (9) corresponds to $r = \infty$).

To generalize this example, we need to reconsider systems of linear equations. Suppose that a system of homogeneous linear equations has m equations and n variables for $n > m$, which means that there are more

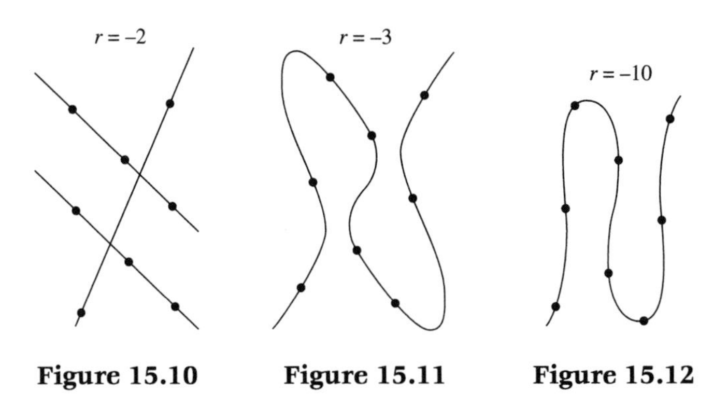

Figure 15.10 Figure 15.11 Figure 15.12

variables than equations. Theorem 14.1 states that there is a nontrivial solution. The next theorem strengthens this by showing that we can use the m equations to express m of the variables in terms of the other $n - m$ when the given equations are not redundant.

For example, consider the system of linear equations in (2) of Section 14, where there are $m = 3$ equations in $n = 5$ variables. We can express the $m = 3$ variables x_1, x_2, x_3 in terms of the remaining $n - m = 5 - 3 = 2$ variables x_4 and x_5. (This method is called *back-substitution* in elementary linear algebra.) Specifically, solving (4) of Section 14 for x_3 gives

$$x_3 = \frac{16}{7}x_4 - \frac{9}{7}x_5. \tag{10}$$

Substituting this expression for x_3 in the first equation in (3) of Section 14 gives

$$-2x_2 - \left(\frac{16}{7}x_4 - \frac{9}{7}x_5\right) - 10x_4 + x_5 = 0,$$

which simplifies to

$$x_2 = -\frac{43}{7}x_4 + \frac{8}{7}x_5. \tag{11}$$

Using (10) and (11) to eliminate x_2 and x_3 from the first equation in (2) of Section 14 gives

$$x_1 - 2\left(-\frac{43}{7}x_4 + \frac{8}{7}x_5\right) + \left(\frac{16}{7}x_4 - \frac{9}{7}x_5\right) + 4x_4 + x_5 = 0,$$

which simplifies to

$$x_1 = \frac{-130}{7}x_4 + \frac{18}{7}x_5. \tag{12}$$

Equations (10)–(12) express x_1, x_2, x_3 as sums of multiples of x_4 and x_5.

Any choice of values for x_4 and x_5 gives a unique solution of system (2) of Section 14. For example, setting $x_4 = 1$ and $x_5 = 2$ in (10)–(12) gives the values for x_1, x_2, x_3 in (6)–(8) of Section 14.

In general, we have the following result:

Theorem 15.2

Consider a system of m homogeneous linear equations in n variables for $m < n$. Assume that none of the equations is a sum of multiples of the others. Then the system can be solved by expressing m of the variables as sums of multiples of the other $n - m$ variables.

The second sentence of the theorem formalizes the idea that the equations are not redundant. When $m = 1$ this sentence means that the one equation in the system is nonzero. When $m > 1$, each equation is nonzero because it is not the sum of zero times the other equations. Thus, in either case, every equation in the system is nonzero.

Proof

Because the first equation in the system is nonzero, it contains a variable x_t with nonzero coefficient. We eliminate x_t from the other equations by adding multiples of the first. This reduces the system to $m - 1$ equations in $n - 1$ unknowns when we set aside the first equation. The solutions of the reduced system correspond to the solutions of the original system by using the first equation of the original system to determine the value of x_t.

Let the equations of the original system be $E_1 = 0, \ldots, E_m = 0$, which each have the form of (1) of Section 14. The equations of the reduced system are $E_2 - r_1E_1 = 0, \ldots, E_m - r_mE_1 = 0$ for real numbers $r_1, \ldots, r_m$. If one equation of the reduced system, say $E_2 - r_2E_1$, were a sum of multiples of the others, we would have

$$E_2 - r_2E_1 = b_3(E_3 - r_3E_1) + \cdots + b_m(E_m - r_mE_1)$$

for real numbers $b_3, \ldots, b_m$. We could rewrite this equation as

$$E_2 = (r_2 - b_3r_3 - \cdots - b_mr_m)E_1 + b_3E_3 + \cdots + b_mE_m,$$

which would contradict the assumption that no equation of the original system is a sum of multiples of the others. Thus, the reduced system also has the property that no equation is a sum of multiples of the others.

We have reduced the system by one equation and one variable. We continue in this way until we have eliminated all m equations. We have also eliminated m variables, one for each equation. As in the discussion accompanying (10)–(12), we can use back-substitution to express each of the eliminated variables as a sum of multiples of the remaining $n - m$ variables. □

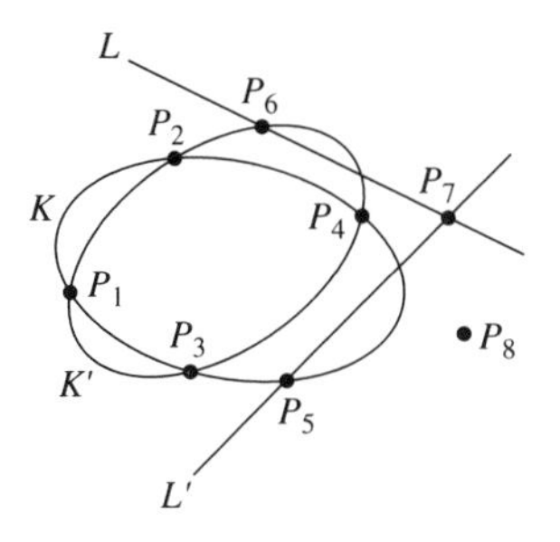

Figure 15.13

As in Example 15.1, requiring a cubic to contain a particular point gives a homogeneous linear equation in the coefficients of the cubic. The next result implies that there are no redundancies in the equations determined by eight points when no four of the points are collinear and no seven lie on a conic. We call two sets of points *disjoint* when they have no points in common.

Theorem 15.3

Let P_1–P_8 be eight points in the projective plane such that no four are collinear and no seven lie on a conic. Then there is a cubic that contains P_1–P_7 but not P_8.

Proof

We divide the proof into three cases, based on the arrangement of the points P_1–P_7.

Case 1

No three of the points P_1–P_7 are collinear (Figure 15.13). Then P_4P_7, P_5P_7, P_6P_7 are three different lines, and so P_7 is the unique point where any two of these lines intersect. Thus, at most one of the three lines contains P_8. By renumbering P_4–P_6, we can assume that neither P_5P_7 nor P_6P_7 contains P_8. The five points P_1–P_5 determine a conic $K = 0$, and the five points P_1–P_4, P_6 determine a conic $K' = 0$ (by Theorem 5.10).

We claim that at least one of the two conics K and K' does not contain P_8. Otherwise, if K and K' both contained P_8, their intersection would include the five points P_1–P_4, P_8, and K and K' would be the same conic (by Theorem 5.10). This conic would contain the seven points P_1–P_6, P_8, which would contradict the hypothesis that no seven of the points P_1–P_8 lie on a conic. This contradiction shows that K and K' cannot both contain P_8.

Let $L = 0$ be the line P_6P_7, and let $L' = 0$ be the line P_5P_7. The cubic $KL = 0$ consists of the conic K and the line L, and the cubic $K'L' = 0$ consists of the conic K' and the line L'. Both of the cubics KL and $K'L'$

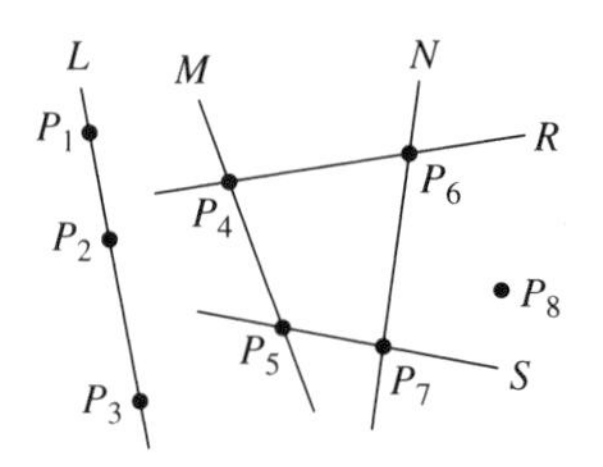

Figure 15.14

contain the seven points P_1–P_7. At least one of these two cubics does not contain P_8, since neither L nor L' contains P_8, and either K or K' does not contain P_8.

Case 2
Three of the points P_1–P_7 are collinear, and one cannot choose a second set of collinear points from P_1–P_7 disjoint from the first. By relabeling P_1–P_7, we can assume that P_1–P_3 lie on a line $L = 0$ and that no three of the points P_4–P_7 are collinear (Figure 15.14). Let $M = 0$, $N = 0$, $R = 0$, and $S = 0$ be the lines P_4P_5, P_6P_7, P_4P_6, P_5P_7, respectively. No two of these lines are equal, because no three of the points P_4–P_7 are collinear. Thus, the two pairs of lines $MN = 0$ and $RS = 0$ do not both contain P_8, since their intersection consists of the four points P_4–P_7. The line L through P_1–P_3 does not contain P_8, since no four of the points P_1–P_8 are collinear. Thus, the two cubics $LMN = 0$ and $LRS = 0$—which consist of the two triples of lines L, M, N and L, R, S—both contain the seven points P_1–P_7, and at least one of them does not contain P_8.

Case 3
Two disjoint sets of three collinear points can be chosen from P_1–P_7. By renaming P_1–P_7, we can assume that P_1, P_2, P_3 lie on a line $L = 0$ and that P_4, P_5, P_6 lie on a line $M = 0$ (Figure 15.15). Neither L nor M contains P_8, since no four of the points P_1–P_8 are collinear. Let $N = 0$ be a line through P_7 that does not contain P_8. The cubic $LMN = 0$ consists of

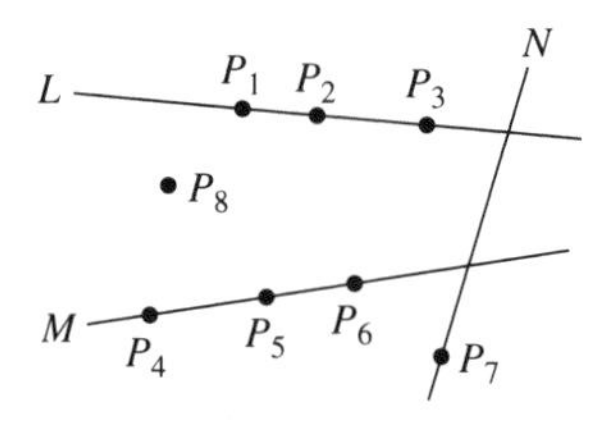

Figure 15.15

the three lines L, M, N, contains the seven points P_1–P_7, and does not contain P_8.

Because Cases 1–3 cover all possibilities, the proof is complete. □

We can now generalize Example 15.1 and determine all cubics through eight points, no four of which are collinear and no seven of which lie on a conic. As stated before Theorem 6.1, we call curves distinct exactly when they are not scalar multiples of each other.

Theorem 15.4

Let P_1–P_8 be eight points in the projective plane such that no four are collinear and no seven lie on a conic. Then the cubics containing P_1–P_8 are $C = 0$ and $rC + D = 0$ for all numbers r, where C and D are distinct cubics.

Proof

The eight points P_1–P_8 give a system of eight homogeneous linear equations in the ten coefficients of the general cubic in (2). None of the equations is a sum of multiples of the others because any seven of the points P_1–P_8 lie on a cubic that does not contain the eighth (by Theorem 15.3). Taking $m = 8$ and $n = 10$ in Theorem 15.2 shows that eight of the ten coefficients of the general cubic can be expressed as a sum of multiples of the other two. If we call the latter two coefficients s and t, we can express all the coefficients of the cubic as sums of multiples of s and t. Collecting the terms with s and those with t shows that the cubics containing P_1–P_8 are $sC + tD = 0$ for real numbers s and t not both zero, where C and D are distinct cubics: The fact that s and t are coefficients of distinct terms of the general cubic implies that C and D are both nonzero and are not scalar multiples of each other. Dividing $sC + tD$ by t when $t \neq 0$ and dividing by s when $t = 0$ (and so $s \neq 0$) shows that the cubics containing P_1–P_8 are $C = 0$ and $rC + D = 0$ for all real numbers r. □

We want an analogue for cubics of Theorem 5.10, which states that five points, no three of which are collinear, lie on a unique conic. We start by extending Theorem 5.10 to include the case where three but not four of the five given points are collinear and the conic is replaced by two lines.

Theorem 15.5

Five points, no four of which are collinear, lie on a unique curve of degree 2.

Proof

Assume first that no three of the five given points are collinear. Then any two lines contain at most four of the points, and so the only curves of degree 2 that contain all five points are conics (by Theorem 5.1). The five points lie on a unique conic (by Theorem 5.10).

On the other hand, assume that three of the given points—say, A, B, C—lie on a line $L = 0$. The two remaining points—say, D and E—lie on a line $M = 0$ (by Theorem 2.2). Then $LM = 0$ is a curve of degree 2 that contains the five points A–E. Conversely, let $Q = 0$ be any curve of degree 2 that contains the five points A–E. The intersection of the line $L = 0$ and the curve $Q = 0$ contains the three points A, B, C. Since Q has degree 2, L is a factor of Q (by Theorem 4.5), and we can write $Q = LN$ for a homogeneous polynomial N of degree 1. Because the line $L = 0$ contains the three points A, B, C, it does not contain D or E (since no four of the points A–E are collinear, by assumption). Since D and E lie on the curve $Q = LN$, they lie on the line $N = 0$. Because D and E determine a unique line (by Theorem 2.2), the lines $N = 0$ and $M = 0$ are the same. Then N and M are scalar multiples of each other, and so are $Q = LN$ and LM. □

The origin $(0, 0)$ satisfies (6) for all real numbers a and d. Thus, Example 15.1 shows that all cubics through the eight points in Figure 15.1 also contain the origin, as Figures 15.2–15.12 suggest. In fact, the next result implies that any two of these cubics intersect exactly nine times, at the eight points in Figure 15.1 and the origin.

Theorem 15.6

Let P_1–P_8 be eight points, no four of which are collinear, and no seven of which lie on a conic. Then all pairs of cubics containing P_1–P_8 intersect in the same nine points, listed by multiplicity. That is, there is a point P_9 such that any two cubics containing P_1–P_8 intersect in exactly the nine points P_1–P_9, listed by multiplicity.

Proof

By Theorem 15.4, the cubics containing P_1–P_8 have the form $C = 0$ and $rC + D = 0$ for all numbers r, where C and D are two cubics that are not scalar multiples of each other. We divide the proof into three claims.

Claim 1

C and D have no common factors other than constants.

Any such factor would have degree 1 or 2, since C and D are not scalar multiples of each other. We consider these two possibilities separately.

Suppose first that C and D have a common factor L of degree 1. We have $C = LQ$ and $D = LR$ for homogeneous polynomials Q and R of degree 2. Since no four of the points P_1–P_8 are collinear, at most three of them lie on L, and at least five of them do not. These points lie on both Q and R, since they lie on C and D. It follows that Q and R are scalar multiples of each other (by Theorem 15.5). This contradicts the assumption that $C = LQ$ and $D = LR$ are not scalar multiples of each other. Thus, C and D have no common factors of degree 1.

Suppose next that C and D have a common factor Q of degree 2. We have $C = LQ$ and $D = MQ$ for homogeneous polynomials L and M of degree 1. Q is not a product of two lines or of one line doubled, by the previous paragraph. If Q is a conic it contains at most six of the points P_1–P_8, by the assumption that no seven lie on a conic. By Theorem 5.1, the only other possibilities are that Q is a single point or the empty set. In any case, at most six of the points P_1–P_8 lie on Q, and so at least two do not. Since the latter points lie on both C and D, they lie on both L and M. Then $L = 0$ and $M = 0$ are the same line (by Theorem 2.2), and so L and M are scalar multiples of each other. This contradicts the assumption that $C = LQ$ and $D = MQ$ are not scalar multiples of each other. Thus, C and D have no common factors of degree 2.

Claim 2

C and D intersect exactly nine times, counting multiplicities, in the real projective plane.

Since C and D have no common factors of positive degree over the real numbers (by Claim 1), the same holds over the complex numbers (by Theorem 11.9). Thus, C and D intersect at exactly nine points, listed by multiplicity, in the complex projective plane (by Bezout's Theorem 11.5). Let P_9 be the ninth point of intersection, in addition to P_1–P_8; P_9 may or may not equal one of the points P_1–P_8.

Consider the map $P \to \bar{P}$ of points in the complex projective plane, which was introduced before Property 12.6. Because C and D have real coefficients, this map interchanges among themselves the points of intersection of C and D in the complex projective plane, listed by multiplicity (by Property 12.6). Because P_1–P_8 have real coefficients, they are all fixed by this map. Thus, the map $P \to \bar{P}$ also fixes P_9. It follows, as in the last paragraph of the proof of Theorem 12.7, that P_9 lies in the real projective plane. Thus, C and D intersect in the nine points P_1–P_9, listed by multiplicity, in the real projective plane, as discussed before Theorem 11.1.

Claim 3

Any two cubics containing P_1–P_8 intersect at the same points, listed by multiplicity, as do C and D.

By Theorem 15.4, any cubic containing P_1–P_8 has the form C or $rC + D$ for a real number r. Let Q be any point. We have

$$I_Q(C, rC + D) = I_Q(C, D)$$

(by Property 3.6(iv)), and so C and $rC + D$ intersect the same number of times at every point as do C and D. If r and s are unequal real numbers, we have

$$I_Q(rC + D, sC + D) = I_Q((r - s)C, sC + D)$$

(subtracting $sC + D$ from $rC + D$, by Property 3.6(iv))

$$= I_Q(C, sC + D)$$

(by the remark after the proof of Theroem 3.6, since $r - s \neq 0$)

$$= I_Q(C, D)$$

(by Property 3.6(iv)). This shows that $rC + D$ and $sC + D$ intersect the same number of times at every point as do C and D.

Claims 2 and 3 establish the theorem. □

In Theorem 15.6, the given points P_1–P_8 are distinct, but P_9 may or may not equal one of them. If P_9 is one of the points P_1–P_8, then any two cubics containing P_1–P_8 intersect twice at this point. If P_9 does not equal any of the points P_1–P_8, then it is a ninth point that lies on all cubics through the eight points P_1–P_8. In either case, the next result shows that every point other than P_1–P_9 lies on a unique cubic through P_1–P_8.

Theorem 15.7

Let P_1–P_8 be eight points, no four of which are collinear and no seven of which lie on a conic. By Theorem 15.6, there is a point P_9 such that the intersection of any two cubics containing P_1–P_8 consists of the points P_1–P_9, listed by multiplicity. If Q is any point other than P_1–P_9, the nine points P_1–P_8 and Q lie on a unique cubic.

Proof

The nine points P_1–P_8 and Q determine a system of nine homogeneous linear equations in the ten coefficients of the general cubic in (2). Because there are more coefficients than equations, the system has a nontrivial solution (by Theorem 14.1), and so P_1–P_8 and Q lie on a cubic. This cubic is unique because Q does not equal any of the points P_1–P_9 where any two cubics containing P_1–P_8 intersect (by Theorem 15.6). □

If we take the points P_1–P_8 in Theorem 15.6 to be the eight points in Figure 15.1, the discussion before Theorem 15.6 shows that P_9 is the origin. If Q is any point other than P_1–P_8 and the origin, the nine points P_1–P_8 and Q lie on a unique cubic (by Theorem 15.7). We can think of Figures 15.2–15.12 as illustrations of this fact.

The general polynomial of degree 2 in (1) of Section 5 has six coefficients. Specifying five points on the curve gives a system of five homogeneous linear equations in the coefficients. If the equations are not redundant, we can use them to express five of the coefficients as multiples of the sixth. Dividing all the coefficients by the sixth shows that the five points lie on a unique curve of degree 2. Theorem 15.5 shows that five points, no four of which are collinear, lie on a unique curve of degree 2, which means that the conditions they impose on curves of degree 2 are not redundant.

Likewise, the general cubic in (2) of this section has ten coefficients. Specifying nine points on the cubic gives a system of nine homogeneous linear equations in the coefficients. If the equations are not redundant, we can use them to express nine of the coefficients as multiples of the tenth. Dividing all the coefficients by the tenth shows that the nine given points lie on a unique cubic.

Theorem 15.7 gives conditions under which nine points P_1–P_8 and Q lie on a unique cubic, which means that the conditions they impose on cubics are not redundant. P_1–P_8 are eight points such that no four are collinear and no seven lie on a conic. The ninth point Q is any point except P_1–P_8 and at most one other: Q cannot be the point P_9 in Theorem 15.6 determined by P_1–P_8, where P_9 may or may not be one of the points P_1–P_8.

Exercises

15.1. Three of the points P_1–P_8 in Example 15.1 lie on the line $x = -1$, two lie on $x = 0$, and three lie on $x = 1$ (Figure 15.2). Why does it follow that no four of these points are collinear and no seven lie on a conic? (This confirms that Theorems 15.6 and 15.7 apply to the given points P_1–P_8.)

15.2. Justify your answers to the following questions algebraically:

(a) What symmetry is exhibited by the graph of (8) for any real number r and by (9)?

(b) What is the geometric relation between the graphs of (8) for a real number r and its negative?

(c) For what values of r does the graph of (8) have a local extremum at a point with x-coordinate 1 or -1?

15.3. Let P_1–P_8 be eight points, no four of which are collinear and no seven of which lie on a conic. Let P_9 be the point determined in Theorem 15.6. Let P_s be one of the points P_1–P_8 other than P_9.

(a) Conclude from Exercise 14.6 or from Exercise 4.26 and Theorem 4.11 that all cubics containing P_1–P_8 are nonsingular at P_s and tangent to different lines there.

(b) Prove that every line on P_s is tangent there to exactly one cubic containing P_1–P_8.

(*Hint*: If C and D are as in Theorem 15.4, one possible approach is to use part (a) to reduce to the case where P_s is the origin, C is tangent at P_s to the x-axis, and D is tangent at P_s to the y-axis.)

15.4. Let P_1–P_8 be eight points, no four of which are collinear and no seven of which lie on a conic. Assume that the point P_9 determined by Theorem 15.6 is one of the points P_1–P_8.

(a) Use Exercise 14.6 or 4.26 to prove that at most one cubic containing P_1–P_8 is singular at P_9.

(b) Prove that the same line is tangent at P_9 to every cubic that contains P_1–P_8 and is nonsingular at P_9.

(c) Conclude from parts (a) and (b) and Theorem 15.4 that P_1–P_8 lie on exactly one cubic singular at P_9.

15.5. Let P_1–P_8 be eight points. Prove that every cubic containing P_1–P_8 is irreducible if and only if no three of the points are collinear and no six lie on a conic.

15.6. Let P_1–P_8 be eight points. If four of the points are collinear, or if seven of the points lie on a conic, prove that any two cubics containing P_1–P_8 have a common factor of degree 1 or 2 and intersect at infinitely many points.

15.7. Let C and D be two cubics that intersect in exactly nine points P_1–P_9, listed by multiplicity, where the points P_1–P_8 are distinct. Prove that any two cubics through P_1–P_8 intersect at exactly the nine points P_1–P_9, listed by multiplicity. (See Exercise 15.6.)

15.8. Consider the following result:

Theorem

Let C be a nonsingular, irreducible cubic. Let K and K' be curves of degree 2 such that C intersects K at points E, F, G, H, W, X, listed by multiplicity, and C intersects K' at points E, F, G, H, Y, Z, listed by multiplicity. Then the lines WX and YZ have the same point as their third intersection with C.

Prove the theorem under the added assumption that the points E, F, G, H, W, X, Y, Z are distinct by using Exercise 15.7, taking D to be the cubic composed of K and line YZ, and considering the cubic composed of K' and line WX. Illustrate the theorem when E–H, W–Z are eight distinct points and both K and K' are conics.

(The theorem includes Theorem 9.6 when K and K' are pairs of lines, and it includes the theorem in Exercise 10.6 when K is a conic and K' is a pair of lines. The theorem is proved in general in Exercise 15.24.)

15.9. Let P_1–P_8 be eight points such that no five are collinear and no curve of degree 2 contains all eight of the points. Prove that there are distinct cubics C and D such that the cubics containing P_1–P_8 are exactly $C = 0$ and $rC + D = 0$ for all real numbers r. (Thus, Theorem 15.4 extends to cases where four of the points P_1–P_8 are collinear or seven lie on a conic. Exercise 15.6 shows that Theorems 15.6 and 15.7 do not extend to these cases.)

15.10. Let P_1–P_8 be eight points. Assume that either five of these points are collinear or all eight lie on a curve of degree 2.

(a) Prove that P_1–P_8 lie on four cubics F_1–F_4 such that $F_1 \neq F_2$, $F_3 \neq F_4$, and the points of intersection of F_1 and F_2 are not the same as those of F_3 and F_4.

(b) Deduce that there do not exist cubics C and D such that the cubics containing P_1–P_8 are $C = 0$ and $rC + D = 0$ for all real numbers r. (This shows that the conditions on the eight points P_1–P_8 in Exercise 15.9 cannot be weakened.)

15.11. Let P_1–P_8 be the eight points $(1,1)$, $(1,0)$, $(1,-1)$, $(0,1)$, $(0,0)$, $(0,-1)$, $(-1,1)$, $(-1,0)$.

(a) As in Example 15.1, find the equations of all cubics through P_1–P_8 and obtain an analogue of (6).

(b) Use part (a) to find polynomials C and D such that the cubics through P_1–P_8 are $C = 0$ and $rC + D = 0$ for all numbers r. Draw graphs of $C = 0$ and $D = 0$ that show P_1–P_8.

(c) Find the coordinates of the point P_9 such that any two cubics through P_1–P_8 intersect at the points P_1–P_9, listed by multiplicity.

(d) As in Figures 15.4–15.12, use appropriate technology to graph the cubics $rC + D = 0$ for a wide range of values of r. Show the points P_1–P_9 on each graph.

15.12. Do Exercise 15.11 for the eight points $(0,3)$, $(0,-3)$, $(2,0)$, $(-2,0)$, $(2,1)$, $(2,-1)$, $(-2,1)$, $(-2,-1)$.

15.13. Do Exercise 15.11 for the eight points $(1,1)$, $(1,0)$, $(1,-1)$, $(-1,1)$, $(-1,0)$, $(-1,-1)$, A, B, where A and B are the points at infinity on lines of slope 1 and -1.

15.14. Do Exercise 15.11 for the eight points $(1,1)$, $(1,0)$, $(1,-1)$, $(-1,1)$, $(-1,0)$, $(-1,-1)$, S, T, where S and T are the points at infinity on lines of slope 2 and -2.

15.15. Do Exercise 15.11 for the eight points $(1,0)$, $(-1,0)$, $(0,1)$, $(0,-1)$, $(1,1)$, $(-1,-1)$, U, V, where U and V are the points at infinity on horizontal and vertical lines.

15.16. Do Exercise 15.11 for the eight points $(1,1)$, $(1,-1)$, $(-1,1)$, $(-1,-1)$, $(2,1)$, $(-2,-1)$, M, N, where M and M are the points at infinity on vertical lines and on lines of slope $\frac{1}{2}$.

15.17. Do Exercise 15.11 for the eight points $(0,0)$, $(1,1)$, $(1,-1)$, $(-1,1)$, $(-1,-1)$, $(2,1)$, $(2,0)$, $(2,-1)$.

15.18. Let $K_1 = 0$, $K_2 = 0$, $K_3 = 0$ be three conics through the same four points A, B, C, D. Let P be a fifth point. Let $L_1 = 0$, $L_2 = 0$, $L_3 = 0$ be three lines through P. Assume that the line $L_i = 0$ intersects the conic $K_i = 0$ at two points Q_i and R_i for $i = 1, 2, 3$, and assume that the eleven points

$$A, B, C, D, P, Q_1, R_1, Q_2, R_2, Q_3, R_3,$$

are distinct. Prove that these eleven points lie on a cubic.

(*Hint*: One posible approach is to note that K_1L_2 and K_2L_1 are cubics that intersect at nine of the eleven points. Deduce that no four of the nine points are collinear and no seven lie on a conic. Note that eight of the nine points lie on the cubic $K_1L_3 + rK_3L_1$ for some real number r. Then apply Theorem 15.6.)

15.19. Let $F = 0$ be nonsingular at a point P. Let n be a positive integer, and let G be the general homogeneous polynomial of degree n with indeterminate coefficients; for example, G is given by (2) of Section 2 for $n = 1$, (1) of Section 5 for $n = 2$, and (1) of Section 8 for $n = 3$. Let d be a nonnegative integer. Prove that there is a system of d linear homogeneous equations in the coefficients of G that is equivalent to the condition that F intersects G at least d times at P. (See Exercises 14.8(a) and 4.23–4.27.)

15.20. Let C be a nonsingular, irreducible cubic. Add points of C with respect to a flex O as in Definition 9.3. Let P_1–P_6 be points of C that are not necessarily distinct. Prove that there is at most one curve of degree 2 that intersects C at P_1–P_6, listed by multiplicity. Prove that such a curve exists if and only if $P_1 + \cdots + P_6 = O$.

(*Hint*: Exercise 15.19 and Theorem 14.1 imply there is a curve of degree 2 whose intersections with C, listed by multiplicity, include P_1–P_5. Then see Exercises 12.16(b), 10.8(b), and 14.12.)

15.21. Let $C = 0$ be a nonsingular, irreducible cubic, and let $D = 0$ be a cubic that intersects C at points P_1–P_9, listed by multiplicity. Let R be the third intersection of line P_1P_2 and C, and assume that R does not equal any of the points P_3–P_9. Let Q_1 be a point of C other than P_1 and P_2, and let Q_2 be the third intersection of line RQ_1 and C. Prove that there is a cubic that intersects C at Q_1, Q_2, P_3–P_9, listed by multiplicity.

(*Hint*: Define lines P_1P_2 and RQ_1 with respect to C. Why can we assume that D is nonsingular at $P_1 - P_9$ by Exercises 4.15 and 4.16? Why does D intersect P_1P_2 at points P_1, P_2, S, listed by multiplicity, for some point S? Why does S lie on a line M that does not contain any of the points $P_3 - P_9, Q_1, Q_2, R$? The desired result follows by using Theorem 6.4 to "peel off" the line P_1P_2 from the intersection of the curve of degree 4 comprised of C and M and the curve of degree 4 comprised of D and the line RQ_1.)

15.22. Let $C = 0$ be a nonsingular, irreducible cubic. Add points of C with respect to a flex O (as in Definition 9.3). Let C intersect a cubic $D = 0$ at points P_1–P_9, listed by multiplicity. Prove that

$$P_1 + \cdots + P_9 = O \tag{13}$$

in the following cases:

(a) There is a line L that intersects C at three of the points P_1–P_9, listed by multiplicity.

(*Hint*: Let Q be a point of L that does not lie on C. We can show that there is a number r such that $rC + D = 0$ contains Q, conclude from Theorems 9.5 and 4.5 that $rC + D$ has L as a factor, and deduce (13) from Exercises 9.2(a) and 10.8(b).)

(b) The general case, where P_1–P_9 are any points of C and are not necessarily distinct.

(*Hint*: By taking Q_1 to be the third intersection of line P_3P_4 and C, we can use Exercise 15.21 to reduce to the case in (a). Part (a) applies directly when the conditions of Exercise 15.21 on R and Q_1 do not all hold.)

15.23. This exercise extends Theorems 15.4, 15.6, and 15.7 to cases where the points P_1–P_8 are not all distinct. Let $C = 0$ be a nonsingular, irreducible cubic. Add points of C with respect to a flex O (as in Definition 9.3). Let P_1–P_8 be points of C that are not necessarily distinct. Set

$$P_9 = -P_1 - \cdots - P_8.$$

(a) Prove that there is a cubic $D = 0$ such that the cubics distinct from C whose intersections with C, listed by multiplicity, include P_1–P_8 are exactly the cubics $rC + D$ for all real numbers r. Prove that these cubics all intersect C at the points P_1–P_9, listed by multiplicity. (See Theorem 15.2 and Exercises 15.19, 12.16(b), 15.22, and the last sentence of the discussion in parentheses at the end of Exercise 14.12.)

(b) If Q is any point in the projective plane other than P_1–P_9, prove that there is exactly one cubic that contains Q and whose intersections with C, listed by multiplicity, include P_1–P_8.

15.24. Use Exercise 15.23(a) to prove the theorem in Exercise 15.8 by taking D to be the cubic composed of K and the line YZ and considering the cubic composed of K' and the line WX.

References

The works listed here provided much of the material in this text. They are recommended sources of further reading.

A fine treatment of synthetic projective geometry and conics appears in

Seidenberg, A., *Lectures in Projective Geometry*, Van Nostrand, Princeton, NJ, 1962.

A wonderful introduction to the number-theoretic study of cubics is provided by

Silverman, Joseph H., and Tate, John, *Rational Points on Elliptic Curves*, Springer-Verlag, New York, 1992.

The following books are devoted primarily to algebraic curves:

Abhyankar, Shreeram S., *Algebraic Geometry for Scientists and Engineers*, American Mathematical Society, Providence, RI, 1990.

Brieskorn, Egbert, and Knörrer, Horst, *Plane Algebraic Curves*, Birkhäuser, Boston, 1986.

Coolidge, Julian Lowell, *A Treatise on Algebraic Plane Curves*, Dover, New York, 1959.

Fulton, William, *Algebraic Curves*, Benjamin/Cummings, Reading, MA, 1969.

Hilton, Harold, *Plane Algebraic Curves*, Oxford University Press, London, 1932.

Kirwan, Frances, *Complex Algebraic Curves*, Cambridge University Press, Cambridge, 1992.

Seidenberg, A., *Elements of the Theory of Algebraic Curves*, Addison-Wesley, Reading, MA, 1968.

Walker, Robert J., *Algebraic Curves*, Springer-Verlag, New York, 1978.

Walker's book is self-contained and incisive. Brieskorn and Knörrer emphasize geometric intuition in their leisurely treatment. Fulton uses abstract algebra to proceed rapidly with the study of curves of all degrees. Kirwan introduces the complex analytic and topological study of complex algebraic curves.

A fine introduction to algebraic curves and higher-dimensional algebraic geometry is given by the following book, which includes a completely accessible proof of Cayley's Theorem that there are exactly 27 lines on every nonsingular cubic surface in complex projective space.

Reid, Miles, *Undergraduate Algebraic Geometry*, Cambridge University Press, Cambridge, 1988.

A more sophisticated, but still concrete, introduction to higher-dimensional algebraic geometry is given by

Harris, Joe, *Algebraic Geometry: A First Course*, Springer-Verlag, New York, 1992.

The following books and articles trace the history of the study of algebraic curves:

Ball, W.W.R., "Newton's Classification of Cubic Curves," *Proceedings of the London Mathematical Society*, Vol. 22, 1890, pp. 104–143.

Bashmakova, Isabella G., "Arithmetic of Algebraic Curves from Diophantus to Poincaré," *Historia Mathematica*, Vol. 8, 1981, pp. 393–416.

Boston, Nigel, "A Taylor-made Plug for Wiles' Proof," *The College Mathematics Journal*, Vol. 26, 1995, pp. 100–105.

Boyer, Carl B., *History of Analytic Geometry*, The Scholar's Bookshelf, Princeton, NJ, 1956.

Coolidge, Julian Lowell, *A History of Geometrical Methods*, Oxford University Press, Oxford, 1940.

Coolidge, Julian, Lowell, *A History of the Conic Sections and Quadric Surfaces*, Oxford University Press, Oxford, 1945.

Dieudonné, Jean, *History of Algebraic Geometry*, Wadsworth, Monterey, CA, 1985.

Kline, Morris, *Mathematical Thought from Ancient to Modern Times*, Oxford University Press, New York, 1972.

Stilwell, John, *Mathematics and Its History*, Springer-Verlag, New York, 1989.

Kline's book is notable among histories of mathematics for its clarity and comprehensiveness.

The proof of the Fundamental Theorem of Algebra given in Section 10 is based on:

Fefferman, Charles, "An Easy Proof of the Fundamental Theorem of Algebra," *American Mathematical Monthly*, vol. 74, 1967, pp. 854–855.

Index

Undergraduate Texts in Mathematics

(continued from page ii)

James: Topological and Uniform Spaces.
Jänich: Linear Algebra.
Jänich: Topology.
Kemeny/Snell: Finite Markov Chains.
Kinsey: Topology of Surfaces.
Klambauer: Aspects of Calculus.
Lang: A First Course in Calculus. Fifth edition.
Lang: Calculus of Several Variables. Third edition.
Lang: Introduction to Linear Algebra. Second edition.
Lang: Linear Algebra. Third edition.
Lang: Undergraduate Algebra. Second edition.
Lang: Undergraduate Analysis.
Lax/Burstein/Lax: Calculus with Applications and Computing. Volume 1.
LeCuyer: College Mathematics with APL.
Lidl/Pilz: Applied Abstract Algebra. Second edition.
Logan: Applied Partial Differential Equations.
Macki-Strauss: Introduction to Optimal Control Theory.
Malitz: Introduction to Mathematical Logic.
Marsden/Weinstein: Calculus I, II, III. Second edition.
Martin: The Foundations of Geometry and the Non-Euclidean Plane.
Martin: Geometric Constructions.
Martin: Transformation Geometry: An Introduction to Symmetry.
Millman/Parker: Geometry: A Metric Approach with Models. Second edition.
Moschovakis: Notes on Set Theory.
Owen: A First Course in the Mathematical Foundations of Thermodynamics.
Palka: An Introduction to Complex Function Theory.
Pedrick: A First Course in Analysis.
Peressini/Sullivan/Uhl: The Mathematics of Nonlinear Programming.
Prenowitz/Jantosciak: Join Geometries.
Priestley: Calculus: A Liberal Art. Second edition.
Protter/Morrey: A First Course in Real Analysis. Second edition.
Protter/Morrey: Intermediate Calculus. Second edition.
Roman: An Introduction to Coding and Information Theory.
Ross: Elementary Analysis: The Theory of Calculus.
Samuel: Projective Geometry. *Readings in Mathematics.*
Scharlau/Opolka: From Fermat to Minkowski.
Sethuraman: Rings, Fields, and Vector Spaces: An Approach to Geometric Constructability.
Sigler: Algebra.
Silverman/Tate: Rational Points on Elliptic Curves.
Simmonds: A Brief on Tensor Analysis. Second edition.
Singer: Geometry: Plane and Fancy.
Singer/Thorpe: Lecture Notes on Elementary Topology and Geometry.
Smith: Linear Algebra. Third edition.
Smith: Primer of Modern Analysis. Second edition.
Stanton/White: Constructive Combinatorics.
Stillwell: Elements of Algebra: Geometry, Numbers, Equations.
Stillwell: Mathematics and Its History.
Stillwell: Numbers and Geometry. *Readings in Mathematics.*
Strayer: Linear Programming and Its Applications.
Thorpe: Elementary Topics in Differential Geometry.
Toth: Glimpses of Algebra and Geometry. *Readings in Mathematics.*

Undergraduate Texts in Mathematics

Troutman: Variational Calculus and Optimal Control. Second edition.
Valenza: Linear Algebra: An Introduction to Abstract Mathematics.
Whyburn/Duda: Dynamic Topology.
Wilson: Much Ado About Calculus.